Department of the Environment

THE UK ENVIRONMENT

Editor: Alan Brown

Assisted by: Chris Groom
Diane Lewis
Christine Rabjohns
David Moir

 A publication of the Government Statistical Service

London: HMSO

The cover photographs are reproduced by the kind permission of
BARNABY'S PICTURE LIBRARY (four urban and rural scenes), and
BRUCE COLEMAN LIMITED (badger photograph by Hans Reinhard).

HMSO publications are available from:

HMSO Publications Centre
(Mail, fax and telephone orders only)
PO Box 276, London, SW8 5DT
Telephone orders 071-873 9090
General enquiries 071-873 0011
(queuing system in operation for both numbers)
Fax orders 071-873 8200

HMSO Bookshops
49 High Holborn, London, WC1V 6HB
(counter service only)
071-873 0011 Fax 071-873 8200
258 Broad Street, Birmingham, B1 2HE
021-643 3740 Fax 021-643 6510
Southey House, 33 Wine Street, Bristol, BS1 2BQ
0272 264306 Fax 0272 294515
9-21 Princess Street, Manchester, M60 8AS
061-834 7201 Fax 061-833 0634
16 Arthur Street, Belfast, BT1 4GD
0232 238451 Fax 0232 235401
71 Lothian Road, Edinburgh, EH3 9AZ
031-228 4181 Fax 031-229 2734

HMSO's Accredited Agents
(see Yellow Pages)

and through good booksellers

Recycled Pa

FOREWORD

The Government's White Paper on the environment, "This Common Inheritance" published in 1990, emphasised its commitment to encourage informed debate on environmental issues, and make information about the environment available to everyone. The environment cannot speak for itself. Careful measurement and observation are the only sure ways of identifying improvement and future problems. We need wide and easy access to data about the environment for everyone if we are to have an informed public debate on the difficult decisions that will face us in ensuring a sustainable future.

This is the first report to collect together environmental information in this easily accessible form. It is the result of a wide consultation on the material to be included, and brings together both the major long term statistical series and also less readily available data from other areas and research programmes. It complements the many other measures that the Government has taken to increase public access to environmental information, including new public registers for environmental data, better information for consumers and the publication of results and reviews of environmental research. It also provides an important contribution to international initiatives, particularly by the OECD, to encourage countries to improve environmental reporting and thereby promote public awareness of environmental issues.

I hope that this report will be used widely and found helpful in public discussion, in schools and as a reference source. We pride ourselves in the UK on the quality of the data which we use to help manage our environment. Further reports will be published at regular intervals and I want them to reflect any suggestions we receive on how it might be improved to increase their usefulness. If you have any suggestions to make please don't hesitate to let us know about them.

Michael Howard

Rt Hon MICHAEL HOWARD QC MP
SECRETARY OF STATE FOR THE ENVIRONMENT

CONTENTS

8   The marine environment   107

INTRODUCTION

This is the first edition of a new statistical report on the environment as promised in the environment White Paper, "This Common Inheritance" (Cmnd 1200) which was published in September 1990.

The aim of the report is to bring together available statistics about the environment in a coherent and comprehensive form and so provide a set of benchmark environmental statistics to inform public debate. The book is intended for the non-technical reader and much of the information is presented in the form of charts, simple tables and explanatory text.

Coverage

Annex B of the White Paper listed the areas considered for inclusion in the report, and over 250 organisations were also consulted. A number of these have provided material for the report.

The environment is a term which covers many features of the world around us. Statistics on the environment, unlike economic statistics, are still in their infancy and there are as yet no universally accepted simple indicators of the "health" of the environment. The range of information available to measure such things as air quality, for example, is considerable, and it is not possible to include all of it in a single report of this kind.

We have therefore selected data for inclusion in the report in an attempt to give a balanced picture. However, more information is available on some aspects of the environment than others, and so the amount of coverage given to particular topics should not necessarily be taken as an indication of their relative importance. At the end of each chapter there are references to sources where further information can be obtained.

The report is divided into seventeen chapters. The first is a very brief introduction to the UK climate. The next nine chapters cover the main environmental resources: air; land; water; and wildlife. Subjects which do not fall neatly into this part of the report are given separate chapters: waste and recycling; noise; radioactivity; environment and health.

These are followed by a chapter which brings together information about some of the pressures on the environment and discusses them in the context of the relevant sectors ie: agriculture; the energy industry; and transport. The next chapter considers public attitudes to the environment and the final chapter deals with environmental expenditure.

The report attempts to cover all the UK, but in many instances this has not been possible, because information is not available (or is not available on a comparable basis) for all the constituent countries. Comparisons are not generally made with other countries outside the UK.

Many topics occur in more than one chapter, reflecting the complex interactions between different parts of the environment. Each topic however, is generally discussed in detail in one chapter where it is most relevant, and there are numerous cross references to other chapters. An index (referring to figure or table numbers) has been included which allows readers to locate major topics where they occur throughout the report. There is also a glossary and a separate listing of abbreviations used in the report.

Information on computer diskette

The report includes around 230 charts and 60 tables. Information on which many of these have been based is available on a computer diskette in a format compatible with most personal computers. This is available on request to purchasers of the report, on application to the Department (on the response form inserted in the report).

Quality of data

As discussed above, not only does the availability of information vary widely between topics, but also the reliability of the figures given is very variable. In some cases, particularly in respect of statistics on waste and on environmental expenditure, only very broad orders of magnitude can be given. The text indicates the likely margins of error, particularly where the level of accuracy has an important bearing on interpretation of the figures presented.

The future

The White Paper recommended publication of a statistical report on the UK environment at regular intervals. This report makes it clear that much further work needs to be done before a comprehensive, accurate assessment can be produced. One of the major contributions of this report is to indicate both the scope, and the limitations, of the information currently available, and to point to areas where data are incomplete or inaccurate. It is intended to be the start of an ongoing process of improvement and development. Readers are therefore encouraged to return the questionnaire included with the volume, indicating which parts of the report they have found useful, and any suggestions they may have for improvements in the future.

Environmental Protection Statistics Division
Department of the Environment
Romney House
43 Marsham Street
London
SW1P 3PY

Telephone enquiry point - 071 276 8052
or 071 276 8425

Acknowledgements

The Editor would like to thank all those who have contributed to the publication, including colleagues in DOE and in other Departments, and the numerous organisations which have provided information, maps and graphical material, and technical guidance. Without the substantial help given by experts in many fields the report would not have been possible.

Special thanks are due in particular to: the Joint Nature Conservation Committee for their very considerable help in drafting the wildlife chapter; the Soil Survey and Land Research Centre and the Macaulay Land Use Research Institute for their contributions on soil, and for providing maps; the British Trust for Ornithology for permission to reproduce charts on bird populations and distributions; the Met Office and the University of East Anglia for their permission to reproduce climate charts and for long term climate data; the National Rivers Authority for permission to reproduce maps and for data underlying a number of figures and tables; ECOTEC Research and Consulting Ltd for their work on environmental expenditure; and the New Scientist for the water cycle diagram. Also, Warren Spring Laboratory; the Institution of Environmental Health Officers and the Royal Environmental Health Institute of Scotland; the Forestry Commission; Ordnance Survey; the Countryside Commission; Natural Environment Research Council; the Royal Society for the Protection of Birds; the National Radiological Protection Board; MORI; and many other organisations which have made contributions either as part of the initial consultation process or during the production of the report.

Thanks also to the production team, in DOE - Chris Groom, Diane Lewis, Christine Rabjohns and David Moir. Chris Morrey and Jock Martin also provided invaluable assistance. Special thanks are also due to colleagues in the Electronic Publishing Unit, particularly Andy Taylor, Roger Jeffs, Claire Parry and Mike Chapman, for preparing camera ready copy and producing the charts, maps and tables.

LIST OF FIGURES AND TABLES

Chapter 4

Figures

Chapter 5

Figures

Chapter 8

Figures

Tables

Chapter 9

Figures

Tables

Chapter 13

Chapter 14

Chapter 15

Chapter 16

Figures

Chapter 17

Figures

SYMBOLS AND CONVENTIONS

In Figures and Tables, where numbers have been rounded to the nearest final digit there may be an apparent slight discrepancy between the sum of the constituent items and the totals.

Percentages in Tables are shown in italics.

See also glossary for items marked (*)

Symbol	Meaning	Symbol	Meaning
..	not available	T	tera $= 10^{12} = 1,000,000,000,000$
-	nil or negligible (less than half the final digit shown)	G	giga $= 10^9 = 1,000,000,000$
		M	mega $= 10^6 = 1,000,000$
p	provisional	k	kilo $= 10^3 = 1,000$
>	greater than	c	centi $= 10^{-2} = 0.01$
<	less than	m	milli $= 10^{-3} = 0.001$
ppm	parts per million	μ	micro $= 10^{-6} = 0.000001$
ppb	parts per billion	n	nano $= 10^{-9} = 0.000000001$
ppmv	parts per million by volume		
t	tonnes	1 billion	= 1 thousand million
ttoe	thousand tonnes of oil or oil equivalent		
		1 tonne	= 1 thousand kilograms
mtoe	million tonnes of oil or oil equivalent	1 kilogram	= 1 thousand grams
kg	kilogram	1 gram	= 1 thousand milligrams
g	gram	1 gram	= 1 million micrograms
mg	milligram	1 kilometre	= 1 thousand metres
μg	microgram	1 square kilometre	= 100 hectares
ng	nanogram	1 square kilometre	= 0.39 square miles
l	litre	1 acre	= 0.40 of a hectare
ml	millilitre	1 hectare	= 10,000 square metres
km	kilometre	1 mile	= 1,609.3 metres
m	metre	1 litre	= 0.001 cubic metres
mm	millimetre		
nm	nanometre		
m^3	cubic metre		
ha	hectare		
dB	decibel*		
pH	measure of acidity/alkalinity		
Bq	becquerel*		
TBq	terabecquerel		
Gy	gray*		
Sv	sievert*		
manSv	man sievert		
mSv	millisievert		
μSv	microsievert		
CO	carbon monoxide*		
CO_2	carbon dioxide*		
HCl	hydrogen chloride*		
CH_4	methane*		
NO	nitric oxide		
NO_2	nitrogen dioxide*		
N_2O	nitrous oxide*		
NO_x	nitrogen oxides		
SO_2	sulphur dioxide*		

1 Climate

1.1 The UK has a cool, moist, temperate, maritime climate with a mild annual average temperature and modest extremes. Nevertheless, the weather can vary widely at different times of the year and between different parts of the country. This variability results from the UK's position at the margin of the principal track of Atlantic depressions. Anticyclones establish themselves around the UK and interrupt the normal south westerly flow, bringing instead airstreams of varied character.

1.2 Figure 1.1 shows the paths of the principal air masses affecting the UK. The most frequent type is a westerly airflow associated with Polar maritime air (ie air originating in polar regions over a water body) and modified by passage over the comparatively warm oceanic current (the North Atlantic Drift, popularly known as the Gulf Stream).

1.3 There is considerable variation in the climate in different parts of the UK. One of the reasons for this is the high ground in the north and west lying across the direction of the prevailing winds. As air rises over high ground its temperature falls, increasing snowfall and snow cover in winter, and the forced uplift of moist air also causes increased cloudiness, more frequent rainfall and fewer hours of bright sunshine. To the lee of high ground, however, the climate is generally drier, warmer and sunnier.

Temperature

1.4 The mild winters mean that the average annual temperature is high for the latitude.

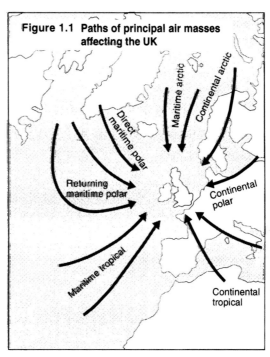

Figure 1.1 Paths of principal air masses affecting the UK

The modest range in temperature is largely due to the North Atlantic Drift and the oceans surrounding the UK which warm up slowly in the summer, storing heat which is released slowly during winter. Occasionally the normal westerly flow over the Atlantic Ocean is interrupted by anticyclonic influences which may last from a day to a season; it is at these times that extremes of temperature are recorded.

1.5 Figure 1.2 shows the mean annual temperature over the UK for the years 1941-70, corrected to mean sea-level. Values range from 11.5 degrees Celsius in the Scillies to 7.5 degrees Celsius in the Shetland Isles. Places on the east coast are generally cooler

---

Temperature information   *Box 1.1*

The temperature information given is based on observations made at about 250 stations for the 30 years 1941-70. This period has been chosen as a reasonable guide to temperatures which might be expected in the short term. To eliminate the altitude effect actual temperatures have been adjusted to mean sea level values. Longer term temperature changes are discussed in Chapter 3 on the global atmosphere.

The *monthly mean daily temperature* is calculated as half the sum of the *mean daily maximum* and the *mean daily minimum*.

The *mean annual temperature* is the arithmetic mean of the mean temperature for each month.

The *annual range* in temperature refers to the difference between the mean temperatures of the warmest and coolest months.

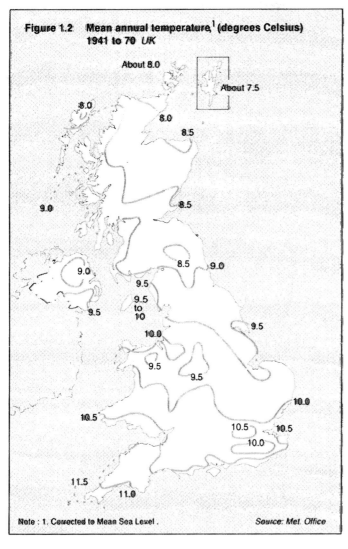

Figure 1.2   Mean annual temperature,[1] (degrees Celsius) 1941 to 70 *UK*

About 8.0

About 7.5

8.0

8.0

8.5

9.0

8.5

9.0

8.5   9.0

9.5

9.5
to
10

9.5

10.0

9.5

9.5

9.5

10.0

10.5

10.5   10.5

10.0

11.5

11.0

Note : 1. Corrected to Mean Sea Level .   Source: Met. Office

temperature at three selected stations in the UK.

1.6 In the winter months temperature differences are greater from east to west than from north to south. Figure 1.4 shows mean temperatures in January, April, July, and September, for the years 1941-70. In spring the pattern is similar to that for the annual mean temperature and in summer the warmest regions are in the south.

Precipitation

1.7 Mean annual rainfall over the UK for the years 1941-70 is shown in Figure 1.5. Annual rainfall varies widely from less than 500 mm in the Thames estuary to more than 2,400 mm over the uplands of Wales, Scotland and the Lake District. There is a similar increase from south east to north west in the total hours of measurable rainfall and in the number of rainy days. The north west is wetter mainly because of the greater frequency of frontal systems and more unstable maritime air crossing Scotland, and because of the mountains.

1.8 Almost all areas have more rain in winter than at other times of the year. The proportion of annual rainfall occurring in summer is higher in eastern areas, decreases towards the north and west, and is lowest over upland areas.

Evaporation and the water balance

1.9 The main determinants of evapotranspiration (loss of water vapour from plants through spores and evaporation from surface water and soil) are wind, temperature and solar radiation. Thick cloud cover and fewer hours of bright sunshine lead to low values of evaporation in upland areas. High coastal values reflect higher wind speeds and more sunshine.

1.10 In general terms, rainfall increases the moisture content of the soil, while evapotranspiration reduces it. When the soil is saturated, the excess water percolates or runs off into lakes, rivers and aquifers. Soil moisture surpluses can occur at any time of the year, but are usually shortlived in summer. When there is not enough rain to keep soil saturated, evapotranspiration withdraws moisture from the soil, and deficits occur. Soil moisture deficits can occur at any time of the year, but are usually small in winter. For more details see Chapter 4 on soil.

Duration of bright sunshine

1.11 Figure 1.6 shows the mean daily duration of bright sunshine for the years

than places on the west coast at the same latitude, because of the presence of the relatively cold North Sea, and generally mean annual temperatures rise from north east to the south west of the country. Figure 1.3 shows fluctuations in mean annual

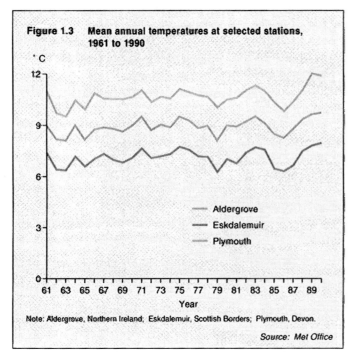

Figure 1.3   Mean annual temperatures at selected stations, 1961 to 1990

°C

12

9

6

3

0

61 63 65 67 69 71 73 75 77 79 81 83 85 87 89

Year

Aldergrove
Eskdalemuir
Plymouth

Note: Aldergrove, Northern Ireland; Eskdalemuir, Scottish Borders; Plymouth, Devon.

Source: Met Office

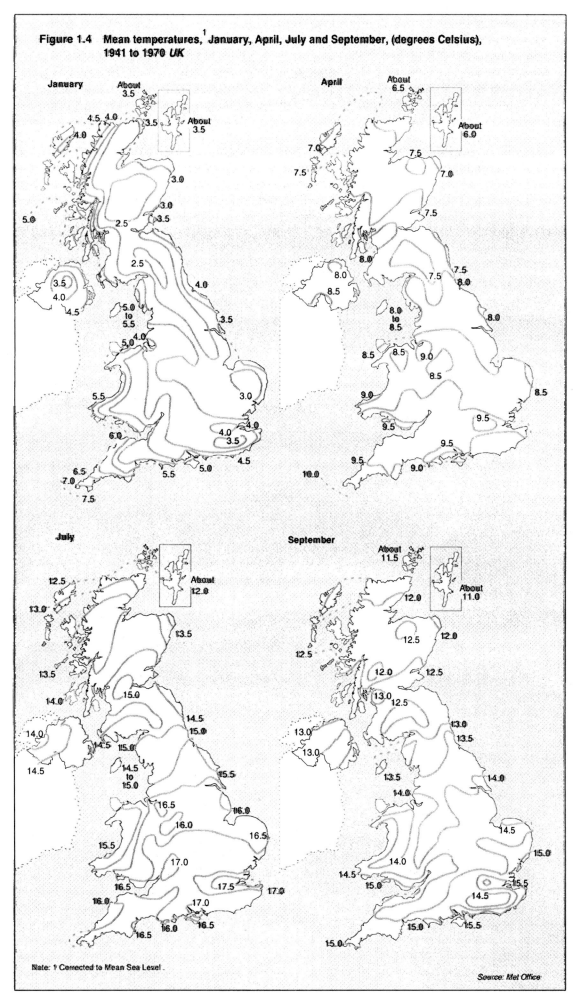

Figure 1.4 Mean temperatures,[1] January, April, July and September, (degrees Celsius), 1941 to 1970 *UK*

Note: 1 Corrected to Mean Sea Level.

Source: Met Office

3

Figure 1.5 Mean annual rainfall, 1941 to 1970 *UK*

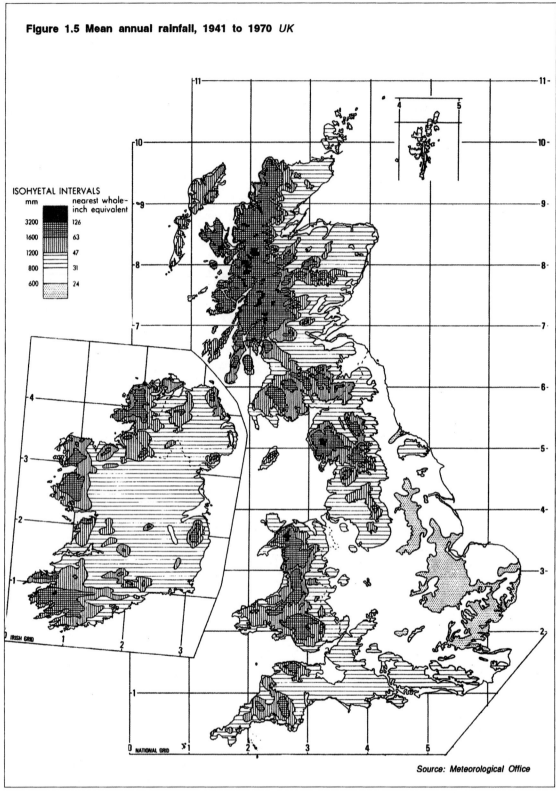

ISOHYETAL INTERVALS

mm	nearest whole-inch equivalent
3200	126
1600	63
1200	47
800	31
600	24

IRISH GRID

NATIONAL GRID

Source: Meteorological Office

1941-70. Average daily sunshine hours vary markedly according to the height of the sun in the sky. The highest values occur on the south coast. Broadly speaking, those places nearer or on the coast receive more sunshine, on average, than those inland; and the south experiences more than the north.

Wind

1.12 Figure 1.7 shows the force and direction of the wind across the UK in an average year. The prevailing wind direction is generally from the south and west and the least frequent winds are from the east and south east. There are, however, regional and seasonal variations. Generally, winds of force five and over are more frequent in the north and west than in the south and east, and winds of force seven are uncommon in England. Calms and light winds are more common in the east than in the west. The prevailing wind patterns in mid-winter are very similar to those for the year as a whole, although winds greater than force 5 and winds from the south west are more frequent and winds from the north east are less frequent. North-easterly winds are more common in the spring than in the other seasons.

1.13 There is considerable variation across the UK in the number of days with gales, as can be seen in Figure 1.8. Strong winds and gales occur most often at coastal stations, particularly those on exposed western coasts. They are least common at inland stations, and on east facing coasts.

Visibility (mist, fog and smog)

1.14 Visibility is broadly related to the concentrations of water droplets and solid particles in the atmosphere. When water droplets are present in suitable concentrations and sizes, mist, fog or thick fog (see Box 1.2) may form if the air cools below its dew point. In large towns and industrial areas visibility may be further reduced by the presence of solid particles: this is known as smoke fog or smog. Significant variations in visibility can occur over short distances due to variations in topography and to man-made influences associated with towns and industrial regions.

1.15 Figure 1.9 shows changes in the incidence of fog and thick fog between 1949 and 1987 for London and Manchester. This period saw the introduction of the Clean Air Acts of 1956 and 1968. See also Chapter 2 on air quality and pollution.

Variability and extremes in the 1980s

1.16 The records show that dramatic climatic variations or extremes are not common in the UK. Generally measurements are within quite a narrow range although exceptions to normal patterns do occur.

1.17 In the 1980s there have been both mild and cold winters and warm and cool summers. The mildest winter was 1988-89 when the mean temperature exceeded the 1941-70 average by 2.7 degrees Celsius. The coldest winter occurred in 1981-82 with a mean temperature 1.2 degrees Celsius below average (mainly because of the coldest December this Century when the mean temperature was 4.7 degrees Celsius below average). The warmest summer was 1983,

Figure 1.6 Mean daily duration of bright sunshine (hours) 1941 to 1970 *UK*

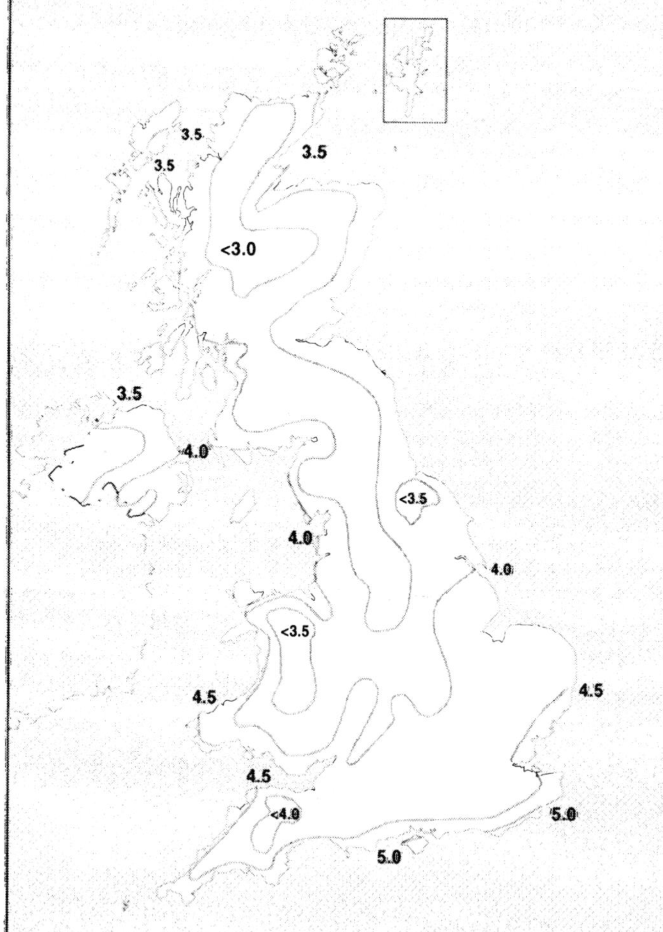

Source: Met Office

with a mean temperature of 17.1 degrees Celsius, 1.7 degrees above the average. July 1983 was the hottest in central England for over 300 years.

1.18 Yearly totals of rainfall in England and Wales reached a peak in the 1930s and declined slowly thereafter. Over the last twenty years, rainfall has continued to be below the 1941-70 average throughout England and Wales, ranging from 2.3 per cent below in Wales, to almost 4 per cent below in south east England. In the latter area two recent winters have been notably drier than normal; 1988-89 when rainfall amounted to 118 mm (55 per cent of the 1941-1970 average) and 1991-92 when 74 mm was recorded. In contrast, the winter of 1989-90 was the second wettest this century

---

Mist, fog and thick fog   *Box 1.2*

Mist visibility equal to or greater than 1000 metres with relative humidity greater than about 95 degrees;

Fog visibility of 1000 metres or less;

Thick fog visibility of 200 metres or less.

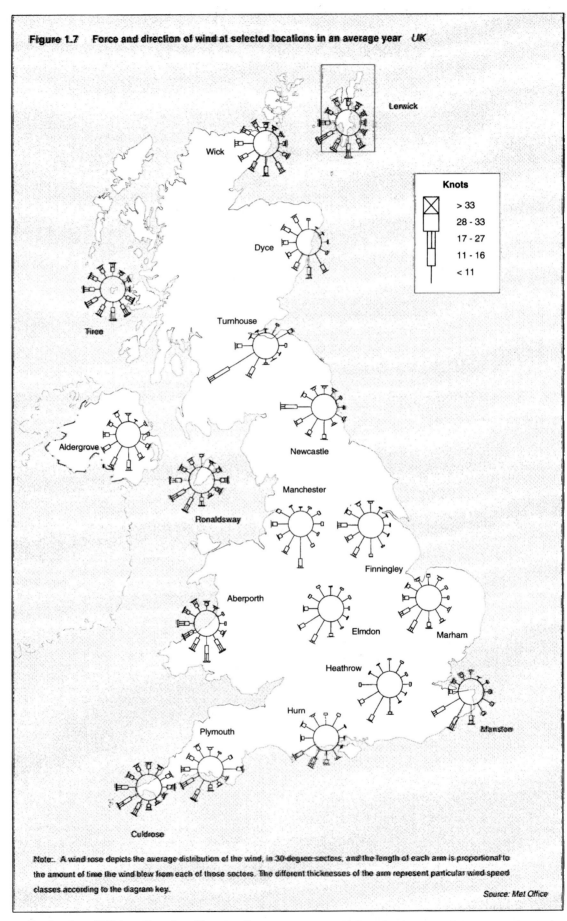

Figure 1.7 Force and direction of wind at selected locations in an average year UK

Note: A wind rose depicts the average distribution of the wind, in 30-degree sectors, and the length of each arm is proportional to the amount of time the wind blew from each of those sectors. The different thicknesses of the arm represent particular wind speed classes according to the diagram key.

Source: Met Office

in England and Wales. See also Chapter 6 on inland water resources and abstraction.

1.19 There is some evidence (at least in the London area) that mean annual wind speeds have declined since reaching a peak in the early 1970s. Nevertheless there have been very strong winds some of which resulted in structural damage and the destruction of many trees. Recent examples are in October 1987 when winds reached hurricane force along parts of the south east coast of England, with gusts up to 115 miles per hour, and January 1990 when storm force winds occurred across much of southern England. See also the section on storm damage in Chapter 5 on land use and land cover.

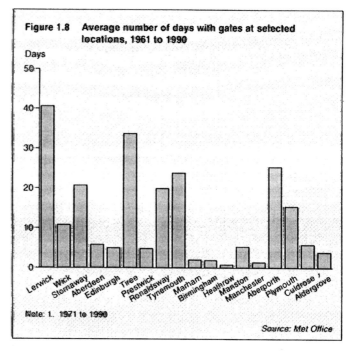

Figure 1.8 Average number of days with gales at selected locations, 1961 to 1990

Note: 1. 1971 to 1990

Source: Met Office

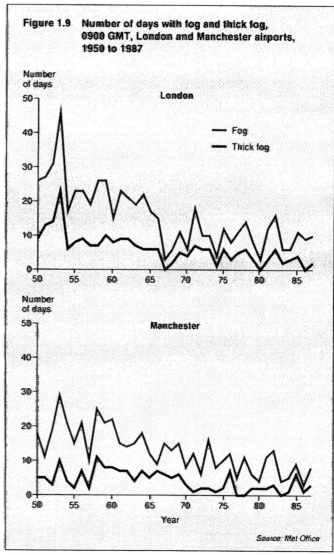

Figure 1.9 Number of days with fog and thick fog, 0900 GMT, London and Manchester airports, 1950 to 1987

Source: Met Office

References and further reading

Chandler, T. J. & Gregory, S., (1976). The Climate of the British Isles. Longman Inc., New York.

2 Air quality and pollution

□ **Sulphur dioxide concentrations have declined greatly in most parts of the UK since 1970. Occasional episodes of "poor" concentrations of sulphur dioxide continue to occur, particularly in Belfast - on 27 days in 1990 (Figures 2.2 and 2.4).**

□ **Black smoke emissions have declined by just under 20 per cent since 1980. Emissions from domestic sources have halved since 1980 whilst those from road transport have doubled (Figure 2.9).**

□ **Surveys in 1986 and 1991 showed that average UK concentrations of nitrogen dioxide increased by 35 per cent over the period, mainly as a result of increased emissions from traffic (2.18).**

□ **Granite-based soils in upland north and west Britain are the most sensitive to acid rain damage. Deposition rates also tend to be high in these areas because of high rainfall, mist and cloud.**

□ **Background levels of ground-level ozone have risen substantially over the last century. The number of hours of "poor" ozone concentrations tends to increase from the north to the south of the country (Figure 2.17).**

□ **Emissions of volatile organic compounds have risen gradually during the 1980s and are now 4 per cent higher than 1980. Emissions from processes and solvents accounted for half of total emissions and road transport just over 40 per cent (Figures 2.19 and 2.20).**

□ **The WHO eight-hour guideline for carbon monoxide concentrations was exceeded on only three occasions in 1990, in Glasgow in November and at Cromwell Road, London in January and December (Figure 2.21). The WHO one-hour guideline has not been exceeded in recent years.**

□ **Carbon monoxide emissions have increased by over 30 per cent since 1980. 90 per cent of emissions are derived from road vehicles (Figures 2.22 and 2.23).**

□ **EC limit values for airborne lead concentrations have been exceeded at only one site since 1985; compliance was achieved at all sites in 1990. Emissions of lead from petrol-engined vehicles have fallen by 70 per cent since the mid 1980s following the reduction in the permitted lead content of petrol and the introduction of unleaded petrol (Figure 2.25).**

2.1 There have long been concerns about damage to human health caused by air pollution, but it is only in the last 20 years that the full extent of its effects on a wide range of environmental features, including soil, insects, fish, wild flowers, crops, forests and buildings have begun to be identified. This chapter gives information about ambient concentrations of a number of atmospheric pollutants in the UK, trends in emissions, and pollutant sources. It covers the mandatory and advisory standards set for the protection of human health and the environment by this country, the European Community (EC) and the World Health Organisation (WHO). Global climate effects of atmospheric gases are dealt with in Chapter 3.

2.2 Pollutants may get into the atmosphere directly from industrial and domestic activities, or they can be formed by chemical processes in the atmosphere. Damage can be caused, too, by pollutants carried through the air for hundreds or thousands of miles from their source, often crossing national boundaries. Air quality in the UK is monitored by networks covering a range of pollutants and a wide variety of environments from urban and industrial centres to the remotest locations in the British Isles (see Box 2.2 on air quality monitoring).

Legislation, standards and guidelines

Box 2.1

Main UK legislation

The main relevant GB provisions currently in force are

- **The Clean Air Acts 1956 and 1968** These acts introduced controls on pollution by smoke, grit and dust from domestic and certain industrial sources, including controls on chimney heights and powers for local authorities to designate smoke control areas.

- **The Health and Safety at Work Act etc 1974** This together with the Alkali Act 1906 places a duty on operators of registered industrial processes to use best practicable means to prevent noxious emissions and render them harmless. The Act is enforced by Her Majesty's Inspectorate of Pollution (HMIP). The provisions relating to air pollution are being replaced by the phased introduction of controls under Part I of the Environmental Protection Act 1990.

- **The Environmental Protection Act 1990** Part I of the Act requires operators of prescribed industrial processes to obtain an authorisation from their local authority (for the less complex processes making emissions mainly to air) or HMIP, or Her Majesty's Industrial Pollution Inspectorate (HMIPI)/River Purification Authorities in Scotland (for more complex ones with significant releases to more than one medium). Best available techniques not entailing excessive cost (BATNEEC) are required to be used to prevent, or where that is not practicable, to minimise polluting releases and render them harmless. Part III of the Act applies statutory nuisance procedures in England and Wales to categories of smoke and other emissions including odours not covered by the Clean Air Acts or Part I of the Act.

- **Road Traffic Acts 1972 and 1974** Regulations under these Acts control emissions from road vehicles.

- **Control of Pollution Act 1974** Regulations under this Act control the composition of motor fuel and fuel oil.

- **The Town and Country Planning Act 1990** This controls the development and use of land and the location of potentially polluting development, as well as controlling other development in proximity to sources of air pollution.

Northern Ireland has comparable legislation except in the case of the Environmental Protection Act 1990 where they plan to introduce similar measures over the next two years.

European Community legislation and other international agreements

Relevant European Community legislation includes

- Directives setting mandatory limit values and guideline values for ambient concentrations of smoke and sulphur dioxide, nitrogen dioxide, and (limit values only) lead. Data on ambient concentrations are given as concentrations in parts per billion (ppb) or micrograms per cubic metre ($\mu g/m^3$) over a range of averaging periods (eg one hour, eight hours, 24 hours are usual);

- a Directive on ground-level ozone was adopted in May 1992, providing for harmonised monitoring and data exchange arrangements and threshold concentrations above which public information and alerts must be issued.

- a framework Directive on industrial emissions, requiring prior authorisation of new or substantially modified plant, the application of BATNEEC, and upgrading of existing plant; daughter Directives imposing more specific requirements in relation to asbestos, large combustion plant, and municipal waste incineration. These set limit values for emissions and, for existing large combustion plant, require specific reductions in national annual emissions;

- Directives on emissions from new cars and heavy diesel vehicles;

- Directives on the sulphur content of gas oil and the lead content of petrol.

The UK is also a party to the UN Economic Commission for Europe (UNECE) convention on Long Range Transboundary Air Pollution and to protocols under the convention on the evaluation and monitoring of such pollution, on oxides of nitrogen and on volatile organic compounds.

World Health Organisation (Europe) air quality guidelines

The World Health Organisation (WHO) published advisory guidelines for ambient concentrations of a wide range of organic and inorganic compounds in 1987. They are set some way below the minimum concentrations at which adverse effects have been observed. Others are set at levels above which adverse ecological effects on health have been observed, on the basis that some plants display a higher sensitivity to air pollutants than humans. For some carcinogenic compounds such as benzene WHO were unable to recommend a safe level.

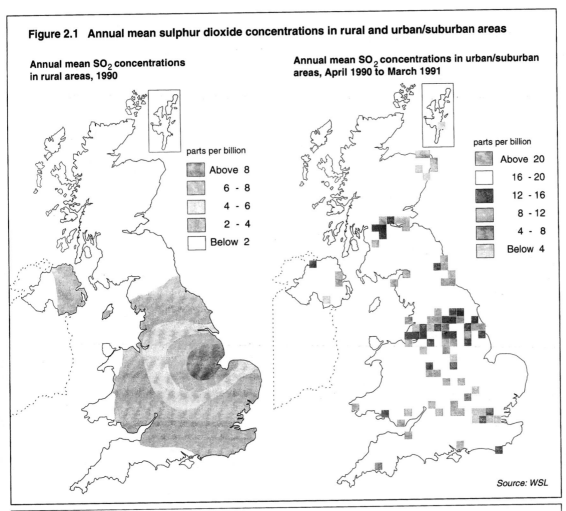

Figure 2.1 Annual mean sulphur dioxide concentrations in rural and urban/suburban areas

Annual mean SO$_2$ concentrations
in rural areas, 1990

Annual mean SO$_2$ concentrations in urban/suburban
areas, April 1990 to March 1991

parts per billion

Above 8
6 - 8
4 - 6
2 - 4
Below 2

parts per billion

Above 20
16 - 20
12 - 16
8 - 12
4 - 8
Below 4

Source: WSL

Air quality monitoring   *Box 2.2*

The purposes of monitoring are to:

- measure trends to help inform and monitor policy;
- check compliance with any formal air quality standards, and adequacy of measures to secure compliance;
- inform the public and the media of day-to-day air quality; and
- understand environmental processes.

Currently there are separate monitoring networks in the UK covering:

   Urban smoke and sulphur dioxide
   Nitrogen dioxide
   Lead particulate
   Acid deposition
   Rural ozone, nitrogen dioxide and sulphur dioxide
   Trace elements
   Carbon monoxide
   Acid waters
   Urban hydrocarbons
   Air toxics

Department of the Environment Air Quality Bulletin system

In October 1990, the Department of the Environment introduced a new system of daily air quality bulletins and forecasts and associated health advice. Bulletins and forecasts categorise air quality as "very good", "good", "poor" or "very poor" for sulphur dioxide, nitrogen dioxide and ozone. Information is given on the health implications of "poor" and "very poor" air quality. This material is available on a free telephone service (0800 556677) and is carried by some national and local newspapers and on CEEFAX (page 196). See Figures 2.4, 2.12 and 2.18 which give information for selected sites, on hourly concentrations classed as "poor" or "very poor".

Sources of air pollutants and trace gases

Emissions

2.3 There are some natural sources of air pollutants, eg sulphur dioxide from volcanoes, dimethyl sulphide from plankton blooms, nitrogen oxides from soils, and hydrocarbons from trees. However, in the urban and industrialised regions of Europe man-made emissions of the most common air pollutants far exceed those from natural sources. Combustion of fuels is a major source, giving rise according to the fuel to emissions of smoke, sulphur dioxide (SO_2), oxides of nitrogen (NO_x), carbon monoxide (CO), carbon dioxide (CO_2), hydrogen chloride (HCl), hydrocarbons and heavy metals. Industrial processes, storage and use of chemicals are also sources of these compounds. Livestock production also gives rise to significant emissions of ammonia and methane.

2.4 Emissions of some pollutants from many sources have decreased greatly for a variety of reasons, including the mandatory use of smokeless fuel in smoke control areas, the widespread use of natural gas for domestic heating, the introduction of new, less polluting,

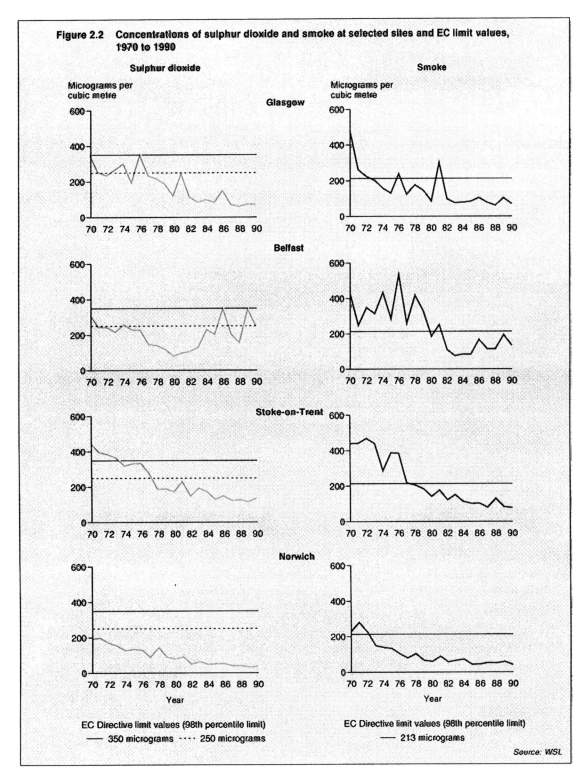

Figure 2.2 Concentrations of sulphur dioxide and smoke at selected sites and EC limit values, 1970 to 1990

Sulphur dioxide — Smoke

Glasgow

Belfast

Stoke-on-Trent

Norwich

EC Directive limit values (98th percentile limit)
—— 350 micrograms ···· 250 micrograms

EC Directive limit values (98th percentile limit)
—— 213 micrograms

Source: WSL

EC limit values for smoke and sulphur dioxide　　　　　*Box 2.3*

(micrograms per cubic metre)[1]

	Smoke	Sulphur dioxide		
Year[a]	80	If smoke less than or equal to	40:	120
		If smoke more than	40:	80
Winter[b]	130	If smoke less than or equal to	60:	180
		If smoke more than	60:	130
Year (peak)[c]	250[2,3]	If smoke less than or equal to	150:	350[2]
		If smoke more than	150:	250[2]

[1] OECD units

[2] Member States must ensure that this value is not exceeded for more than three consecutive days and must try to prevent and reduce such instances where the value has been exceeded.

[3] Equivalent to 213 British Standard units.

[a] Median of daily mean values

[b] Median of daily mean values: October to March

[c] 98 percentile daily mean concentration - ie the level exceeded by the highest 2 per cent of daily mean values during the year.

---

industrial processes and the installation of abatement equipment on existing ones.

2.5 The growth of road traffic however, has led to large increases in emissions from motor fuel combustion, despite continuing progress in reducing emission levels from individual vehicles. Some new processes and products, too, have given rise to problems not previously encountered, such as increasing emissions of volatile organic compounds, the main precursors, with oxides of nitrogen, of ground level ozone.

Factors affecting concentrations and deposition

2.6 Atmospheric processes exert an important influence on the behaviour of pollutants and on their dispersion. Wind disperses pollutants from their sources and can transport them over long distances. Calm weather conditions and temperature inversions contribute to the localised build up of pollutants at ground level particularly in wintertime. Concentrations of pollutants are also subject to seasonal and year-on-year variability due to climatic conditions, eg sulphur dioxide and nitrogen dioxide concentrations tend to be higher in years with cold winters because more heating fuel is burnt, and long periods of sunshine and high temperatures generate more secondary pollutants such as ozone. Tall chimneys help to reduce ground-level concentrations of pollutants close to industrial plants and fossil fuel burning power stations.

The pollutants
Sulphur dioxide and black smoke

2.7 Sulphur dioxide (SO_2) is released by the combustion of sulphur-containing fuels such as coal, smokeless fuel and oil. Black smoke consists of fine particles (mainly carbon) from incomplete combustion of fossil fuels. SO_2 and smoke can cause temporary breathing problems for some people, and for sensitive groups of people long term exposure to smoke can bring general respiratory difficulties. SO_2 is also one of the principal contributors to acid rain; and oily smoke from diesel vehicles soils buildings and makes the urban environment grimy.

2.8 Concentrations of SO_2 are measured at sites in separate urban and rural monitoring networks. Smoke concentrations are measured

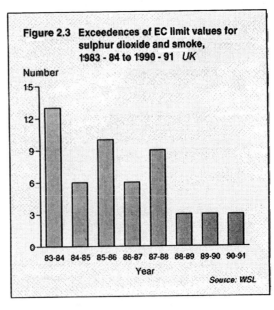

Figure 2.3 Exceedences of EC limit values for sulphur dioxide and smoke, 1983 - 84 to 1990 - 91 *UK*

Number

Year

Source: WSL

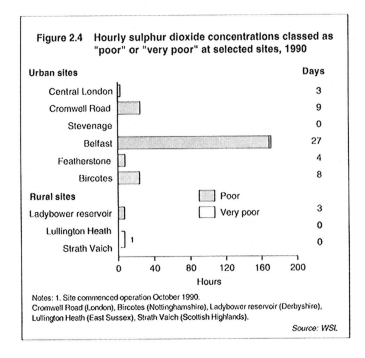

Figure 2.4   Hourly sulphur dioxide concentrations classed as "poor" or "very poor" at selected sites, 1990

Urban sites	Days
Central London	3
Cromwell Road	9
Stevenage	0
Belfast	27
Featherstone	4
Bircotes	8
Rural sites	
Ladybower reservoir	3
Lullington Heath	0
Strath Vaich	0

Legend: Poor, Very poor

Notes: 1. Site commenced operation October 1990.
Cromwell Road (London), Bircotes (Nottinghamshire), Ladybower reservoir (Derbyshire), Lullington Heath (East Sussex), Strath Vaich (Scottish Highlands).

Source: WSL

at urban sites only. The right hand map at Figure 2.1 shows average SO_2 concentrations during 1990-91 at urban and suburban locations in the UK. There are fewer rural measurement sites so information on the distribution of SO_2 in rural areas is less precise. However, the left hand map, which is based on measurements made in 1990, gives an idea of rural concentrations.

2.9 Air quality standards for ground level atmospheric concentrations of SO_2 and smoke are set in EC Directive 80/779 which came into force in 1983. The standards (or limit values) are based on WHO criteria for safe levels of the pollutants. The limit values are expressed in micrograms per cubic metre ($\mu g/m^3$) over a range of averaging periods (see Box 2.3 for mean daily limit values). Derogations were agreed for 29 local authority areas in the UK which were allowed up to a further 10 years (ie up to 1 April 1993) to reduce ground level concentrations of sulphur dioxide and smoke to comply with the Directive's limit values. In 1989 the number of derogation areas in the UK was reduced to 22 because seven areas were no longer breaching the limit values. Trends in annual concentrations of both SO_2 and smoke at four sites in the UK and a comparison with the EC limit values are given in Figure 2.2. Figure 2.3 shows where UK monitoring sites exceeded the limit values between 1983 and 1991. This happened in seven local authority areas in the UK in 1983-84, but in Belfast alone in 1990-91. Belfast is a derogation area and has until April 1993 to comply. Daily information on SO_2 concentrations is made available through the Air Quality Bulletin system referred to in Box 2.2. Figure 2.4 shows for selected bulletin sites the number of days when hourly SO_2 concentrations were classed as "poor" or "very poor".

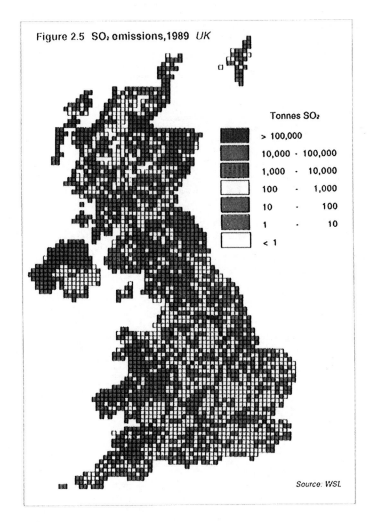

Figure 2.5 SO_2 emissions, 1989 *UK*

Tonnes SO_2

> 100,000	
10,000 - 100,000	
1,000 - 10,000	
100 - 1,000	
10 - 100	
1 - 10	
< 1	

Source: WSL

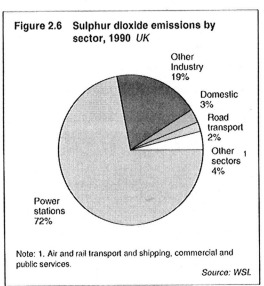

Figure 2.6   Sulphur dioxide emissions by sector, 1990 *UK*

Other Industry 19%
Domestic 3%
Road transport 2%
Other sectors 4% 1
Power stations 72%

Note: 1. Air and rail transport and shipping, commercial and public services.

Source: WSL

2.10 Figure 2.5 shows estimated SO_2 emissions across the UK for 10 km grid squares. This shows the highest emissions around power stations and major industrial areas. The estimated relative contributions of the main emission producing sectors are shown in Figure 2.6. SO_2 emission estimates are probably accurate to within 10 to 15 per cent.

2.11 Since the 1960s SO_2 emissions have declined to under 4 million tonnes in 1983 and subsequent years (see Figure 2.7). The increasing use of sulphur-free fuels such as natural gas, lower industrial energy demand and energy conservation have contributed to the decline, particularly in urban areas. Emissions by industry other than power stations have declined from 27 per cent of the total in 1980 to 19 per cent in 1990. The UK is committed (under the EC Large Combustion Plants Directive) to reducing total emissions of SO_2 from existing large combustion installations, which includes power stations, by 20 per cent by 1993, by 40 per cent by 1998 and by 60 per cent by 2003, taking 1980 as the baseline.

2.12 Black smoke emissions are produced mainly from road transport and from homes. Road transport accounted for almost 46 per cent of total emissions in 1990, with buses and heavy goods vehicles contributing most. Homes accounted for 33 per cent, mostly from coal burning. The relative contributions of each sector and fuel type to total emissions are shown in Figure 2.8.

2.13 The trends in total smoke emissions and by sector from 1970 to 1990 are shown in Figure 2.9. The domestic sector has contributed most to the reduction, mainly due to a decline in the use of coal. The contribution to emissions from coal combustion fell from 59 per cent to 37 per cent between 1980 and 1990. However, emissions from diesel-fuelled vehicles have increased steadily. The contribution to smoke emissions from diesel fuel rose from 19 per cent in 1980 to 42 per cent in 1990. Currently the level of total emissions is about the same as it was in 1984.

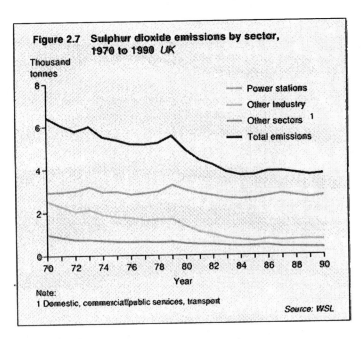

Figure 2.7 Sulphur dioxide emissions by sector, 1970 to 1990 UK

Thousand tonnes

Power stations
Other Industry
Other sectors [1]
Total emissions

Year

Note:
1 Domestic, commercial/public services, transport

Source: WSL

soon after the enabling legislation but orders are still being made. The number of smoke control orders and premises affected rose by nearly 14 per cent and 18 per cent respectively between 1980 and 1990. The area covered increased by nearly 27 per cent.

Smoke from crop residue burning

2.15 Stubble burning which can create a major public nuisance, will be banned in England and

Table 2.1 Smoke Control Orders, 1980 to 1990, UK			
	1980	1985	1990
Total number of orders	5,581	6,025	6,342
Total area covered (thousand hectares)	759	869	962
Total number of premises[1] affected (thousands)	8,080	8,899	9,519

Note: 1. Domestic and industrial premises.

Source: DOE, WO, SOEnD, DOE(NI)

2.14 Although there is evidence of earlier reductions in the use of coal arising from underlying structural changes in the economy, the Clean Air legislation of the 1950s and 1960s reinforced this trend by forcing people away from using coal to smokeless fuel and continues to do so. Table 2.1 shows the number of smoke control orders in operation, and the area and number of premises covered. The majority of smoke control orders were made

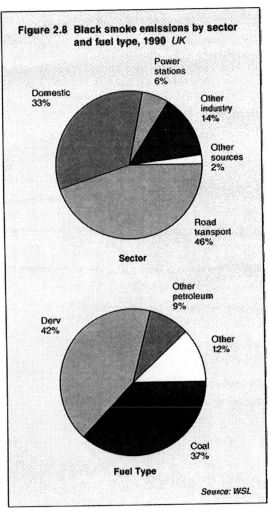

Figure 2.8 Black smoke emissions by sector and fuel type, 1990 UK

Power stations 6%
Domestic 33%
Other industry 14%
Other sources 2%
Road transport 46%

Sector

Other petroleum 9%
Derv 42%
Other 1.2%
Coal 37%

Fuel Type

Source: WSL

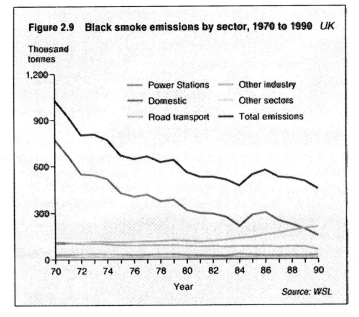

Figure 2.9 Black smoke emissions by sector, 1970 to 1990 *UK*

Thousand tonnes

— Power Stations ⋯ Other industry
— Domestic ⋯ Other sectors
— Road transport — Total emissions

Source: WSL

undertakes a survey of local authorities on the problems and complaints caused by stubble burning. The number of complaints recorded has varied considerably; 234 in 1988, 2,487 in 1989, 1,124 in 1990 and 1,153 in 1991.

Nitrogen oxides

2.16 Nitric oxide (NO) and nitrogen dioxide (NO_2) (jointly termed NO_x) are gases formed in combustion processes both from the nitrogen present in fuels and from the oxidation of nitrogen in air. The effects of oxides of nitrogen are varied. High concentrations can reduce plant growth and cause visible damage to sensitive crops, add to acid deposition, and play a part in the formation of ground level ozone (see section on ground level ozone later in this chapter). Acute exposure to high concentrations of nitrogen dioxide can cause breathing problems. Figure 2.10 shows mean nitrogen dioxide concentrations in rural and urban areas. Estimates are less accurate than those for SO_2 but are probably accurate to plus or minus 30 per cent.

2.17 There are EC and WHO air quality standards and guidelines for ground level concentrations of nitrogen dioxide (see Box 2.4). Compliance with the EC Directive for nitrogen dioxide is monitored continuously at seven sites in urban and industrial locations

Wales after the 1992 harvest. New regulations were introduced in 1991 to control the conditions under which burning may be carried out in the interim. Between 1983 and 1991 the proportion of cereal straw burnt fell from 38 per cent to 15 per cent. The amount of straw baled and removed during this period has remained constant while the straw ploughed in or cultivated has risen from 2 per cent to 27 per cent. The National Society for Clean Air each year

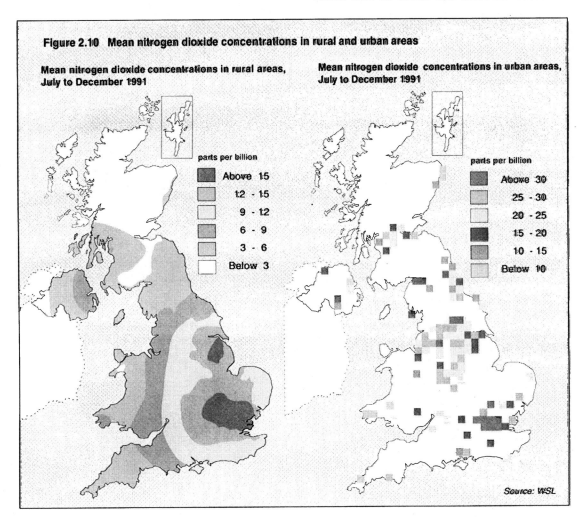

Figure 2.10 Mean nitrogen dioxide concentrations in rural and urban areas

Mean nitrogen dioxide concentrations in rural areas, July to December 1991

parts per billion

Above 15
12 - 15
9 - 12
6 - 9
3 - 6
Below 3

Mean nitrogen dioxide concentrations in urban areas, July to December 1991

parts per billion

Above 30
25 - 30
20 - 25
15 - 20
10 - 15
Below 10

Source: WSL

EC and WHO standards and guidelines for levels of nitrogen dioxide concentrations — **Box 2.4**

EC Directive 85/203 set the limit value for ambient nitrogen dioxide at 105 ppb hourly mean concentration as a 98 percentile. The Directive also gives guide values of 70 ppb on the same basis or 26 ppb on a 50 per cent basis.

WHO air quality guidelines (for Europe) relate to short term nitrogen dioxide exposure. These are expressed in terms of two time periods for which maximum values have been set ie

0.21 ppm	1 hour	- hourly guideline
0.08 ppm	24 hours	- daily guideline.

where the risk of population exposure to high concentrations is expected to be greatest. It is also monitored continuously at a kerbside and a suburban site and a number of rural locations.

2.18 Extensive surveys in 1986 and in 1991 showed that average UK concentrations of nitrogen dioxide increased by 35 per cent over the period, mainly as a result of increased emissions from traffic. High concentrations usually occur in winter, particularly in calm cold weather when pollutants are trapped close to the ground by temperature inversions, and the highest ever in the UK were recorded in London in December 1991. Figure 2.11 shows concentrations of NO_2 at selected sites from 1976 to 1990 compared with the EC limit value. Concentrations in London are higher than in other UK towns and cities because there is more road traffic there than anywhere else.

2.19 The hourly and daily mean concentration WHO guidelines for nitrogen dioxide are most often exceeded at busy kerbside and roadside locations. Away from the kerbside, the WHO 1-hour guideline has been exceeded in London, Manchester and Glasgow in some years. Figure 2.12 shows for selected sites the number of hours in 1990 when NO_2 concentrations were classed as "poor". No recordings were classed as "very poor" at these sites in 1990. Information on nitrogen dioxide concentrations is made available through the Air Quality Bulletin system referred to in Box 2.2.

2.20 Figure 2.13, showing the distribution of NO_x emissions, highlights the relationship between emission levels and major conurbations and power stations. Total NO_x emissions were 2.7 million tonnes in 1990, over three quarters of which came from coal, petrol and diesel fuel. Figure 2.14 shows the contributions to emissions by fuel type and by sector in 1990.

2.21 Figure 2.15 shows that over the last two decades total estimated NO_x emissions were around 2.2 to 2.4 million tonnes a year until 1986 but then rose steadily to the 1990 level of 2.7 million tonnes. Over this period emissions from road transport rose by nearly 800 thousand tonnes, from 27 per cent of all emissions in 1970 to 51 per cent in 1990. Power stations' contributions fell from 37 per cent in 1970 to 28 per cent in 1990.

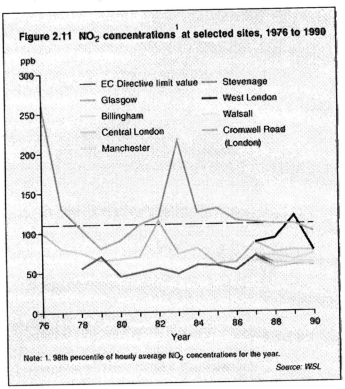

Figure 2.11 NO_2 concentrations[1] at selected sites, 1976 to 1990

Note: 1. 98th percentile of hourly average NO_2 concentrations for the year.

Source: WSL

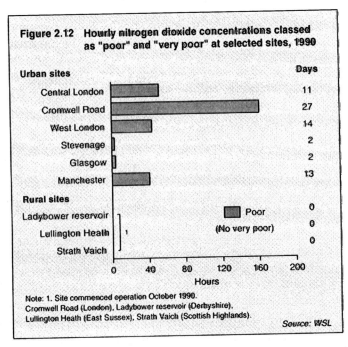

Figure 2.12 Hourly nitrogen dioxide concentrations classed as "poor" and "very poor" at selected sites, 1990

Urban sites	Days
Central London	11
Cromwell Road	27
West London	14
Stevenage	2
Glasgow	2
Manchester	13

Rural sites	
Ladybower reservoir	0
Lullington Heath	0
Strath Vaich	0

Poor (No very poor)

Note: 1. Site commenced operation October 1990.
Cromwell Road (London), Ladybower reservoir (Derbyshire),
Lullington Heath (East Sussex), Strath Vaich (Scottish Highlands).

Source: WSL

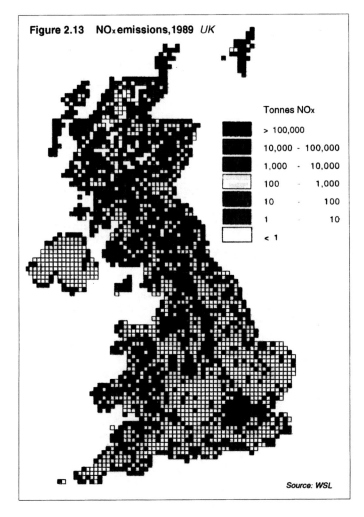

Figure 2.13 NOx emissions, 1989 *UK*

Tonnes NOx

> 100,000

10,000 - 100,000

1,000 - 10,000

100 - 1,000

10 - 100

1 - 10

< 1

Source: WSL

at long distances from the sources of the pollution. Intensive research has greatly increased the understanding of its effects, in recent years. It is now known that through acidification of soils in geologically sensitive areas, acid rain can inhibit plant nutrition and restrict the range of flora and fauna. Freshwaters in such areas can be made toxic to aquatic plants, invertebrates and fish. Through the effects on the food chain, this can lead to losses of other fauna such as natterjack toads, dippers and otters. Acid rain also damages building materials such as stone, concrete and metals. See also references to acid deposition in Chapter 4 on soil, Chapter 7 on inland water quality and Chapter 10 on wildlife.

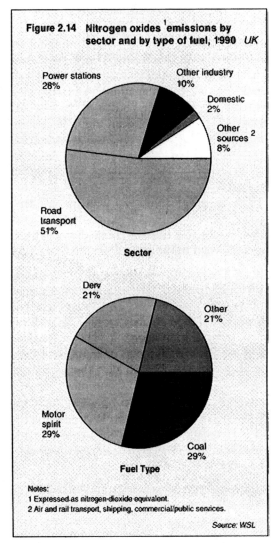

Figure 2.14 Nitrogen oxides [1] emissions by sector and by type of fuel, 1990 *UK*

Power stations 28%

Other industry 10%

Domestic 2%

Other sources [2] 8%

Road transport 51%

Sector

Derv 21%

Other 21%

Motor spirit 29%

Coal 29%

Fuel Type

Notes:
1 Expressed as nitrogen dioxide equivalent.
2 Air and rail transport, shipping, commercial/public services.

Source: WSL

2.22 The UK is committed (under the EC Large Combustion Plants Directive) to reducing total emissions of nitrogen oxides from existing combustion installations by 15 per cent by 1993 and by 30 per cent by 1998 taking 1980 emissions as the baseline. To meet the Directive targets, low NO_x burners will be fitted in all 12 of the major coal-fired power stations in England and Wales (73 per cent of UK coal fired capacity). EC Directive 91/441/EEC also sets specific standards for car emissions for all new passenger vehicles. In effect, this makes catalytic converters compulsory for all new cars coming onto the road from 1 January 1992. The converters should reduce harmful emissions, including nitrogen oxides, by around 80 per cent.

Acid deposition

2.23 "Acid rain" is the term used to refer to the various processes by which acidic gases and particles are deposited on land and water, often

2.24 Sulphur dioxide (SO_2), nitrogen oxides (NO_x) and hydrogen chloride (HCl), are the main

Critical loads

Box 2.5

A critical load is a quantitative estimate of exposure to one or more pollutants below which significant harmful effects on specified sensitive elements of the environment do not occur, according to present knowledge. The critical load approach has been developed to show up those areas in which damage is being caused by deposition of acid pollution and to allow more effective abatement strategies to be developed. Critical loads maps of acidity in soils and freshwaters have been produced and Figure 2.16 shows areas in the UK where critical loads for acidity of soils and freshwater are exceeded [1,3,5].

Review Group Reports
Box 2.6

A number of expert Review Groups have been established and sponsored by the Department of the Environment to examine various aspects of acid deposition. A summary of the more recent reports of these Groups is given below.

The UK Review Group on Acid Rain (UKRGAR) published its third report [2] in 1990 which summarised the understanding of acid deposition and the available network measurements over the UK for the years 1986 to 1988.

The UK Critical Loads Advisory Group (UKCLAG) published two reports [3,4] in 1991 giving critical load maps for soils in the UK and for standing freshwaters in Scotland respectively. A third report [5] was published in March 1992 giving provisional critical load maps for British freshwaters superseding that for freshwaters in Scotland published in 1991. The most vulnerable soils and freshwaters are to be found in upland areas of north and west Britain.

The UK Acid Waters Review Group (UKAWRG) published its second report [6] in 1988 which concluded that parts of Scotland, Wales and England, especially northern England, possess moderately severe acid waters leading, in some cases, to depletion of fish stocks. Acid deposition appears to be the major causal factor. In some regions of the UK, afforestation has exacerbated freshwater acidity.

The UK Photochemical Oxidants Review Group (UKPORG) published its second report [7] in 1990 and summarised the understanding of the oxides of nitrogen, together with their role in acid rain and ozone formation.

The UK Terrestrial Effects Review Group (UKTERG) acknowledged in its first report [8], published in 1988, that acid deposition was accelerating acidification of some soils and changes in soil biology might result ultimately. Such changes were likely to alter plant nutrition and change the chemistry and biology of freshwaters. Major agricultural crops were unlikely to be damaged directly by current concentrations of sulphur dioxide and nitrogen oxides but few data existed on the impact of air pollution on natural vegetation other than lichens. There was no direct proof of pollution-related forest decline in the UK, although some forests were subjected to pollution climates which might be expected to cause stress. The Forestry Commission have also reported on forest health surveys - see Chapter 5 on land use and cover.

The UK Buildings Effects Review Group (UKBERG) published its first report [9] in 1989, and concluded that:

- current rates of weathering of stone on historic buildings are higher than natural rates and weathering on all buildings was greater in urban areas;
- there is no unequivocal evidence that current rates of weathering of stone and most metals in the structure of historic buildings are significantly different from those in the recent past despite significant decreases in acidic emissions over the past decades;
- natural stone buildings face a greater risk than modern buildings.

contributors to acid rain. Acidic pollutants reach the earth's surface in dry weather by absorption from the atmosphere by plants, trees, soils and by water surfaces ("dry deposition"). In wet weather, they may fall to the ground in rain drops or snow ("wet deposition"). In fog and mist droplets may be captured by plants and trees ("occult deposition").

2.25 Acidic pollutants travel great distances from their sources. International agreements on the control of acidic emissions and depositions are being developed through the UNECE Convention on Long Range Transboundary Air Pollution, signed in 1979. This work is paying increasing attention to the concept of critical loads of pollutants. This entails looking at sensitive elements of the environment to see how small the load has to be to avoid significant harm (see Box 2.5 on critical loads). The Department of the Environment has formed a

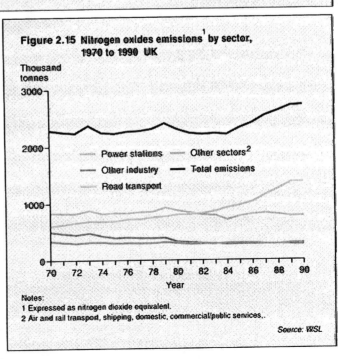

Figure 2.15 Nitrogen oxides emissions[1] by sector, 1970 to 1990 UK

Thousand tonnes

Legend:
- Power stations
- Other industry
- Road transport
- Other sectors[2]
- Total emissions

Year

Notes:
1 Expressed as nitrogen dioxide equivalent.
2 Air and rail transport, shipping, domestic, commercial/public services.

Source: WSL

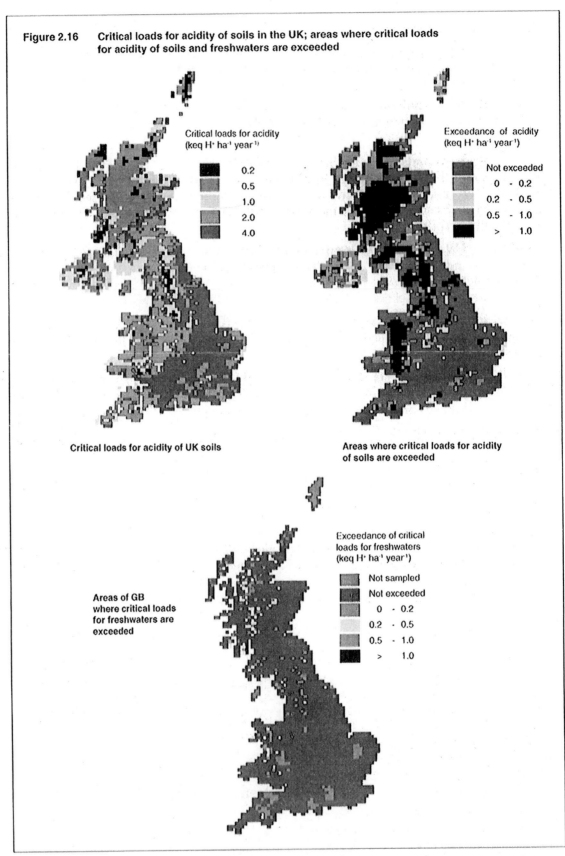

Figure 2.16 Critical loads for acidity of soils in the UK; areas where critical loads for acidity of soils and freshwaters are exceeded

Critical loads for acidity
(keq H⁺ ha⁻¹ year⁻¹)

0.2
0.5
1.0
2.0
4.0

Critical loads for acidity of UK soils

Exceedance of acidity
(keq H⁺ ha⁻¹ year⁻¹)

Not exceeded
0 - 0.2
0.2 - 0.5
0.5 - 1.0
> 1.0

Areas where critical loads for acidity of soils are exceeded

Exceedance of critical loads for freshwaters
(keq H⁺ ha⁻¹ year⁻¹)

Not sampled
Not exceeded
0 - 0.2
0.2 - 0.5
0.5 - 1.0
> 1.0

Areas of GB where critical loads for freshwaters are exceeded

UK Critical Loads Advisory Group (UKCLAG) to prepare maps of critical loads for freshwaters, soils, and plants (see Box 2.6 on Review Group reports) and critical levels for vegetation (trees, crops and natural ecosystems) and buildings and materials.

2.26 The UK has ratified a protocol on nitrogen oxides under the UNECE Convention on Long Range Transboundary Air Pollution. This requires national emissions of nitrogen oxides to be brought back to 1987 levels by 1994.

Ground level ozone

2.27 Ozone occurs naturally throughout the atmosphere. Maximum concentrations are found in the stratosphere (15 to 50 km from the surface of the earth) where ozone acts as a shield to harmful ultraviolet radiation (see also the section on the depletion of the ozone layer in Chapter 3, and Chapter 14 on environment and health). In the troposphere, extending from ground level to about 15 km, ozone occurs naturally in small quantities but concentrations can be increased by chemical reactions involving other pollutants. There are no significant direct emissions of ozone into the atmosphere so it is referred to as a secondary pollutant. Ozone is formed by a complex series of reactions between nitrogen oxides and volatile organic compounds (VOCs), in the presence of sunlight. The levels encountered in the UK can have small effects on people's lungs and irritate eyes. Ozone can also reduce the yield of some sensitive crops and damage natural vegetation and organic materials and it contributes to the damage caused by acid rain by promoting the oxidation of air pollutants to acidic species.

2.28 There is evidence that the pre-industrial near ground-level concentrations of ozone were typically 10 to 15 ppb. Since industrialisation, increased atmospheric pollution as a result of human activity appears to have doubled the background concentrations over the past 100 years so that current annual mean concentrations are approximately 30 ppb over the UK.

2.29 Concentrations can rise substantially above background levels in summer heat waves when there are continuous periods of bright sunlight with temperatures above 20 degrees Celsius, and light winds. Once formed, ozone can persist for several days and can be transported long distances.

2.30 Figure 2.17 shows the distribution over the UK of episodes that have exceeded 100 ppb hourly mean concentrations in the summer of 1990. The contours indicate the number of hours of high ozone episodes and show a marked gradient increasing in a north/ south direction. From this distribution it is clear that pollution transported with continental air masses plays a significant role in UK ozone episodes. The number of hours exceeding specified concentration levels can, however, vary substantially from year to year. Figure 2.18 shows for selected sites the periods during 1990 when hourly ozone concentrations caused the air quality to be classed as "poor". No hourly concentrations were recorded as "very poor" at these sites in 1990 (see Box 2.2 on monitoring).

2.31 The EC Directive on ozone adopted in May 1992 commits member states to ozone monitoring, information exchange and providing information and guidance to the general public.

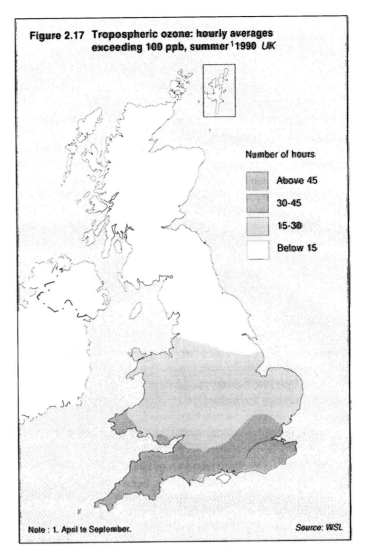

Figure 2.17 Tropospheric ozone: hourly averages exceeding 100 ppb, summer [1] 1990 UK

Number of hours

Above 45

30-45

15-30

Below 15

Note: 1. April to September. Source: WSL

Volatile organic compounds (VOCs)

2.32 Volatile organic compounds (VOCs) include a large number of chemical compounds which are able to evaporate into a gas and take part in chemical reactions. Some VOCs can cause drowsiness, eye irritation and coughing,

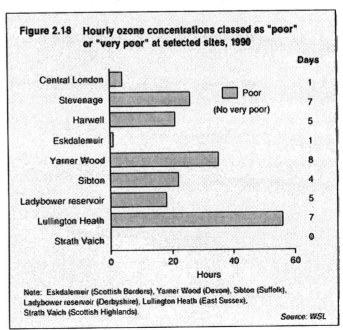

Figure 2.18 Hourly ozone concentrations classed as "poor" or "very poor" at selected sites, 1990

Poor (No very poor)

Site	Hours	Days
Central London		1
Stevenage		7
Harwell		5
Eskdalemuir		1
Yarner Wood		8
Sibton		4
Ladybower reservoir		5
Lullington Heath		7
Strath Vaich		0

Hours

Note: Eskdalemuir (Scottish Borders), Yarner Wood (Devon), Sibton (Suffolk), Ladybower reservoir (Derbyshire), Lullington Heath (East Sussex), Strath Vaich (Scottish Highlands).

Source: WSL

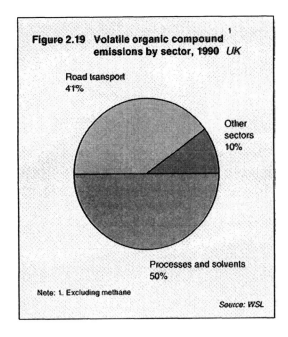

Figure 2.19 Volatile organic compound[1] emissions by sector, 1990 UK

Road transport 41%

Other sectors 10%

Processes and solvents 50%

Note: 1. Excluding methane

Source: WSL

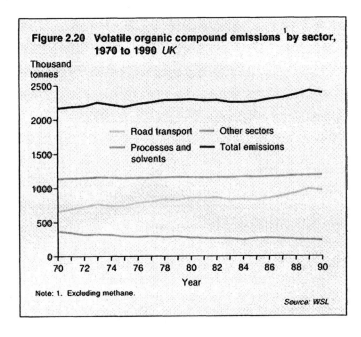

Figure 2.20 Volatile organic compound[1] emissions by sector, 1970 to 1990 UK

Thousand tonnes

Road transport — Other sectors
Processes and solvents — Total emissions

Year

Note: 1. Excluding methane.

Source: WSL

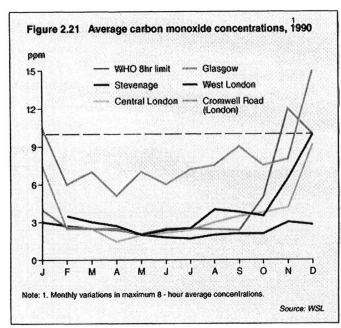

Figure 2.21 Average carbon monoxide concentrations, 1990[1]

ppm

WHO 8hr limit — Glasgow
Stevenage — West London
Central London — Cromwell Road (London)

Note: 1. Monthly variations in maximum 8 - hour average concentrations.

Source: WSL

while others such as benzene are carcinogenic. All, to varying degrees, are capable of producing photochemical oxidants including ground-level ozone, by reactions with nitrogen oxides in the presence of sunlight. Some also deplete the ozone layer.

2.33 Sources of VOC emissions are many and varied. They include a great range of industrial processes, particularly those using solvent-containing coatings, adhesives and cleaning agents (around 50 per cent of total emissions). Evaporation of petrol and emissions in vehicle exhausts are other major sources: emissions from road transport accounted for 41 per cent of the total in 1990 and have increased by around 12 per cent since 1980. Estimates of aggregate UK VOC emissions are however highly uncertain, with an estimated accuracy of only plus or minus 30 per cent. Figure 2.19 shows VOC emissions by sector in 1990 and Figure 2.20 shows the trends in total emissions and by source over the last two decades.

2.34 The UK signed in November 1991 a protocol on VOCs under the UNECE Convention on Long Range Transboundary Air Pollution. This will require national VOC emissions to be reduced by 30 per cent from 1988 to 1999. Specific control measures already taken or in hand include a 5 per cent by volume limit on the benzene content of petrol (EC Directive 85/210/EEC); the introduction of emission standards for new cars (EC Directive 91/441/EEC), which will reduce emissions of VOCs and other emissions from new cars from the end of 1992 by about 80 per cent; and the progressive implementation of air pollution controls over a wide range of processes under Part I of the Environmental Protection Act 1990.

Carbon monoxide (CO)

2.35 Carbon monoxide (CO) is derived from the incomplete combustion of petrol. It is toxic at high concentrations and affects physical co-

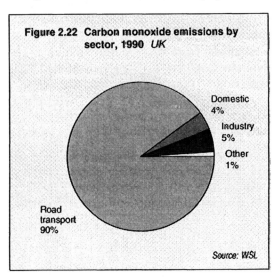

Figure 2.22 Carbon monoxide emissions by sector, 1990 UK

Domestic 4%

Industry 5%

Other 1%

Road transport 90%

Source: WSL

ordination, vision and judgement. CO can also contribute indirectly to global warming (see also Chapter 3 on the global atmosphere).

2.36 The WHO air quality guideline recommends that concentrations of carbon monoxide should not exceed 10 ppm over an eight-hour period, and 25 ppm over a one hour period. Figure 2.21 shows that for the five monitoring sites operating in 1990, the WHO eight-hour guideline was exceeded in November 1990 in Glasgow and in January and December 1990 at Cromwell Road, London. The WHO one-hour guideline has not been exceeded in recent years.

2.37 UK estimates of CO emissions are currently accurate to plus or minus 30 per cent. Estimates are based on emission factors derived from relatively few measurements and they vary widely with combustion conditions and driving patterns. Figure 2.22 shows emissions by sector, and trends over the last two decades are shown in Figure 2.23. Total estimated emissions in 1990 were 6.7 million tonnes of which 90 per cent were caused by road transport. This was more than 2 million tonnes higher than in 1970. The EC vehicle Directives will reduce future emissions of carbon monoxide from road transport.

Lead

2.38 Lead is a trace element which can get into the atmosphere from leaded petrol, coal burning and from metal works. Concentrations can accumulate in the environment and in the body and can affect health, particularly children's.

2.39 A monitoring network of eleven sampling sites around three works measures compliance with EC Directive 82/884/EEC which established limit values for airborne lead concentrations of 2.0 µg per cubic metre of air expressed as an annual average. Since 1985, only one site (in Walsall) has exceeded the limit value. This area had a derogation from compliance with the Directive until the end of 1989. The annual mean for 1990, the first year of compliance, was 1.3 µg/m³. Surveys have also been carried out on the effect on blood lead concentrations of the reduction in the lead content of petrol and these are discussed in Chapter 14 on environment and health.

2.40 Limits on the lead content of petrol (EC Directives 78/611/EEC and 85/210/EEC) have progressively reduced lead concentrations. Unleaded petrol has become widely available and successive increases in the duty differential between leaded and unleaded petrol to over 5p per litre have encouraged sales. Virtually all petrol stations in the UK now sell unleaded petrol compared with only around 10 per cent as recently as late 1988. Deliveries of unleaded

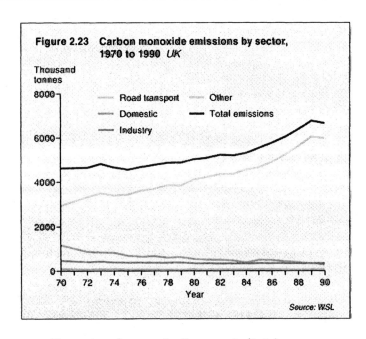

Figure 2.23 Carbon monoxide emissions by sector, 1970 to 1990 UK

Thousand tonnes

Road transport   Other
Domestic   Total emissions
Industry

Source: WSL

petrol have risen from under 3 per cent of total petrol deliveries to petrol stations in November 1988 to 46 per cent in April 1992.

2.41 Figure 2.24 shows average lead concentrations at selected sites between 1980

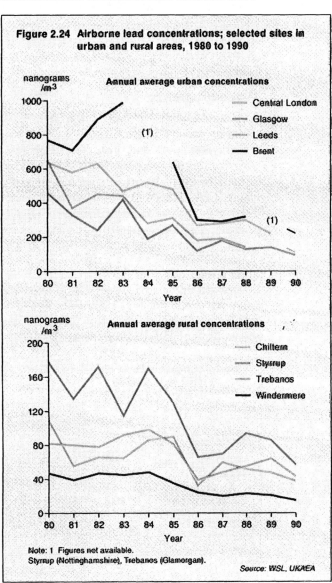

Figure 2.24 Airborne lead concentrations; selected sites in urban and rural areas, 1980 to 1990

nanograms /m³   Annual average urban concentrations

Central London
Glasgow
Leeds
Brent

nanograms /m³   Annual average rural concentrations

Chiltern
Styrrup
Trebanos
Windermere

Note: 1 Figures not available.
Styrrup (Nottinghamshire), Trebanos (Glamorgan).

Source: WSL, UKAEA

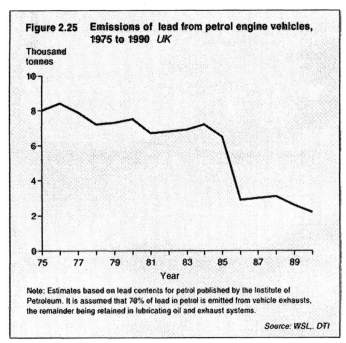

Figure 2.25   Emissions of lead from petrol engine vehicles, 1975 to 1990 *UK*

Note: Estimates based on lead contents for petrol published by the Institute of Petroleum. It is assumed that 70% of lead in petrol is emitted from vehicle exhausts, the remainder being retained in lubricating oil and exhaust systems.

Source: WSL, DTI

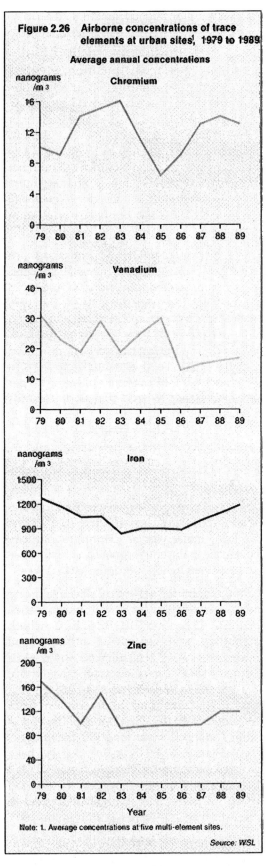

Figure 2.26   Airborne concentrations of trace elements at urban sites[1], 1979 to 1989

Average annual concentrations

Chromium

Vanadium

Iron

Zinc

Note: 1. Average concentrations at five multi-element sites.

Source: WSL

and 1990. Monitoring shows that concentrations halved at urban sites between 1985 and 1986, and there has been a further reduction since then as unleaded petrol became more widely available and used. The downward trend in emissions from petrol engine vehicles is shown in Figure 2.25. However, the increased use of unleaded petrol is to some extent offset by increases in volumes of traffic.

Trace elements

2.42 In addition to lead, other elements normally present in trace quantities, mainly arsenic, selenium, zinc, iron, vanadium and chromium are also monitored and recent trends in concentrations at urban sites are given in Figure 2.26. Trends in the concentrations of the principal elements are generally downwards as coal is used less in homes and factories.

Bad smells

2.43 There were over 13,500 complaints to Environmental Health Officers (EHOs) from the public about bad smells in 1989-90 in England and Wales. These are shown in Table 2.2 by source, for the last three years. Most complaints

Table 2.2 Complaints about odours, by source, 1987-88 to 1989-90, England and Wales			
	1987-88	1988-89	1989-90
Industrial processes			
No. of complaints[1]	9714	9649	9833
Complaints per million population	259	264	292
Agricultural practices			
No. of complaints[1]	3756	3973	3705
Complaints per million population	99	108	109

Note: 1 From reporting local authorities only

Source: IEHO

(about three quarters) concern smells from industrial processes; there were around 10,000 such complaints in 1989-90 relating to around 2,000 premises. A similar number of complaints were recorded in the previous two years.

2.44 Complaints about farm smells totalled 3,700 in 1989-90, about a quarter of the total number recorded. EHOs classed as "justifiable" around 40 per cent of agricultural complaints. Figure 2.27 shows the relative proportions of justifiable complaints about agricultural practices by source. Almost half of the justifiable complaints related to spreading slurry on land and a further 18 per cent related to slurry storage. Nearly 30 per cent related to smells from agricultural buildings. In July 1992, the Agriculture Departments of England and Wales issued a Code of Good Agricultural Practice for the Protection of Air giving practical advice to farmers on, amongst other things, how to minimise farm smells and emissions of ammonia. Similar advice has been issued in Scotland.

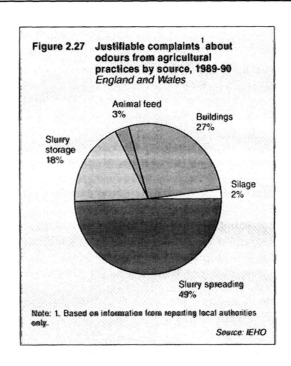

Figure 2.27 Justifiable complaints[1] about odours from agricultural practices by source, 1989-90 England and Wales

Animal feed 3%
Buildings 27%
Slurry storage 18%
Silage 2%
Slurry spreading 49%

Note: 1. Based on information from reporting local authorities only.

Source: IEHO

References and further reading

1. Department of the Environment, (1991). Acid Rain; Critical and Target Loads Maps for the United Kingdom. DOE.

2. Department of the Environment, (1990). Acid Deposition in the United Kingdom 1986-1988. Third Report of the UK Review Group on Acid Rain. Department of the Environment and Department of Transport Publication Sales Unit.

3. Department of the Environment, (1991). Critical Load Maps for the United Kingdom. I Soils. A Report to the Department of the Environment by the UK Critical Loads Advisory Group. DOE.

4. Department of the Environment, (1991). Critical Load Maps for the United Kingdom. II Freshwaters. A Report to the Department of the Environment by the UK Critical Loads Advisory Group. DOE.

5. Department of the Environment, (1992). Critical and Target Loads Maps for Freshwaters in Great Britain. A Third Report to the Department of the Environment by the UK Critical Loads Advisory Group. DOE.

6. Department of the Environment, (1988). Acidity in United Kingdom Fresh Waters. Second report of the UK Acid Waters Review Group. HMSO.

7. Harwell, (1990). Oxides of Nitrogen in the United Kingdom. Second Report of the UK Photochemical Oxidants Review Group. Harwell Laboratory.

8. Department of the Environment, (1988). The Effects of Acid Deposition on the Terrestrial Environment in the United Kingdom. UK Terrestrial Effects Review Group. HMSO.

9. Department of the Environment, (1989). The Effects of Acid Deposition on Buildings and Building Materials. UK Building Effects Review Group Report. HMSO.

Harwell, (1987). Ozone in the United Kingdom. First Report of the UK Photochemical Oxidants Review Group. Harwell Laboratory.

Department of Health, (1991). Ozone. First Report of the Advisory Group on Medical Aspects of Air Pollution Episodes. HMSO.

3 The global atmosphere

☐ The atmospheric concentrations of the major long-lived greenhouse gases (carbon dioxide, methane, chlorofluorocarbons and nitrous oxide) are increasing because of human activities (3.3). The Intergovernmental Panel on Climate Change estimates that without controls on greenhouse gas emissions, this will lead to global average temperature increasing at a rate between 0.2 and 0.5 degrees Celsius per decade over the next century with the possibility of significant regional climatic changes (3.26).

☐ Man-made emissions of carbon dioxide contribute most to the enhanced greenhouse effect, with methane, CFCs and nitrous oxide emissions having a further significant influence (Figure 3.3). Increases in tropospheric ozone as a result of emissions of nitrogen oxides and hydrocarbons are also thought to be important but are currently difficult to quantify. The contribution of CFCs to global warming may be offset by a global cooling effect due to the stratospheric ozone depletion with which they are associated (3.6, 3.21).

☐ UK carbon dioxide emissions from fossil fuel combustion have remained relatively constant over the last 10 years (Figure 3.7). Power stations account for 34 per cent of current emissions, other industrial sources 26 per cent and road transport 19 per cent (Figure 3.4). Emissions from power stations and other industrial sources have declined in relative importance over the last 10 years while the relative importance of road transport emissions has increased (3.12).

☐ UK carbon dioxide emissions are around 2.7 per cent of the global total from fossil fuel combustion. UK per capita emissions are currently about 2.8 tonnes of carbon a year, close to the European average, compared with around 5 tonnes for the USA and 0.2 tonnes for typical developing countries (3.11).

☐ There is now firm evidence that the increase in atmospheric concentrations of halocarbons (such as CFCs) cause the formation of the "ozone hole" over Antarctica each Austral spring. There is also increasing evidence linking these compounds with the more general depletion of the stratospheric ozone layer, which has reached 8 per cent in northern mid-latitudes during the spring (3.30).

☐ The current and prospective phase-out schedules for ozone depleting substances imply that atmospheric concentrations will peak in the stratosphere between 1995 and 2005 (Figure 3.17).

☐ UK consumption of CFCs fell by over 50 per cent between 1986 and 1989, mainly as a result of reduced use in aerosols (3.32).

3.1 Changes in the global atmosphere, resulting in climate warming and ozone depletion, are two of today's greatest environmental challenges. Since the ozone layer protects us from harmful solar radiation, significant depletion could have serious consequences for life on earth. It was first proposed in the mid-1970s that ozone in the stratosphere could be depleted by chemical reactions involving chloro-fluorocarbons (CFCs), which led to a US government ban on the use of CFCs in aerosols in 1978. It was not until 1985 however, that unequivocal evidence of ozone depletion came to light with the discovery of the "ozone hole" above the Antarctic by the British Antarctic Survey which had been monitoring the vertical ozone column since

---

The atmosphere

Box 3.1

The climate system

The atmosphere is a layer of gases extending up to 500 km above the earth (see Figure 3.1), with half of its mass concentrated in the lowest 6 km. It interacts strongly with the surface of the earth - oceans, icecaps, land, and vegetation - in the global climate system. Heat and water are carried by the atmosphere, redistributing them across the surface from areas that are hot or wet to areas that are cold or dry. These transfers help make the planet habitable.

Greenhouse gases and climate

Important components of the atmosphere are the so-called greenhouse gases. They are minor constituents of the atmosphere but have a profound effect. The atmosphere is largely transparent to radiation from the sun. Some solar radiation is reflected by clouds or the surface, but a large proportion heats the earth's surface. The warmed earth heats the atmosphere by convection, and radiates heat as long wave infra-red radiation which is absorbed by the greenhouse gases in the atmosphere, also warming the air. Some of the heat of the atmosphere is then re-radiated to the ground, making the earth's surface some 33 degrees Celsius (C) warmer than it would otherwise be. The gases are referred to as greenhouse gases because their behaviour is similar in effect to glass in a greenhouse; glass allows solar radiation in, which heats the interior, but reduces the outward emission of heat radiation (see Figure 3.2). Without the greenhouse effect the earth would be virtually uninhabitable. However, by increasing the levels of greenhouse gases in the atmosphere (for example by destroying forests and through the use of fossil fuels) man may be inadvertently affecting the climate.

The ozone layer

The ozone layer is found between approximately 15 and 50 km above the earth's surface and is a region of the atmosphere which has a relatively high concentration of ozone. This absorbs some solar radiation, and consequently is a relatively warm layer of the atmosphere, providing a stable "cap" to the more turbulent lower atmosphere (troposphere). The ozone layer protects the earth from harmful ultraviolet radiation which is emitted by the sun and any significant reduction in ozone concentration may have serious consequences for life on earth. Climate can also be affected by changes in the amount of stratospheric ozone, affecting stratospheric temperatures and wind patterns.

Ground level ozone

Although stratospheric ozone has a beneficial effect as a shield against ultraviolet radiation, it can have harmful effects when it appears in high concentrations at ground level where it can be regarded as a pollutant. The effects of "ground level ozone" are considered in more detail in Chapter 2 on air quality and pollution.

1957. Also in the mid 1980s the environmental implications for global warming, of the steady increase in concentrations of the so-called greenhouse gases in the atmosphere (see Box 3.1) became more widely known and a topic of major concern.

3.2 This chapter discusses greenhouse gases and ozone depletion and considers how global climate and the ozone layer have changed in the past and may change in the future.

The greenhouse gases

3.3 Since the industrial revolution man has been adding increasing amounts of greenhouse gases to the atmosphere mainly by the release of carbon dioxide, firstly through the use of fossil fuels for heating, transportation and the generation of electricity, and secondly through increasing destruction of forests. The effect of this is expected to be a warming of the global climate and changes in the patterns of atmospheric heat and water transport. Significant local climatic changes may result (see Box 3.3).

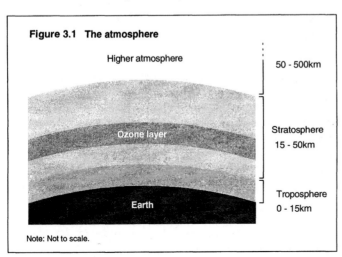

Figure 3.1 The atmosphere

Higher atmosphere — 50 - 500km

Ozone layer

Stratosphere 15 - 50km

Troposphere 0 - 15km

Earth

Note: Not to scale.

3.4 The greenhouse gases have a common physical property; they are strong absorbers of heat radiation (infra-red). Atmospheric concentrations of the most important greenhouse gas, water vapour, are not directly affected by emissions from human activities. An exception is the water vapour released into the high troposphere or low stratosphere by aircraft which may have an effect on climate by encouraging the formation of high, thin cirrus clouds. However, the amount of water vapour in the atmosphere is limited by temperature which means that any warming of the atmosphere as a result of increased concentrations of other greenhouse gases, will be amplified due to an associated increase in water vapour.

3.5 The main greenhouse gases emitted as a result of human activity are carbon dioxide (CO_2), nitrous oxide (N_2O), methane (CH_4) and CFCs. Tropospheric ozone is also a greenhouse gas. Its natural concentration is determined by incursions of stratospheric ozone into the troposphere and by natural chemical reaction cycles involving methane, carbon monoxide, nitrogen oxides and volatile organic compounds in the presence of sunlight. By adding to or altering the concentrations of these naturally occurring precursors, man can directly influence the concentration of tropospheric ozone. There is some evidence that background ozone levels in the troposphere have doubled during the last century [1].

3.6 Table 3.1 shows the properties of key greenhouse gases. The first two columns give the average atmospheric concentrations of greenhouse gases and how they are changing. The direct global warming potentials (GWP) of these gases (given in the third column) show their relative importance for global climate change due to radiative properties of the molecules themselves. The figures show the likely warming contribution of each gas relative to an equal weight of CO_2 over a period of 100 years. The fourth column shows the lifetime of each greenhouse gas (the period during which about two thirds of each greenhouse gas will be lost to the atmosphere after its emission). Some greenhouse gases also have indirect radiative effects through their interaction with atmospheric chemical processes, and the final column indicates the current estimate of whether the indirect effect of each greenhouse gas is positive or negative. Following the recent Supplementary Report in 1992 of the Intergovernmental Panel on Climate Change [2] (IPCC), earlier quantitative estimates of indirect effects have been withdrawn as

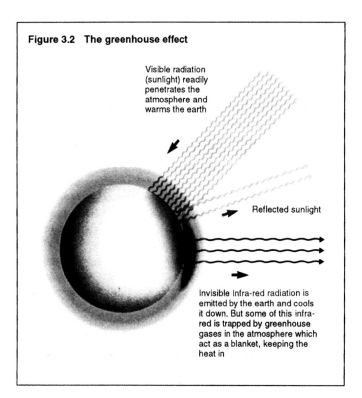

Figure 3.2 The greenhouse effect

Visible radiation (sunlight) readily penetrates the atmosphere and warms the earth

Reflected sunlight

Invisible Infra-red radiation is emitted by the earth and cools it down. But some of this infra-red is trapped by greenhouse gases in the atmosphere which act as a blanket, keeping the heat in

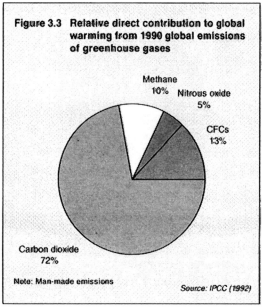

Figure 3.3 Relative direct contribution to global warming from 1990 global emissions of greenhouse gases

Methane 10%
Nitrous oxide 5%
CFCs 13%
Carbon dioxide 72%

Note: Man-made emissions

Source: IPCC (1992)

Table 3.1 Properties of key greenhouse gases

	Current (1994) average atmospheric concentration (ppmv)	Current rate of change % per annum	Direct Global Warming Potential (GWP)	Lifetime (years)	Type of indirect effect
Carbon dioxide	355	0.5	1	120	none
Methane	1.72	0.6 - 0.75	11	10.5	positive
Nitrous oxide	0.31	0.2 - 0.3	270	132	uncertain
CFC 11	0.000255	4	3,400	55	negative
CFC 12	0.000453	4	7,100	116	negative
CO[1]				months	positive
NMHC[1]				days/months	positive
NOx[1]				days	uncertain

Note: 1. Emissions affecting tropospheric ozone concentration.
Carbon monoxide (CO)
Non methane Hydrocarbon (NMHC)
Nitric oxide and nitrogen dioxide (NOx)

Source: IPCC

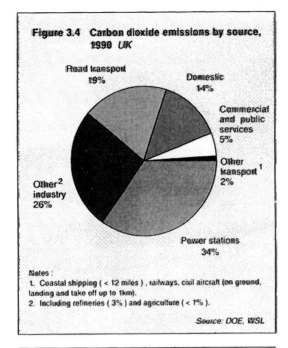

Figure 3.4   Carbon dioxide emissions by source, 1990 UK

Road transport 19%

Domestic 14%

Commercial and public services 5%

Other transport[1] 2%

Power stations 34%

Other[2] industry 26%

Notes :
1. Coastal shipping (< 12 miles), railways, civil aircraft (on ground, landing and take off up to 1km).
2. Including refineries (3%) and agriculture (< 1%).

Source: DOE, WSL

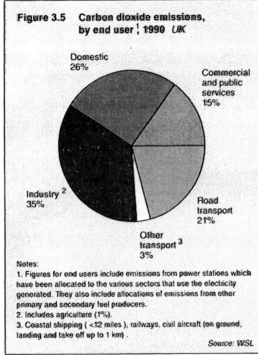

Figure 3.5   Carbon dioxide emissions, by end user[1] 1990 UK

Domestic 26%

Commercial and public services 15%

Road transport 21%

Other transport[3] 3%

Industry[2] 35%

Notes:
1. Figures for end users include emissions from power stations which have been allocated to the various sectors that use the electricity generated. They also include allocations of emissions from other primary and secondary fuel producers.
2. Includes agriculture (1%).
3. Coastal shipping (<12 miles), railways, civil aircraft (on ground, landing and take off up to 1 km) .

Source: WSL

additional factors of uncertainty have been discovered. In addition, it is now believed that stratospheric ozone depletion due to CFCs produces a global cooling effect that to some extent counters the high global warming potential of the CFC molecules themselves.

3.7 Figure 3.3 shows the relative direct greenhouse warming effect over the next 100 years of world greenhouse gas emissions in 1990. Although, on an equal weight basis, CO_2 is the least potent greenhouse gas (see GWPs in column 3 in Table 3.1), it contributes most to global warming because the volume of emissions is so large. When the indirect radiative effect of CFCs is taken into account, the relative importance of CO_2 is increased.

3.8 Under the Framework Convention on Climate Change the UK will be committed to take action aimed at returning emissions of CO_2 and of other greenhouse gases (not controlled by the Montreal Protocol) to 1990 levels by the year 2000.

3.9 The following sections discuss each of the greenhouse gases resulting from human activity.

Carbon dioxide

3.10 Carbon dioxide is a key natural component of the carbon cycle and is the means whereby carbon is exchanged between large reservoirs in the land and ocean biospheres and the atmosphere (see Box 3.2). Man-made CO_2 emissions, largely from the burning of fossil fuels and destruction of forests have perturbed the natural cycle. The steady increase in CO_2 concentrations from about 280 parts per million by volume (ppmv) before the industrial revolution, to 355 ppmv in 1991 indicates that about half the man-made emissions are not absorbed into the natural reservoirs, but remain in the atmosphere.

3.11 IPCC estimate man's global emissions of CO_2 from combustion in 1990 as around 6,000 million tonnes of carbon (ranging from 5,500 to 6,500 million tonnes), of which the UK's contribution amounts to 160 million tonnes (2.7 per cent) [2]. On a per capita basis, UK emissions are currently about 2.8 tonnes of carbon per capita per year, close to the European average, compared with around 5 tonnes per capita per year for the USA and 0.2 tonnes per capita per year for typical developing countries. IPCC estimate that a further 1,600 million tonnes of carbon (plus or minus 1,000 million) is released from land use changes, primarily deforestation. The sources of UK emissions are given in

Forests and the carbon cycle   Box 3.2

Forests are important recyclers of carbon dioxide. Growing trees and other plants trap free carbon from carbon dioxide in the atmosphere, in a process called photosynthesis. Although there is no net addition of carbon once forests are mature, they continue to store the carbon already absorbed. This carbon returns to the atmosphere slowly as trees decay, and immediately if they are burnt; but it stays locked up if the timber is put to long lasting use.

Figure 3.4. Power stations account for 34 per cent of UK emissions and other industrial sources for 26 per cent. Road transport accounts for 19 per cent. Power station and other fuel processing industry emissions such as refineries can also be divided up amongst the end users of delivered energy. On this basis in 1990 some 35 per cent of CO_2 emissions were traceable to industry, 21 per cent to road transport, 26 per cent to the domestic sector, and 15 per cent to the commercial and public sector (see Figure 3.5). The burning of coal (mainly in coal fired power stations) accounts for over 40 per cent of emissions, petroleum fuels 35 per cent and natural gas 19 per cent (see Figure 3.6).

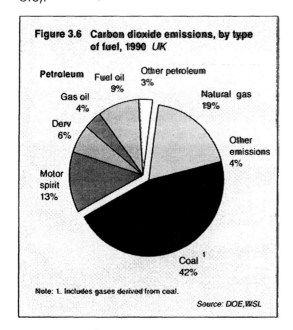

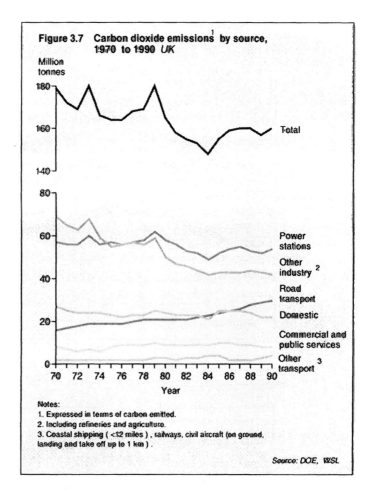

3.12 Figure 3.7 shows recent trends in UK emissions by source. In recent years, total UK emissions of CO_2 (expressed as carbon) peaked in 1979 at 181 million tonnes, declined in the early 1980s, rising again to reach 160 million tonnes in 1990. Individual sources show some levelling out of trends since the mid 1980s, but road transport emissions continue to increase. Figure 3.8 shows trends in emissions by end user (after dividing emissions from power stations and other fuel processing industries amongst end users of delivered energy).

3.13 Background CO_2 concentrations have been measured at remote locations such as Hawaii and polar regions since 1957. The long lifetime of CO_2 in the atmosphere and the relatively rapid mixing of gases throughout the troposphere within a few years, means that it is only necessary to take measurements at a relatively few remote sites to determine background trends. Long term changes in concentrations over 250 years are shown in Figure 3.9. The information is derived from

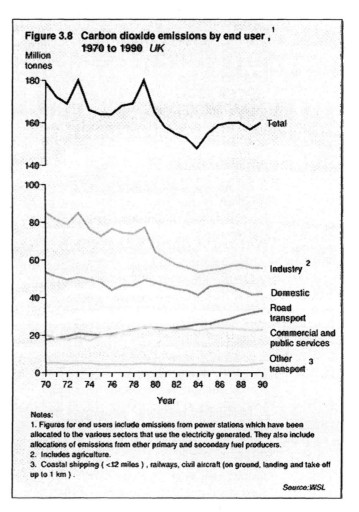

31

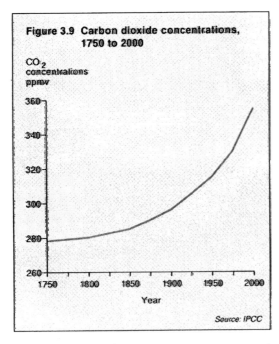

Figure 3.9 Carbon dioxide concentrations, 1750 to 2000

CO_2 concentrations ppmv

Year

Source: IPCC

a combination of these recent instrumental measurements and ice core data. Concentrations from tens of thousands of years ago can also be estimated from deep samples of ice in Greenland and Antarctica. These show that there were significant fluctuations in CO_2 levels between the ice ages and the interglacials. These have been shown to be correlated with changes in temperature on this time scale, although cause and effect are not easy to disentangle. The Department of the Environment (DOE) has recently commissioned CO_2 measurements at Mace Head in the Republic of Ireland in order to monitor the emissions of CO_2 from the European continent. This can be carried out by analysis of the difference in concentrations of CO_2 between clean and polluted air.

Nitrous oxide

3.14 Atmospheric concentration of this gas is about 8 per cent greater than in pre-industrial times, and is increasing at between 0.2 and 0.3 per cent per year. Main UK sources are industry and emissions from soil processes, with smaller contributions from road transport and other fuel combustion.

Methane

3.15 Methane is generated from a number of diverse biological and industrial sources. It is destroyed in natural chemical cycles in the atmosphere and plays an important role in the production of tropospheric ozone.

3.16 Global methane emissions through man's activities are estimated to be of the order of 360 million tonnes per year, of which the UK's share is around 4.4 million tonnes per year (1.2 per cent). The main sources of methane emissions in the UK are

shown in Figure 3.10. Animals account for 26 per cent of emissions (mainly from cattle), landfill sites 23 per cent, deep mined coal 19 per cent, offshore oil and gas production 21 per cent, and leakage from gas distribution 8 per cent.

3.17 Recent trends in estimated UK methane emissions by source are given in Figure 3.11. Total emissions increased gradually in the 1970s, levelled off in the early eighties, fell substantially in 1984 (due to the miners' strike) and increased again to pre-strike levels in 1986. Since 1986, emissions have been declining mainly due to a fall in emissions from coal mining and off-shore oil and gas production.

3.18 Global methane concentrations have more than doubled since the industrial revolution. Concentrations of atmospheric methane are measured for DOE at Mace Head, in the Republic of Ireland, as part of the international Global Atmospheric Gases Experiment (GAGE).

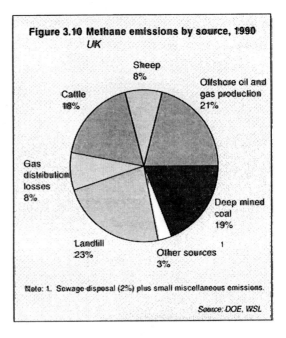

Figure 3.10 Methane emissions by source, 1990 UK

Sheep 8%
Cattle 18%
Offshore oil and gas production 21%
Gas distribution losses 8%
Deep mined coal 19%
Landfill 23%
Other sources 3%

Note: 1. Sewage disposal (2%) plus small miscellaneous emissions.

Source: DOE, WSL

Ozone

3.19 Ozone occurs both in the stratospheric ozone layer where it is generated by the direct action of ultraviolet light from the sun on oxygen, and in the troposphere due to intrusion from the stratosphere and photochemical reactions between nitrogen oxides and hydrocarbons. Background levels of tropospheric ozone have about doubled since the industrial revolution, probably as a result of increasing emissions of nitrogen oxides from power stations, cars and industry, and man-made hydrocarbons including methane. Stratospheric ozone depletion is discussed later in this chapter.

Chlorofluorocarbons

3.20 Chlorofluorocarbons are wholly man-made substances, the use of which grew rapidly during the 1960s and 1970s. Being non-inflammable, of low toxicity, odourless, cheap and convenient, they have found uses in diverse fields such as refrigeration, insulation, air conditioning, foam blowing, aerosol sprays and degreasing. They are however, powerful greenhouse gases. In general they have a long atmospheric lifetime, breaking down primarily in the stratosphere. CFCs are now being phased out because of their important role in depleting the ozone layer in the stratosphere.

3.21 The contribution CFCs make to global warming is complicated because although they are powerful greenhouse gases themselves they lead to the depletion of stratospheric ozone, which is also a greenhouse gas. The reduction in stratospheric ozone leads to cooling of the atmosphere. The extent to which direct CFC warming is countered by ozone depletion cooling may be substantial and the net impact of CFCs in climate warming is still being evaluated.

Climate change

3.22 Recent climate observations and trends for the UK are discussed in Chapter 1. In this section long term trends in the UK and global climate are considered in the context of the increase in greenhouse gases. Box 3.3 summarises some of the possible impacts of climate change in the UK.

Temperature

3.23 Global temperatures have been increasing over the past century, as shown in Figure 3.12 (in terms of differences from the 1951-80 average). Some of this rise might be expected to be due to the increasing greenhouse effect. However, other factors also affect global temperatures such as changes in ocean circulation, solar output, volcanoes and air pollution products such as aerosols. Volcanoes and pollution tend to reduce temperatures. The global temperature change observed over the past 100 to 130 years of 0.45 degrees Celsius (plus or minus 0.15 degrees Celsius) is consistent with the expected increase in temperature estimated to result from increasing greenhouse gases, taking into account the negative effect of ozone depletion and aerosols.

3.24 Temperatures over the UK show a similar long term trend during the last 130 years. Temperatures have been recorded in England for the last three centuries and the mean annual temperatures for four sites in

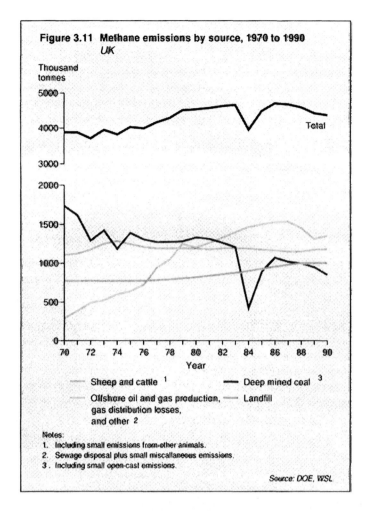

Figure 3.11 Methane emissions by source, 1970 to 1990 UK

Thousand tonnes

Sheep and cattle [1] Deep mined coal [3]
Offshore oil and gas production, gas distribution losses, and other [2] Landfill

Notes:
1. Including small emissions from other animals.
2. Sewage disposal plus small miscellaneous emissions.
3. Including small open-cast emissions.

Source: DOE, WSL

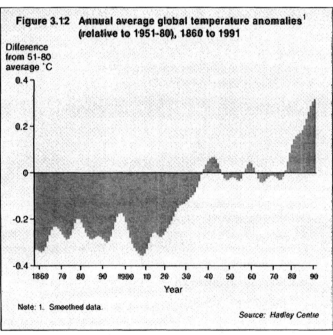

Figure 3.12 Annual average global temperature anomalies[1] (relative to 1951-80), 1860 to 1991

Difference from 51-80 average °C

Note: 1. Smoothed data.

Source: Hadley Centre

central England are shown as differences from the 1951-80 average in Figure 3.13. The global temperature differences from the 1951-80 average are also shown for comparison. The central England temperatures show greater fluctuations as would be expected for a small area. The recent upward trend in central England temperatures appears to have occurred primarily in the winter half of the

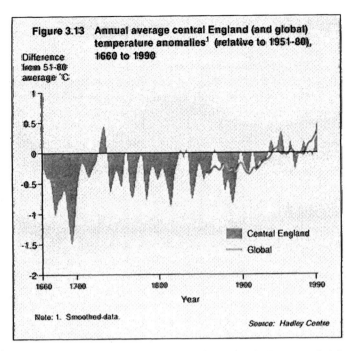

Figure 3.13 Annual average central England (and global) temperature anomalies[1] (relative to 1951-80), 1660 to 1990

Difference from 51-80 average °C

Central England
Global

Year

Note: 1. Smoothed data.

Source: Hadley Centre

year as shown in Figure 3.14. Figures 3.13 and 3.14 also show a generally cold period during the 17th and 18th centuries. Although the recent warming over England is probably a reflection of global changes, other fluctuations are known, in part, to be due to changes in atmospheric circulation, possibly related to changes in sea surface temperature.

Rainfall

3.25 The annual average rainfall for England and Wales over the last two centuries (shown as differences from the 1951-80 average in Figure 3.15) has fluctuated considerably but no significant long term trends are evident. However there have been some long term changes in the seasonal distribution of rainfall. Figure 3.16 provides evidence of a drying tendency in summer and an increase in rainfall in winter. Again, changes in atmospheric circulation are likely direct causes of these changes but the reasons for the circulation variations are not known. At present it is not possible to identify the relationship, if any, between changes in

---

Climatic change impacts in the UK   *Box 3.3*

The UK Climate Change Impacts Review Group (CCIRG) was set up by the Department of the Environment in 1990 to inform Government Ministers of the potential impacts of climate change in the UK, to identify research needed and to highlight the policy issues to be addressed. CCIRG has produced climate change scenarios to consider potential impacts of climate change in the UK [3]. These scenarios (which are not predictions) assume that, globally, by 2030, world surface air temperature will increase by 0.7 to 2.0 degrees Celsius, with a best estimate of 1.4 degrees Celsius. Summer temperature changes in the UK were considered to be comparable to the global mean value using the best model predictions available at that time. However, it was suggested that winter temperatures would warm faster than the global average. By 2030, winters in the UK were estimated to be approximately 1.5 to 2.1 degrees Celsius warmer than now. Under this scenario, winter precipitation in the UK is considered likely to increase by about 5 per cent on average by 2030, although the changes in summer precipitation were considered to be much more uncertain.

If they occurred, such climatic changes are likely to have appreciable effects in the UK. There could be significant movement of species northwards and to higher elevations; many sensitive species might be lost to the UK, although migration and invasion could increase the overall number of species. Some habitats may diminish or disappear eg wetlands and peatlands.

The mean global sea level could increase by 20 cm relative to today by the year 2030; this estimate can be applied broadly to the UK. There might be an increased frequency of storm events and coastal flooding. Areas particularly vulnerable to changes in sea level, unless action is taken, include parts of the coasts of East Anglia, Lancashire and the Yorkshire/Lincolnshire area, the Essex mudflats, the Somerset levels, the Sussex coastal towns and the Thames estuary. The north Wales coast, the Clyde and Forth estuaries, and Belfast Lough would also be vulnerable. (See also Chapter 9 which gives more details about changing sea levels).

The water content of soils would be likely to decrease in response to increased evaporation. Such changes would have a major effect on the types of crops, trees and other land uses that soils in a particular area can support. The pattern of UK land use might change as a result. If summers were drier and warmer, many soils would shrink more than usual, with important implications for structural stability. The areas most affected would be central, eastern and southern England where there are clayey soils of large shrink-swell potential.

Increases in the frequency of hot dry periods would lead to decreases in water availability but increases in water demand. Groundwater levels (groundwater provides about 20 per cent of the water supply in England and Wales - see also Chapter 6 on inland water resources and abstraction) could be reduced in the south and the east.

greenhouse gas concentrations and rainfall in the UK.

Climate predictions

3.26 Global climate change and its possible impacts have been considered by IPCC. In its 1992 Supplementary Report [2], IPCC confirmed that emissions resulting from human activities are substantially increasing the atmospheric concentrations of greenhouse gases. They estimate that without controls on greenhouse gas emissions, this will lead to global average temperature increasing at a rate between 0.2 and 0.5 degrees Celsius per decade over the next century with the possibility of significant climate changes. However, there are many uncertainties in their predictions, particularly with regard to the timing, magnitude and regional patterns of climate change.

3.27 Predicting the future course of the UK's climate is particularly difficult. Although the world temperatures are expected to increase, the climate of the UK is strongly influenced by the surrounding ocean. How the ocean and atmospheric circulations, and consequently temperatures, will change is not clear. The Meteorological Office (Hadley Centre) general circulation model of the ocean and atmosphere shows less warming in the north Atlantic and greater warming on the continent with higher levels of greenhouse gases. Such changes may mean that future rates of warming will be slightly less over the UK than globally but to what extent is still uncertain and will depend also on any changes in atmospheric circulation which occur.

3.28 The UK Climate Change Impacts Review Group (CCIRG) used scenarios (not predictions) of climate change for the UK, to consider the potential impacts [3]. The conclusions are summarised in Box 3.3.

Depletion of the ozone layer

3.29 The ozone layer helps prevent very harmful short wavelength (less than 280 nm) ultraviolet-C radiation from the sun penetrating to the surface of the earth, and substantially reduces the amount of harmful ultraviolet-B (wavelength 280-320nm) radiation reaching the earth's surface. Ultraviolet-B can cause skin cancer and can have harmful effects on plants (including agricultural crops) and marine organisms.

3.30 A number of natural and man-made trace gases are known to cause ozone destruction. However, it is the increase in the concentrations of man-made gases since the 1950s that has led to the very significant

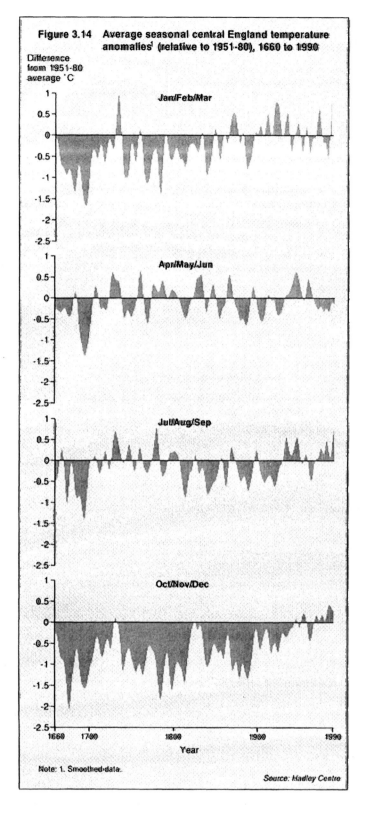

Figure 3.14   Average seasonal central England temperature anomalies[1] (relative to 1951-80), 1660 to 1990

Difference from 1951-80 average °C

Jan/Feb/Mar

Apr/May/Jun

Jul/Aug/Sep

Oct/Nov/Dec

Year

Note: 1. Smoothed data.

Source: Hadley Centre

depletion in ozone which has been observed over Antarctica (the so-called "ozone hole"). The ozone hole covers an area of some 18 million sq km between about early September and mid November. Ozone almost disappears entirely at heights between 15 and 25 km resulting in a total ozone column reduction of up to 60 per cent. Such an event is well outside the normal range of variations in ozone levels between seasons and at different latitudes. In addition, total ozone measured

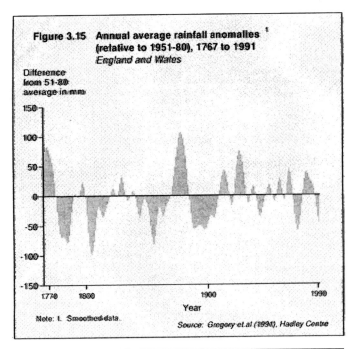

Figure 3.15 Annual average rainfall anomalies [1] (relative to 1951-80), 1767 to 1991 *England and Wales*

Difference from 51-80 average in mm

Year

Note: 1. Smoothed data.

Source: Gregory et al (1991), Hadley Centre

annually, has fallen by more than 12 per cent south of 60°S since late 1978. There is also evidence of a more general but less severe decline in ozone concentrations elsewhere. In northern latitudes in the early months of the year, ozone levels have fallen by 6 to 8 per cent and a 4 to 6 per cent decline has been observed in the same latitudes in April and May since late 1978. In this section the properties of the main ozone depleting gases, their emissions and atmospheric concentrations are discussed.

Production and consumption of CFCs and halons

3.31 Information about the production and consumption of CFCs and halons in the European Community (EC) and in the world, is given in Table 3.2 for the latest years for which information is available.

3.32 Some information is also available for UK consumption of CFCs and halons and, for CFCs, the areas of application. Consumption of CFCs fell by over 50 per cent between 1986 and 1989 (from 63,000 tonnes to 31,000 tonnes) mainly as a result of reduced CFC use in aerosols [4]. UK consumption of halons fell from about 2,000 tonnes in 1986 to 1,500 tonnes in 1990, a reduction of 25 per cent.

3.33 The Montreal Protocol on substances that deplete the ozone layer, which was agreed in 1987, introduced limits on the production and consumption of CFCs and halons. Agreement was reached by the parties to the Protocol in June 1990 to

Table 3.2 CFC and Halon production [1] and consumption [1] in the EC and world

Thousand tonnes

	1986	1987	1988	1989	1990	1991
EC						
CFCs						
Production	429	435	413	355	269	246
Consumption	304	317	299	223	175	154
Halons						
Production	70	74	..	..	..	..
Consumption	44	55	..	..	..	..
World						
Production/						
Consumption	932	1,012	1,020	907	618	..

Note: 1. Production and consumption figures have been multiplied by a factor (the ozone depletion factor, or ODP) to reflect the properties damage to the ozone layer.

Source: DTI (1990) for EC figures and OECD for world figures

The ozone depleters
Box 3.4

The (man-made) ozone depleters are the halogenated hydrocarbons or halocarbons - the hydrogen in hydrocarbon molecules such as methane and ethane is replaced either wholly or in part by the halogens fluorine, chlorine and bromine. Their breakdown in the stratosphere releases chlorine and bromine which actively take part in the destruction of ozone. The most well known halogenated hydrocarbons are the chlorofluorocarbons (CFCs). These consist of carbon, chlorine and fluorine.

More recently, partially halogenated hydrocarbons have been developed, partly as replacements for CFCs. They have similar properties but are significantly less active ozone depleters, mainly because they decay in the lower atmosphere, thus shortening their lifetime in the atmosphere and reducing the volume reaching the stratosphere. There are two classes of partially halogenated carbons: the hydrochlorofluorocarbons (HCFCs), containing carbon, hydrogen, chlorine and fluorine; and the hydrofluorocarbons (HFCs) which are similar to HCFCs but do not contain chlorine and are not therefore ozone depleters.

The brominated halocarbons, including the halons, are very powerful ozone depleters (molecule for molecule about 40 times more effective at destroying ozone). The man-made halons, used mainly as fire extinguishants, are controlled by the Montreal Protocol. However the main source of bromine in the stratosphere is still methyl bromide which is emitted naturally from the oceans and is widely used as a fumigant. About 25 per cent of emissions are considered to be from human activities.

Other widely used substances which result in destruction of stratospheric ozone are carbon tetrachloride, 1,1,1 trichloroethane and methyl chloroform (a widely used solvent). These are to be phased out under the Montreal Protocol.

phase out CFCs, halons and carbon tetrachloride by the year 2000 (with exemptions for production of halons for any essential uses) and 1,1,1 trichloroethane by 2005. The EC agreed subsequently to phase out CFCs by mid 1997.

Concentrations of CFCs

3.34 Global concentrations of CFCs in the Global Atmospheric Gases Experiment and other halogenated hydrocarbons are measured at remote locations and concentrations (monthly mean values) over the last two decades are shown in Figure 3.17 for two of the more common substances (CFC 11 and 12). Concentrations have been rising rapidly, but the rise is expected to slow and then reverse over the next few years as the Montreal Protocol takes effect [5] as shown by the broken lines in Figure 3.17.

Chlorine and bromine loading

3.35 Chlorine and bromine are released into the atmosphere from the breakdown of the ozone depleters (see Box 3.4), and catalyse chemical reactions which convert ozone into oxygen. A measure of the potential damage to the ozone layer is given by the total amount of chlorine in the atmosphere, called chlorine loading (ozone depletion increases as chlorine loading increases). Figure 3.18 shows the calculated chlorine loading in recent decades and projections of future loading, assuming that there is global compliance within the current provisions of the Montreal Protocol. A separate case in which controls similar to the current European regulation are applied globally is also shown. Chlorine loading is expected to peak towards the end of the century and decline thereafter. Since there is a lag of about five years before CFCs reach the stratosphere, maximum ozone depletion is expected to occur between the years 1995 and 2005, depending on the phase out date.

Ozone measurements

3.36 Direct stratospheric measurements are not easy to make since the stratosphere extends from between 15 and 50 km above the earth. Most modern aircraft can just reach the lower stratosphere and balloons can attain heights of around 35 km. Measurement methods are also expensive.

3.37 As a result, ground based remote sensing techniques have been developed to observe stratospheric conditions. These techniques make it possible for example to measure the total column of ozone above a specific location on the earth's surface. The British Antarctic team at Halley Bay first observed the "ozone hole", which occurs over the Antarctic for a few months each

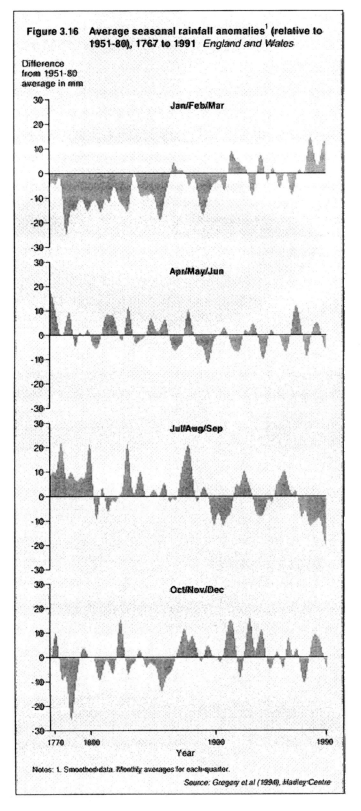

Figure 3.16 Average seasonal rainfall anomalies[1] (relative to 1951-80), 1767 to 1991 England and Wales

Notes: 1. Smoothed data. Monthly averages for each quarter.
Source: Gregory et al (1994), Hadley Centre

year, in 1984 using a remote sensing instrument, the Dobson spectrophotometer. Figure 3.19 shows the change in total ozone in October at Halley Bay over the past 35 years. This shows the substantial loss of ozone which has occurred in recent years.

3.38 Satellites have provided a global coverage of ozone measurements since 1978. These have revealed that there has been an underlying decline in ozone outside the tropics, which has reached 8 per cent in

37

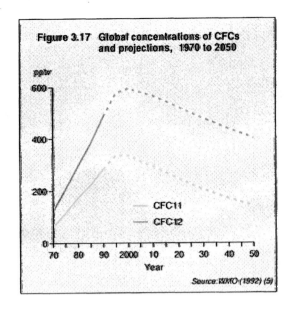

Figure 3.17 Global concentrations of CFCs and projections, 1970 to 2050

pptv

CFC11
CFC12

Year

Source: WMO (1992) (5)

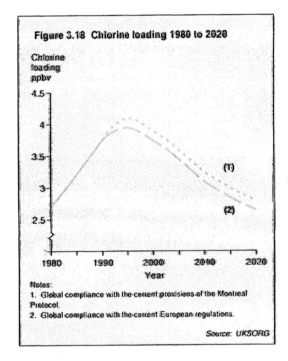

Figure 3.18 Chlorine loading 1980 to 2020

Chlorine loading ppbv

(1)

(2)

Year

Notes:
1. Global compliance with the current provisions of the Montreal Protocol.
2. Global compliance with the current European regulations.

Source: UKSORG

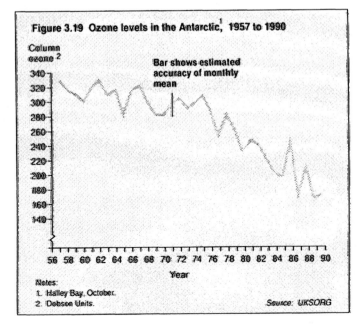

Figure 3.19 Ozone levels in the Antarctic,[1] 1957 to 1990

Column ozone [2]

Bar shows estimated accuracy of monthly mean

Year

Notes:
1. Halley Bay, October.
2. Dobson Units.

Source: UKSORG

middle latitudes of the northern hemisphere in spring.

Ultraviolet radiation

3.39 Depletion of stratospheric ozone is expected to lead to an increase in the amount of ultraviolet-B radiation reaching the surface of the earth. Shortwave ultraviolet radiation (mainly ultraviolet-B) can be harmful to life. High doses can retard plant growth, cause skin cancers and cataracts, and reduce immunity to disease. The Department of the Environment has funded the measurement of ultraviolet radiation spectra since 1989 at Reading University. Also, the National Radiological Protection Board has made broad band ultraviolet-B measurements at three sites over the last three and a half years. These measurements have not been made for long enough to reveal any trends since many other factors, such as changes in low level ozone, pollution, and cloudiness, affect the amount of ultraviolet radiation received at the surface and it may take some years before any trends can be detected.

References and further reading

1. Houghton, J.T., Jenkins, G.J., & Ephraums, J.J. (eds), (1990). Climate Change. The IPCC Scientific Assessment. Cambridge University Press.

2. Houghton, J.T., Callander, B.A., & Varney, S.K., (eds), (1992). Climate Change 1992. The Supplementary Report to the IPCC Scientific Assessment. Cambridge University Press.

3. UK Climate Change Impacts Review Group, (1991). The Potential Effects of Climate Change in the United Kingdom. HMSO.

4. Department of Trade and Industry, (1990). CFCs and Halons. Alternatives and the Scope for Recovery and Recycling and Destruction. A Report by Coopers and Lybrand Deloitte in association with Mott Macdonald and C.S. Todd & Associates. HMSO

5. World Meteorological Organisation, (1992). Scientific Assessment of Ozone Depletion, 1991. Report No 25. WMO.

UK Stratospheric Ozone Review Group, (1991). Stratospheric Ozone 1991. HMSO.

Department of the Environment and the Meteorological Office, (1989). Global Climate Change. HMSO.

Department of the Environment, (1991). The Ozone Layer. DOE.

Gregory, J.M., Jones, P.D. & Wigley, T.M., (1991). International Journal of Climatology, 11, 331-345.

4 Soil

☐ **Soil is not an unlimited resource and can be lost, degraded or improved by natural processes and by human activities (4.4).**

☐ **The National Soils Inventory and national soil maps at 1:250,000 scale provide an overall view of the range and distribution of soils in GB (4.10).**

☐ **The risk of leaching of nitrate and pesticides into aquifers is greatest on shallow or sandy soils particularly where they overlie permeable substrates (eg limestone and sandstone) (4.12).**

☐ **In the last 50 years inorganic fertiliser applications have increased by between 5 and 10 times to improve agricultural crop yields, and the area of grassland converted to arable use has increased (4.46, 4.56).**

☐ **In years of drought the shrinkage of some clays causes structural damage (4.14). Increasingly dry years may result in greater damage. Clayey soils are found predominantly in the south and east of England (Figure 4.3).**

☐ **About a third of arable land in England and Wales (20,500 sq km) has been identified as being at risk from wind or water erosion (Figure 4.5).**

☐ **Increasing acidity in soils is thought to be a contributory factor in forest decline in central Europe, although there is little clear evidence in the UK (4.35).**

☐ **Disposal of sewage sludge on land has proved a significant source of heavy metal contamination in some soils. Currently about half of all domestically produced sewage sludge is used for agricultural purposes, but this can be expected to increase with the implementation of the Urban Waste Water Treatment Directive and the banning of disposal at sea by the end of 1998 (4.42).**

☐ **An increasing awareness of the slow degradation of some organic chemicals has led to the development of strict controls on the use of pesticides and the introduction of less polluting alternatives (4.52).**

4.1 Soils are an essential component of most terrestrial ecosystems. They have long been recognised for their importance for food production and forestry, but less so for their contribution to the landscape or in providing habitats for wildlife.

4.2 Soils can also play an important role in preventing pollutants which may be present in rainwater, from entering surface waters (ie rivers, lakes etc) and groundwater in aquifers (underground water bearing rocks), by filtering out fine particles and adsorbing some of the chemicals dissolved in the water. Their role as a buffer and as a source of many potential environmental pollutants is increasingly being appreciated and understood.

4.3 Most soils have taken thousands of years to develop. Almost all UK soils have at some stage been cultivated, manipulated and managed by man principally for agricultural purposes in the lowlands and for forestry and other uses in the uplands.

4.4 Soil is not an unlimited resource and can be lost, degraded or improved by natural processes and by human activities. In recognition of the threat to soil there have been moves in western Europe since the mid 1970s towards the development of soil protection policies aimed at the prevention, or the reversal, of adverse changes.

4.5 This chapter discusses the types and distribution of soils in the UK and their key characteristics, and assesses the range of concerns regarding damage to soils as a result of natural and man-made factors.

Soil formation and distribution

4.6 Soils develop from the weathering of geological parent material (eg bedrock or glacial till), into progressively finer particles. Rainwater containing dissolved oxygen, carbon dioxide and acids derived from decomposing plant residues and atmospheric pollutants chemically weather the fine particles, releasing nutrients required for plant growth such as calcium, potassium and magnesium. Biological processes in the soil result in further development. The nature of the parent material, climate, local topography, vegetation and land-use, influence the extent of weathering and the soil type that develops. The nature and properties of soils can therefore vary significantly over relatively small areas (for instance within single fields) depending on, for example, local changes in parent material, slope and aspect.

4.7 There are a number of soil classification systems used throughout the world. The system developed by the Soil Survey of England and Wales (now the Soil Survey and Land Research Centre - SSLRC) is used to

Table 4.1	Occurrence of the most common soil types	
	England and Wales, Scotland	
		percent
Soil type	England and Wales	Scotland
Lithomorphic soils	7	10
Brown soils (including pelosols)	45	19
Podzols	5	24
Gley soils (including surface water and groundwater)	40	23
Peat soils	3	24

Source: Avery (1990), MLURI

classify soils in England and Wales, according to broad differences in the composition or origin of the soil material and the presence or absence of certain diagnostic features. At the highest level soils are divided into several categories known as major soil groups. Within each major group, soils are progressively subdivided into soil groups, soil subgroups and then into soil series. A slightly different system was developed at the Macaulay Land Use Research Institute for Scottish soils. However, the maps produced include essentially the same kind of information.

4.8 The availability and accuracy of soil maps of the UK are very varied. Detailed maps (1:63,360, 1:50,000 and 1:25,000) exist for about 20 per cent of the total land area of England and Wales and 50 per cent of Scotland. However, soil maps, based on less detailed sampling, are available for the whole of England, Wales and Scotland at a scale of 1:250,000. In Northern Ireland, a soil survey is currently underway and is expected to be completed in 1995. The survey aims to map soil types at 1:10,000 scale and to produce a comprehensive database of Northern Ireland soils. The map in Figure 4.1 shows the UK part of the 1:1,000,000 scale soil map of Europe.

4.9 Box 4.1 provides a brief description of the main characteristics of the major soil groups and Table 4.1 shows the relative occurrence of the different soil types.

4.10 Data from the National Soils Inventory for England and Wales, based on a 5 km sampling grid were published in 1992 and data from the Scottish Soils Inventory have been available since 1988. The National Soils Inventory and national soil maps at 1:250,000 scale provide a useful overall picture of soils in GB. Mapping at finer resolution has also been conducted for many representative landscapes.

Soil texture

4.11 Most soils are composed of sand, silt and clay particles which may be bound

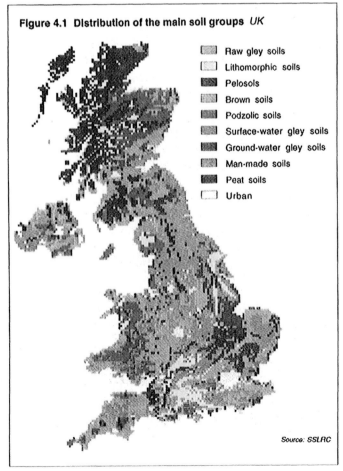

Figure 4.1 Distribution of the main soil groups *UK*

- Raw gley soils
- Lithomorphic soils
- Pelosols
- Brown soils
- Podzolic soils
- Surface-water gley soils
- Ground-water gley soils
- Man-made soils
- Peat soils
- Urban

Source: SSLRC

Description of the major soil groups   Box 4.1

Raw gley soils

Soils which occur in mineral material that has remained waterlogged since deposition. They are chiefly confined to intertidal flats or saltings.

Lithomorphic soils

Shallow soils usually formed directly over bedrock in which the only significant soil forming process has been the formation of an organic or organic-enriched mineral surface.

Pelosols

Slowly permeable clayey soils. They crack deeply in dry seasons.

Brown soils

Generally free draining brownish or reddish soils overlying permeable materials.

Podzolic soils

Soils with dark brown, black or ochreous subsurface layers resulting from the accumulation of iron, aluminium or organic matter leached from upper layers. They normally develop as a result of acid weathering conditions.

Surface water gley soils

Seasonally waterlogged slowly permeable soils.

Ground water gley soils

Soils with prominently mottled or grey subsoils resulting from periodic waterlogging by a fluctuating groundwater table.

Peat soils

Predominantly organic soils derived from partially decomposed plant remains that accumulate under waterlogged conditions.

Man made soils

Soils formed in material modified or created by human activity eg earth containing manures or refuse, soils which result from unusually deep cultivation or soil forming materials for use in land restoration following mining or quarrying.

---

together with organic matter or other cementing agents to form aggregates, which in turn join together to form larger structural units. The degree of binding and structural development is strongly influenced by the relative proportions of the different sized particles to give a range of soil textures. Many soils contain a roughly equal mix of the three particle size categories and are known as loams.

4.12 Many of the properties of a soil are strongly influenced by texture. Sandy soils, for example, usually have weak or no structure. The spaces between the coarse particles are relatively large and allow rapid drainage of water. This limits the amount of water that is retained in the soil and available for plant growth. It also increases the risk of leaching of nitrate and pesticides into surface waters, and into aquifers where soils overlie permeable substrates (eg limestone and sandstone). Figure 4.2 shows areas of GB with predominantly sandy soils.

4.13 Soils with a high clay content have very different properties. The spaces between individual particles are considerably smaller than between sands and silts, which greatly impedes the drainage of water. Some clays are, however, well structured as a result of cycles of shrinking and swelling as the soils dry out and re-wet. In these soils individual soil blocks are separated by relatively large fissures which act as the predominant drainage routes. Water moves much more slowly through the main soil mass. This can result in rapid drainage under some conditions, but very poor drainage following closure of fissures during prolonged wet

---

National Soils Inventory   Box 4.2

The National Soils Inventory comprises descriptions of soil profiles at sites located at 5 km grid intersects together with analytical data for soils at the 10 km intersect. Information held within the database is available from the Soil Survey and Land Research Centre for data on England and Wales, and from the Macaulay Institute for Scottish soils.

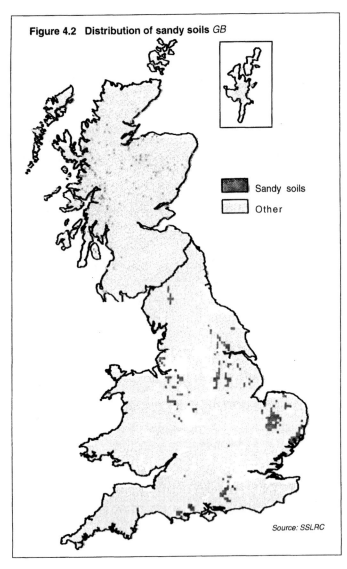

Figure 4.2 Distribution of sandy soils *GB*

Sandy soils

Other

Source: SSLRC

periods. Because of the physical and chemical properties of many clays, they are generally very good at buffering percolating water before it enters water courses. Problems can occur where dry clay soils are exposed to sudden heavy bursts of rainfall, as much of the rain tends to flow across the soil surface rather than through it. This can limit the extent to which a soil buffers the water, and increase the risk of flash floods as water courses recharge more quickly.

4.14 Under very dry summer conditions, excessive shrinkage of soils with a high clay content can cause structural damage to houses built on them. Under average climatic conditions clay shrinkage is not a major problem, but in major dry periods such as 1971 to 1976, 1984 and in recent years, extensive subsidence damage was caused (see also the section on droughts in Chapter 6). Figure 4.3 indicates the distribution of clayey soils in GB. Problems are likely to increase if droughts become more common and severe.

4.15 Silty soils, particularly those with low organic matter contents, often have a very weak structure and are susceptible to waterlogging. This can restrict seedling germination, and increase the risk of erosion. The physical and chemical properties of loamy soils will depend on the relative proportions of sand, silt and clay. Where the soil depth is sufficient they are generally the better quality soils (at least for agriculture and forestry).

Soil water

4.16 An important feature of soil is its ability to provide a reservoir of water for plant growth. This depends on a number of factors including texture, organic matter content and density. During winter, soils wet up to a wetness state known as field capacity. With free drainage, any excess water drains through the soil. Impeded drainage results in waterlogging. There comes a point in spring where the loss of water by plant uptake and evaporation is greater than rainfall and the soil begins to dry out. A soil moisture deficit develops which is measured as the amount of rainfall required to return the soil to field capacity.

4.17 The demand placed on soil water supplies by growing plants can be estimated using published meteorological data. The Maximum Potential Soil Moisture Deficit (PSMD) gives a measure of the potential use of water by plants. This figure can be adjusted for individual crops which may have different annual growth patterns. It is possible to estimate for individual soil series the amount of water available to individual crops. By comparing the PSMD of an area with the plant available water it is possible to estimate the droughtiness of particular soils for different crops. Figure 4.4 provides an indication of soil droughtiness in England and Wales for grass. This indicates that drought is most severe in eastern England in general and parts of East Anglia in particular. In contrast, drought is far less of a problem in the west of England and Wales. Similar maps are available for other common crops such as winter wheat, sugar beet and potatoes.

4.18 Soils which lie within the range of a rising groundwater table, and those that are not free draining may become waterlogged for a part of the year. Extended waterlogging results in the depletion of oxygen and the development of anaerobic (oxygen deficient) conditions. This can have significant effects on plants and animals growing on and living in soils, and the chemical form and availability of certain elements, including some which are potentially toxic (especially when present in abnormal amounts). Soils can be classified according to their duration and degree of waterlogging. One such system has six categories ranging from well drained (Wetness Class I) to almost permanently waterlogged (Wetness Class VI).

4.19 The patterns of water movement through soils are complex and are related to several soil and site factors. A hydrological classification of soil types has recently been developed by the Institute of Hydrology, SSLRC and the Macaulay Land Use Research Institute, called HOST (Hydrology of Soil Types). This identifies eleven basic patterns of water movement, which can be further subdivided based on additional factors in the soil or underlying rock. This system can be related to specific soil types and when linked to other information, can be used to evaluate the fate of fertilisers and agrochemicals applied to the soil.

Soil nutrients

4.20 Plants require a number of elements to ensure healthy growth. A fundamental role of soil is to provide a steady supply to meet plant requirements. These essential elements are commonly divided into two groups, major and minor nutrients, depending on the relative amounts needed by most plants. The major nutrients include nitrogen, phosphorus, potassium, sulphur, magnesium and calcium. Minor nutrients include iron, manganese, copper, zinc and boron. It should be noted that some minor nutrients, essential for the development of plants are toxic to both plants and animals and are considered as pollutants if present in excessive amounts.

4.21 Sources of the different nutrient elements and the rate of availability vary and involve complex interactions within the soil. In the absence of fertilisers, nitrogen and sulphur are mainly derived from the breakdown of soil organic matter and inputs from acidic deposition. Elements such as phosphorus, potassium and calcium on the other hand are predominantly provided by the weathering of soil particles or parent material.

4.22 Although nutrient elements are generally present in high concentrations in most soils almost all are in forms unavailable to plants. The lack of availability can limit plant growth and influence both the composition of individual plants and the range of species in a community. Many wild flowers for example, characteristic of some natural and semi natural systems, thrive because low available nutrient levels combined with particular ranges of acidity (pH) prevent excessive colonisation by more invasive species characteristic of nutrient rich soil. In intensive agricultural systems, the rate of

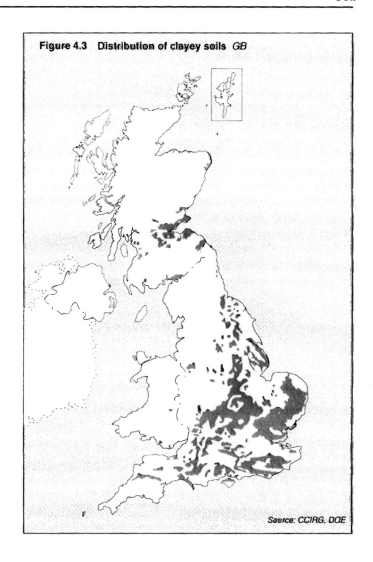

Figure 4.3 Distribution of clayey soils GB

Source: CCIRG, DOE

availability of nutrients is often too slow for optimum plant productivity and the shortfall is made up with additions of fertiliser and manures.

Soil Quality

4.23 The concept of soil quality has been the focus of much discussion. The quality of soil can only be assessed in relation to its use. Soil considered as highly suitable for one purpose, such as providing a surface for building on, may be unsuitable for other purposes such as agriculture or forestry. A number of existing systems used to classify land in terms of the suitability for particular uses (especially agriculture and forestry) such as the agricultural land classification, implicitly involve an assessment of soil quality as well as other factors such as climate and topography. More recently the concept of soil quality has been used to address the significance of soil contamination and degradation. This is a complex issue, as a particular soil contaminant can have varying degrees of significance on both the soil which it affects and on other systems influenced directly or indirectly by the soil (humans, plants, animals, surface waters etc).

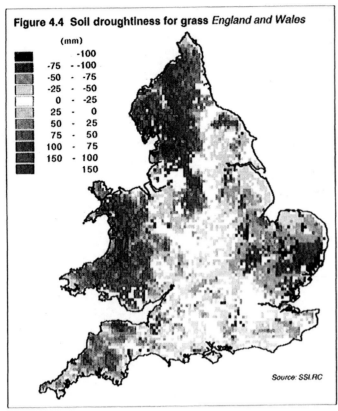

Figure 4.4 Soil droughtiness for grass *England and Wales*

(mm)

	-100
-75	- -100
-50	- -75
-25	- -50
0	- -25
25	- 0
50	- 25
75	- 50
100	- 75
150	- 100
	150

Source: SSLRC

4.24 The Interdepartmental Committee on the Redevelopment of Contaminated Land (ICRCL) have published threshold concentrations of some organic and inorganic contaminants in soils, above which action may be needed to avoid adverse effects for a particular site end-use. These values which are currently being reassessed are aimed for use in redeveloping land which is considered to be contaminated. As published they do not relate to land which has already been developed.

4.25 An important step in identifying the effects on soil quality of a potential contaminant is an understanding of the complex interactions that occur within the soil. This is considered further with respect to certain pollutants later in this chapter.

Sensitivities/threats to soils

4.26 The main sensitivities/threats to soils are:

- Increasing urban areas, motorway and road building and industrial development
- erosion
- acidification
- accumulation of pollutants
- organic matter loss and deteriorating soil structure

and associated problems:

- nitrate and phosphates in water resources
- organic contaminants in water resources

More information on many of these issues is given in Chapters 2, 5 and 7 on air quality, land and inland water quality.

Increasing urban areas

4.27 Urban and industrial development can result in the permanent loss of land for uses such as agriculture, forestry or nature conservation. Around 10 per cent of land in the UK is classified as urban and this is gradually increasing (see also the section on urban areas in Chapter 5). In many cases where building is planned, soils are removed, and can be used to make environmental improvements in other areas. The restoration of large areas of former dereliction using imported soils is a prime example. Despite this, much soil is lost or buried as a result of urban expansion. See also the development control section in Chapter 5.

Erosion

4.28 Soil erosion is the loss of soil particles by the action of wind or water. In general, soils with a good permanent vegetation cover are unlikely to suffer from erosion problems. The problem is most significant where the vegetation is removed for periods of time such as for arable agriculture. It has been estimated that a third of the arable area in England and Wales (20,500 km²) is at risk from erosion [1]. Figure 4.5 shows the areas most at risk. Erosion of blanket peats in the uplands can also be severe as a result of grazing, burning and drainage.

4.29 Wind erosion is confined largely to sandy and peaty soils. Water erosion is more prevalent in the main arable areas, in upland areas and in areas of intensive recreation (for example footpaths within National Parks).

4.30 The risk of water erosion is greatest during periods of heavy rainfall when the soil is saturated, the soil surface is bare and movement of water across the soil surface increases. The resulting runoff can quickly remove large volumes of soil which bury or destroy crops in localised areas, and leave channels (rills and gullies) which in the worst cases can inhibit agricultural machinery cultivating the land. In the long term average soil depth is reduced. Once erosion channels are formed these act as the main route for water, which exacerbates the problem.

4.31 Other indirect problems include sediment deposition on roads and in water courses and reservoirs etc, and the loss of seeds and fertiliser from the soil surface. Phosphate applied as fertiliser is generally concentrated at the soil surface and may be washed into surface waters bound to soil particles. Agrochemicals may also enter water courses bound to soil particles.

4.32 Control measures include the retention or planting of strips of permanent vegetation (trees or hedges) to form shelter belts which reduce the effects of wind and tillage techniques to reduce the rate at which water is able to move across the soil surface.

Acidification

4.33 The distribution of acid soils is closely related to climate and parent material. Soil acidification is a natural process and in areas of high rainfall leaching of nutrient elements (eg calcium, potassium, magnesium) from the soil leads to increasing soil acidity. Even unpolluted rainwater is naturally slightly acidic because it contains dissolved carbon dioxide. Acid soils therefore dominate large areas of the UK. However, man's activities can accelerate the process. Changes in land use, the application of fertilisers and the increased inputs of acidic precipitation from the release of oxides of nitrogen and sulphur from the burning of fossil fuels, have been identified as potentially major threats. See also the section on sources of air pollutants in Chapter 2.

4.34 The wide range of soil acidity (as measured by pH values) found in natural soils is reflected in variations in the vegetation they can support. Plants are sensitive to the pH of a soil both through direct effects on plant roots and indirectly through the availability of nutrient elements at different pHs. In acid soils increasing amounts of metals become available (eg aluminium and manganese) which can be toxic to plants. Most plants show satisfactory growth at a pH between 5 and 8. At a pH less than 5, the number of plant species able to tolerate the conditions rapidly decreases and specialised plants dominate. For agricultural purposes, pH can economically be controlled by liming, to reverse the effect of the acid.

4.35 Increasing acidity in soils is thought to be a contributory factor in forest decline in central Europe, although there is little clear evidence in the UK. It also has the effect of increasing the solubility of metals in the soil including aluminium, lead, cadmium and zinc. This increases the risk of leaching into, and contamination of, surface and groundwater.

4.36 The concept of "critical loads" has been used to assess the impact of acidic deposition. This is a quantitative estimate of the exposure to one or more pollutants below which significant changes in soil do not occur (see also references to critical loads in Chapter 2). For acidification the critical load calculation

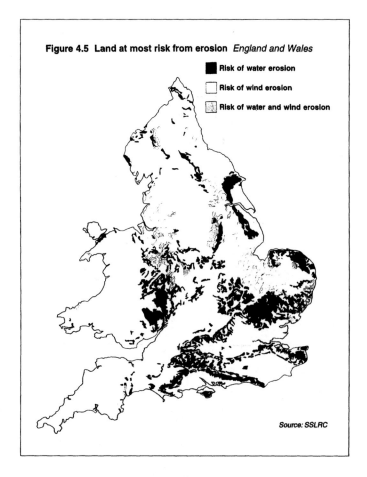

Figure 4.5 Land at most risk from erosion *England and Wales*

- ■ Risk of water erosion
- □ Risk of wind erosion
- ▨ Risk of water and wind erosion

Source: SSLRC

is based on the total acidic inputs, soil mineralogy, texture and soil moisture regime.

4.37 The most sensitive soils to acidification are generally those with low organic matter and clay contents. Soils containing carbonates or with a higher clay content have a greater capacity to buffer acidification.

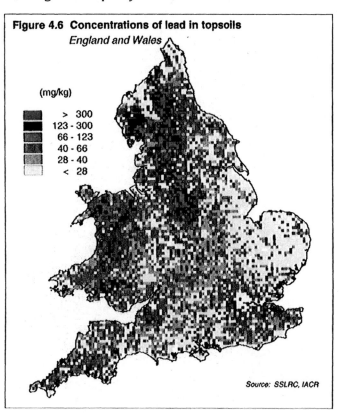

Figure 4.6 Concentrations of lead in topsoils
England and Wales

(mg/kg)

| > 300 |
| 123 - 300 |
| 66 - 123 |
| 40 - 66 |
| 28 - 40 |
| < 28 |

Source: SSLRC, IACR

4.38 Figure 2.16 (in Chapter 2) shows critical loads for acidity of soils in the UK and areas where critical loads are exceeded. These include areas of Scotland, Wales and Northern Ireland.

Heavy metal contamination

4.39 Soil contamination by heavy metals can originate from a number of sources including geological parent material, industrial processes (atmospheric emissions, waste disposal, effluent disposal) and farming practices, (eg applications of sewage sludge, farm wastes and fertilisers). There may be high local concentrations in and around areas of historic metal mining dating back in some cases to pre Roman times in south west England (south Devon and Cornwall), North Wales, the Pennines and north west England. Figure 4.6 shows concentrations of lead in topsoils and Figure 4.7 concentrations of cadmium in topsoils in England and Wales. Similar maps exist for other metals.

4.40 Recent studies have shown that levels of certain heavy metals, including lead, zinc and copper can vary more widely in urban and industrial areas, than those normally found in agricultural land. There is also concern in areas with elevated levels of metals that reworking or disturbing sites may mobilise the metals with adverse effects on water sources.

4.41 Once in a soil many heavy metals bind to organic matter or clays and are largely unavailable to plants. The availability of most metals increases as a soil becomes more acid,

increasing the risk to plants and animals through uptake, and metal leaching into surface or groundwater under extreme conditions. Heavy metals have also been shown to affect the microorganisms within a soil sometimes influencing the rate of nutrient cycling and organic matter breakdown within the soil. This can affect soil structural development and the nutrient availability to plants.

4.42 The disposal of increasingly large amounts of sewage sludge on land, which is likely to follow the banning of disposal at sea by the end of 1998 and the implementation of the Urban Waste Water Treatment Directive, has been identified as a potentially significant source of heavy metal contamination in soils. Currently about half of all domestically produced sewage sludge is used for agricultural purposes. Sludges from different sources can be very variable. The risk of contamination is dependent on the metal content of the sludge which will therefore depend on its source, the rate of application and the long term pattern of disposal on a particular field. See also Chapter 7 on inland water quality and Chapter 11 on waste.

4.43 Guidance on the implications for the use of land arising from the presence in soil of some elements and compounds is given in notes produced by ICRCL.

4.44 Heavy metal pollution through industrial emissions to the atmosphere are controlled through air pollution legislation (see Chapter 2 on air quality and pollution) whilst landfilling wastes are controlled by the Environmental Protection Act 1990, Control of Pollution Act 1974 and the Pollution Control and Local Government (Northern Ireland) Order 1978 (see Chapter 11 on waste).

4.45 The Department of the Environment (DOE) has published guidelines for the use of sewage sludge on agricultural land (Sludge (Use in Agriculture) Regulations 1989 as amended by the Sludge (Use in Agriculture) Regulations 1990). Maximum acceptable levels of metal contaminants have been set on the basis of predicted rates of uptake by growing crops or ingestion by livestock. These maximum acceptable levels are currently under review.

Figure 4.7 Concentrations of cadmium in topsoils England and Wales

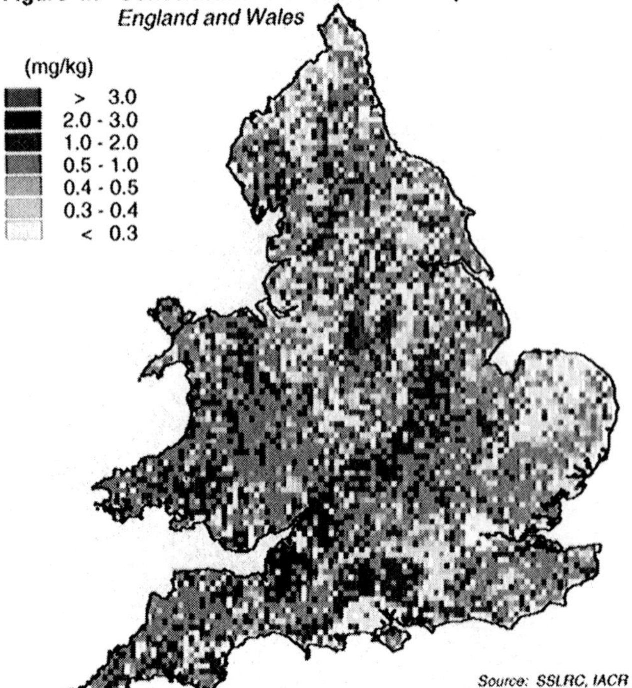

(mg/kg)
> 3.0
2.0 - 3.0
1.0 - 2.0
0.5 - 1.0
0.4 - 0.5
0.3 - 0.4
< 0.3

Source: SSLRC, IACR

Nitrate and phosphates in water resources

4.46 Soils used for intensive agriculture require additional nutrients (particularly nitrogen, phosphorus and potassium) to maintain optimum plant productivity. Most soils contain large reserves of these nutrients held within the mineral or organic fractions. In the last 50 years inorganic fertiliser applications have increased by between 5 and 10 times to increase agricultural crop yields. However, if amounts in excess of plant requirements are applied, eg as inorganic fertilisers or organic farm wastes, there is the risk of these materials polluting water resources. Some farming practices can lead to leaching of nitrate out of the rooting zone. This nitrate can arise from excessive applications of nitrogen applied either as inorganic fertiliser or as manures, or from the release of nitrate during the natural decomposition of some of the soil organic matter. The risk of pollution from inorganic sources of phosphorus is less common because this element is quickly bound to the soil. However, there are risks of phosphate pollution of water courses where excessive quantities of farm wastes, containing high levels of phosphate, are applied to soil, or where eroded soils enter a river.

4.47 Leaching of nitrate is most pronounced on free draining soils during winter and early spring. The overall pattern of loss is complex and difficult to quantify. The actual loss ultimately depends on the soil type, land use, rainfall patterns, drainage characteristics, the amount and availability of nitrogen applied, the timing of applications in relation to crop growth, and for grassland the intensity of grazing by livestock. Figure 4.8 shows the potential soil leaching risk under given land use and culivation conditions.

4.48 Steps have been taken to limit the amount of nitrate leaching, including research into the optimum timing and application rates of fertiliser and manure. Research has shown that as crop growth is increased by nitrogen made available from fertilisers, organic manures or soil reserves, nitrogen uptake and removal by the crop also increase. However if nitrogen is available in amounts larger than needed to further stimulate growth, then nitrogen uptake will reach a plateau and residue will accumulate in the soil, which may be lost by leaching. If nitrogen is made available at times of year when little or no crop growth is taking place, this nitrogen may also be lost by leaching if rainfall exceeds evaporation. Policies, such as the pilot nitrate scheme, incorporate measures which can

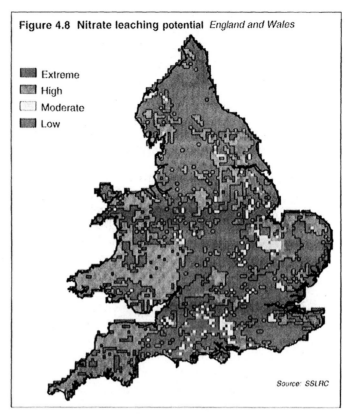

Figure 4.8 Nitrate leaching potential *England and Wales*

Extreme
High
Moderate
Low

Source: SSLRC

locally reduce the risk by maintaining a plant cover and regulating the application of fertilisers and manures. However, even under these conditions, nitrate which has already leached below the rooting zone in earlier years, will continue to move through the deeper substrate into the groundwater. The NRA are producing maps which identify areas of particular risk of groundwater contamination from point sources and from diffuse pollution including both nitrate and pesticides. See also sections on nitrate in Chapter 7 on inland water quality.

Pesticides and organic solvents

4.49 Modern agricultural production systems rely on the use of pesticides for crop protection and disease control purposes which may be applied either directly or indirectly to soil.

4.50 Pesticides can affect a soil directly by adsorption onto clays and organic matter, by affecting soil microorganisms and plant growth, and by transport through the soil to surface and groundwater and thereby drinking water supplies. This has stimulated interest into their behaviour and toxic effects in both the soil and water environment. See section on pesticides in Chapter 7.

4.51 Pesticides are not only used in agriculture. The control of weeds along railways, motorways, footpaths and amenity areas for example has been identified as a significant source of runoff, especially as many of the areas treated are adjacent to artificial drainage systems leading to direct contamination of surface and groundwaters.

4.52 An increasing awareness of the slow degradation of some organic chemicals has led to the development of strict controls on the use of pesticides, and the introduction of less polluting alternatives. Pesticides are controlled by regulations made under the Food and Environment Protection Act 1985.

4.53 The behaviour of a pesticide in soil is influenced by a number of factors including its chemical properties, climate and the soil type (particularly the soil texture and drainage characteristics). As with nitrate, leaching is highest in sandy soils, and except where surface runoff or flow down cracks is significant, lower in clay soils.

4.54 Organic solvents can derive from a range of industrial, urban or agricultural activities. The majority of pollution incidents involve point sources and are usually concerned with much higher initial concentrations of the chemical. These can have significant long term effects on the soil, and on pollution of water.

Soil organic matter loss and deteriorating soil structure

4.55 Soil organic matter is a vital component of productive and stable soils. It is an important source of plant nutrients, improves water retention and soil structure, and is important in terms of the soil's buffering capacity against many of the threats discussed earlier.

4.56 Many areas of permanent grassland have been ploughed up and used for arable agriculture over the last 50 years. Repeated cultivations, removal of the vegetation (the crop), and a reduction in the use of organic manures has resulted in a rapid reduction of the organic matter content of these soils. A threshold level of 2 per cent organic carbon has been suggested, below which a soil is increasingly vulnerable to structural instability.

4.57 The loss of organic matter on a global scale has also been identified as a contributory factor towards global warming as the organic carbon is ultimately converted to carbon dioxide increasing the levels in the atmosphere. See also references to the greenhouse effect in Chapter 3.

4.58 As the organic matter content of a soil decreases, the factors that maintain soil structural stability are reduced, and soils become more prone to loss of structure and compaction. Compaction is most common and severe where heavy loads such as wheeled vehicles and animal hooves are repeatedly run over clayey soils under wet conditions. Lighter soils are less prone to this form of compaction and ameliorative techniques are generally easier.

References and further reading

1. Graziani C.A., (1987). Report drawn up on behalf of the Committee on the Environment, Public Health and Consumer Protection on the Erosion of Agricultural Soils and on Wetlands in the European Community. European Parliament Session Document 1987-88. Doc A2-20/87.

Avery, B.W., (1990). Soils of the British Isles. CAB International.

Institute of Terrestrial Ecology, Soil Survey and Land Research Centre, (1989). An Assessment of the Principles of Soil Protection in the UK. Volume 1 Concepts and Principles. Institute of Terrestrial Ecology.

Institute of Terrestrial Ecology, Soil Survey and Land Research Centre, (1989). An Assessment of the Principles of Soil Protection in the UK. Volume 2 Review of Current Major Threats. Institute of Terrestrial Ecology.

Soil Survey of England and Wales, (1984). Soils and their Use in Northern England. Soil Survey of England and Wales Bulletin No. 10. SSLRC.

Soil Survey of England and Wales, (1984). Soils and their Use in Wales. Soil Survey of England and Wales Bulletin No. 11. SSLRC.

Soil Survey of England and Wales, (1984). Soils and their Use in Midland and Western England. Soil Survey of England and Wales Bulletin No. 12. SSLRC.

Soil Survey of England and Wales, (1984). Soils and their Use in Eastern England. Soil Survey of England and Wales Bulletin No. 13. SSLRC.

Soil Survey of England and Wales, (1984). Soils and their Use in South West England. Soil Survey of England and Wales Bulletin No. 14. SSLRC.

Soil Survey of England and Wales, (1984). Soils and their Use in South East England. Soil Survey of England and Wales Bulletin No. 15. SSLRC.

The Macaulay Land Use Research Institute, (1982). The Scottish Soil and Land Capability for Agriculture, Handbooks 1 to 7 and associated maps. Macaulay.

The Macaulay Land Use Research Institute, (1984). The Scottish Soil and Land Capability for Agriculture, Handbook 8; Organisation and Methods. Macaulay.

Agricultural Advisory Council on Soil Structure and Soil Fertility, (1970). Modern Farming and the Soil (The Strutt Report). HMSO, London.

5 Land use and land cover

□ **About three quarters of the UK is covered by agricultural land, around one tenth by forest and woodland, and just over one tenth by urban land. Since the late 1940s (in England and Wales) there has been an increase in woodland, an increase in urban land, a small decrease in farmland and a large decrease in semi-natural vegetation (Figure 5.1 and Table 5.1).**

□ **Between 1984 and 1990, 52,000 km of hedgerows in GB were removed and 26,400 km of new hedges appeared (5.7).**

□ **In 1989, Green Belts in England covered 1.5 million hectares, about 12 per cent of the country and more than double the designated area in 1979. In Scotland there are five Green Belts covering nearly 145,000 hectares (Figures 5.2 and 5.3).**

□ **Nearly 20 per cent of the UK is covered by National Parks, Areas of Outstanding Natural Beauty and National Scenic Areas (Figure 5.4).**

□ **Since 1947, the area of GB covered by forest and woodland has increased from 1.4 million hectares to over 2.3 million hectares in 1991. Nearly all of the increase is in coniferous forest, most of it in upland areas (5.46). Coniferous forest increased from 380,000 hectares to 1.5 million hectares over this period (Figure 5.8).**

□ **The total stock of vacant urban land in England in 1990 was estimated at 60,000 hectares of which 25,000 hectares had previously been developed (5.61). In 1988 there were just over 5,000 hectares of vacant urban land in Scotland (5.63).**

□ **The area of derelict land in England declined by 11 per cent between 1982 and 1988. Derelict land declined in all regions except Yorkshire and Humberside (5.70, 5.71).**

□ **In 1988, 96,100 hectares of land in England were affected by permissions for surface mineral workings. Between 1982 and 1988, 20,600 hectares were reclaimed, largely for agricultural and amenity use (Figures 5.23 and 5.25).**

□ **Most listed buildings in England are in reasonable condition and almost 93 per cent are secure. But 37,000 listed buildings may be at risk from neglect and a further 73,000 are vulnerable (5.96).**

5.1 The way land is used is continually changing, affecting both the rural and urban environments. Changes are shaped by a framework of controls and policies for protection in order to balance competing needs for new development, conservation, agriculture, etc. This chapter provides an outline of the main uses and changes and the way changes are controlled, and it also deals with the built heritage. There are four main sections: the first discusses land uses in the UK and land use changes; the second covers the ways changes in use are regulated; the third looks at the main categories of land use individually, and some of the trends, features and measures affecting them; and the fourth describes the background to heritage protection in the UK.

Land use and land cover

5.2 The following paragraphs contain information about the nature of the land in the UK, described here as land use or land cover. Land use means the types of use to which the land is put, for example, agricultural, forestry, industrial, or residential.

Table 5.1 Land cover and use: 1947, 1969 and 1980 *England and Wales*								
	Monitoring landscape change survey [1] % cover						Forestry Census	Agricultural Census
	Woodland	Semi-natural vegetation [2]	Farmed land [3]	Water and wetlands [4]	Other land [5]	Total	Forest and woodland [6]	Agricultural land [7]
1947	7.0	12.6	72.7	1.3	6.4	100.0	6.3	80.5
1969	7.9	10.1	72.1	1.1	8.8	100.0	7.2	78.5
1980	7.9	9.2	71.8	1.1	9.9	100.0	7.9	77.6

1. Photographs were interpreted for a range of dates around those shown. Relative standard errors are approximately: woodland (7%), semi-natural vegetation (10%), farmed land (1.5%), water/wetland (17%), other land (1.5%).
2. Includes heath , bracken, gorse, heather and grassland not included under farmed land.
3. Cultivated land, and improved, neglected and rough pasture.
4. Coastal and inland waters, freshwater marsh, salt marsh and peat bog.
5. Includes built-up land, urban open space, major transport routes, bare rock, sand, shingle, mineral works and derelict land. Urban land can be defined in many ways. For example, Professor Best estimated that urban land defined as covering all land under urban uses covered the following proportion of England and Wales: 6.7% in 1930, 8.8% in 1950, 10.8% in 1970 and 11.6% in 1980. There is evidence to suggest the MLC survey underestimated urban land.
6. All forest land and woodland including woodland on agricultural holdings.
7. All agricultural land including crops, fallow, grasses, rough grazing woodland and other land on farms, eg farm roads, yards, buildings, ponds, etc.

Land cover describes what the land looks like, the different features and characteristics such as roads and buildings, woodlands, crops, and areas of water. The main sources of information are given and summarised in Box 5.1.

Land Use

5.3 About three quarters of the land area of the UK is in agricultural use. The remaining quarter comprises mainly forest and woodland, and urban areas. Figure 5.1 shows these main land uses in the UK as a whole and also by individual country. Over two thirds of land in Wales, Scotland and Northern Ireland is grass and rough grazing and under 10 per cent is crops and fallow, whereas in England a third is crops and fallow and only 39 per cent is grass and rough grazing. Although agriculture has been declining slightly, forest and woodland has been

increasing at an average rate of 22,000 hectares a year over the last decade, and now accounts for more than 10 per cent of the land area. Urban land use has also increased and accounts for about 10 per cent.

5.4 Since 1985, the Department of the Environment (DOE), Scottish Office and Welsh Office have collected information about changes in land use (see Box 5.1). The Ordnance Survey (OS) collect information in the course of their map revision work. They record details of site size, location (grid reference), previous land use, new land use and the year they believe change happened. Changes involving physical development tend to be recorded sooner than changes between other uses (for example, between agriculture and forestry), some of which may not be recorded for some years. Results are published in DOE Statistical Bulletins[1].

Land Cover

5.5 Information about land cover is available from the sample survey Monitoring Landscape Change in England and Wales (MLC), published in 1986[2] (see Box 5.1). This provided estimates of countryside change in England and Wales since the late 1940s. Table 5.1 shows the distribution of broad categories of landscape features from the MLC survey, together with Agricultural and Forestry Census figures, for 1947, 1969 and 1980. Between the late 1940s and 1980 there was an increase in woodland; a small decrease in farmed land and a large decrease in semi-natural vegetation. The losses of farmed land to built-up land were mainly from improved grassland; gains to farmed land were mainly from semi-natural vegetation and broadleaved woodland; cultivated land increased mainly at the

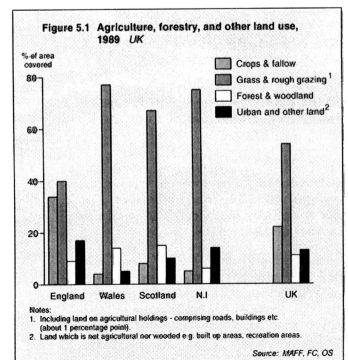

Figure 5.1 Agriculture, forestry, and other land use, 1989 *UK*

%·of area covered

Crops & fallow
Grass & rough grazing [1]
Forest & woodland
Urban and other land [2]

England   Wales   Scotland   N.I   UK

Notes:
1. Including land on agricultural holdings - comprising roads, buildings etc. (about 1 percentage point).
2. Land which is not agricultural nor wooded e.g. built up areas, recreation areas.

Source: MAFF, FC, OS

expense of improved grassland; and whilst some new coniferous woodlands replaced old broadleaved or coniferous woods, most were previously upland areas (largely marginal agricultural land).

5.6 Change statistics alone, however, without assessment of the quality of the features that have changed, do not indicate the significance of change for wildlife habitat, landscape quality or management requirements. An important new source of information is the 1990 Countryside Survey, which will provide statistics on the current status of the land cover and vegetation in GB and changes over the last decade. (See Box 5.1).

5.7 Changes in land cover have mainly followed changes in land use and management practices. In particular, bigger fields mean

Table 5.2 Survey to monitor landscape change: length [1] of linear features: 1947, 1969, 1980 and 1985
England and Wales

					Thousand kilometres	
	Hedgerows	Fences	Banks	Open ditches	Walls	Wood-land fringe
1947 [2]	796	185	151	122	117	241
1969 [2]	703	193	140	116	114	241
1980 [2]	653	199	132	111	111	243
1985	621	210	128	112	108	243

1. Relative standard errors are approximately: fences (5%), hedgerows and woodland fringe (8%), banks (18%), walls (26%), open ditches (29%).
2. Photographs were interpreted for a range of dates around this year.

Source: DOE, CC

fewer hedges. The MLC survey showed a substantial net loss of hedgerows of about 4,000 to 5,000 km a year over the period 1947 to 1980. Table 5.2 sets out survey information on changes to linear landscape features (hedges, walls, etc) between 1947 and 1985. The most recent information on hedgerows comes from the 1990 Countryside Survey (see Box 5.1). This survey found that in 1984 hedgerows in GB totalled 549,000 km and that by 1990 some 52,000 km (10 per cent) had been removed, but that new hedges totalling 26,400 km appeared in this period. The survey also reports a decline in the quality of some hedgerows, which indicates that a lack of long-term management as well as deliberate removal is the cause of some losses. The ecological effects of changes to land cover including the linear features are now widely recognised. See also Chapter 10 on wildlife.

5.8 A separate study has looked at landscape change in the National Parks and the Broads (see Box 5.1). The findings show the overall extent of change in the last decade or so, in the Parks as a whole and in individual Parks. There has been an overall 13 per cent increase in cultivated land and a 2 per cent

increase in improved pasture over the period. Coniferous forest, including felled and planted land, increased by 11 per cent. However, much of the substantial increase in young and mature coniferous forest resulted from planting before the 1970s. The survey results suggested that a decrease in new coniferous planting may have occurred during the period studied. Fences increased by 5.4 per cent, often replacing hedges and walls, which decreased by 4.6 per cent and 1.3 per cent respectively. Other categories which decreased were rough pasture (11 per cent), heath (2 per cent) and grass moor (2 per cent). These overall figures do not reveal the wide variation in change from Park to Park.

5.9 In Scotland, a land cover map, based on interpretation of air photographs, has been compiled. The land cover data are being digitised and the database will be available in the autumn of 1992.

5.10 In Northern Ireland, separate but comparable surveys are being undertaken to provide statistics on the current status of land cover and vegetation. The Northern Ireland Countryside Survey uses a similar methodology to the 1990 Countryside Survey to provide baseline statistics for Areas of Outstanding Natural Beauty (AONBs) and the wider countryside in Northern Ireland. Information from the survey is available for some areas but, as this is the first such survey, there are no data on trends.

Controls on how land is used

Development plans and development control

5.11 The framework for controlling land use change is largely provided for by the town and country planning system. Planning decisions on proposals to build on land, or change its use, are usually made by local authorities. Decisions must accord with the development plans prepared by local planning authorities unless other considerations indicate otherwise, eg government guidance on Green Belt protection.

5.12 Development plans comprise local plans (in which district councils set out development control policies and proposals for their area) and structure plans (in which county councils set out strategic policies as a framework for district councils); or unitary plans (which combine the function of structure and local plans in London and metropolitan boroughs). Environmental considerations must be taken into account in preparing the plans, which must contain policies on conservation of the natural beauty

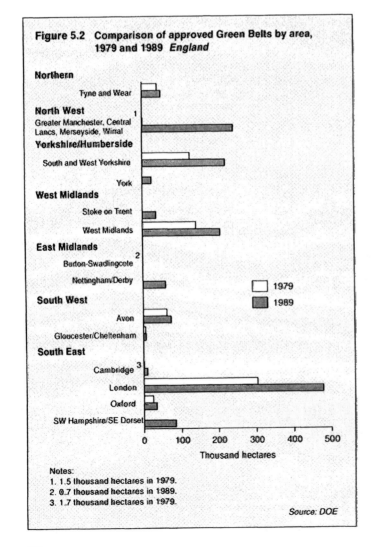

Figure 5.2 Comparison of approved Green Belts by area, 1979 and 1989 *England*

Thousand hectares

Notes:
1. 1.5 thousand hectares in 1979.
2. 0.7 thousand hectares in 1989.
3. 1.7 thousand hectares in 1979.

Source: DOE

and amenity of land, and on the improvement of the physical environment. The Planning and Compensation Act 1991 makes mandatory the preparation of district wide local plans in England and Wales. In Scotland, planning authorities have been legally required to prepare local plans for their areas since 1975.

5.13 Environmental assessment (EA) (also known as environmental impact assessment or EIA) has formed a part of the development control process in GB since 1988[6]. Similar arrangements apply in Northern Ireland. A series of Regulations implement EC Directive 85/337 and apply to certain types of project which are likely to have significant effects on the environment by virtue of their nature, size or location. Developers are required to assemble in an environmental statement relevant information about the likely environmental effects of projects. Planning authorities must take the environmental statements and public comments into account when reaching their decisions.

5.14 Local authorities must also be able to respond quickly when planning rules are broken, for example, where a site is being developed without consent or in ways which breach a planning condition. The Planning and Compensation Act 1991 has strengthened planning authorities' enforcement powers.

Designated and protected areas

Green Belts

5.15 By 1989, Green Belts in England totalled 1.5 million hectares, about 12 per cent of the country and more than double the designated area in 1979 (see Figure 5.2). Planning permission would normally not be granted for most proposals for development in the Green Belt. They are established to check the unrestricted sprawl of built-up areas; to safeguard the surrounding countryside from further encroachment; to prevent neighbouring towns from merging into one another; to preserve the special character of historic towns; and to assist in urban regeneration. The largest Green Belt is around London and large areas have also been designated in the North West, South and West Yorkshire and the West Midlands. In Scotland, there are five Green Belts covering nearly 145,000 hectares. Figure 5.3 shows the areas covered by Green Belts in England and Scotland. There are no Green Belts in Wales or Northern Ireland.

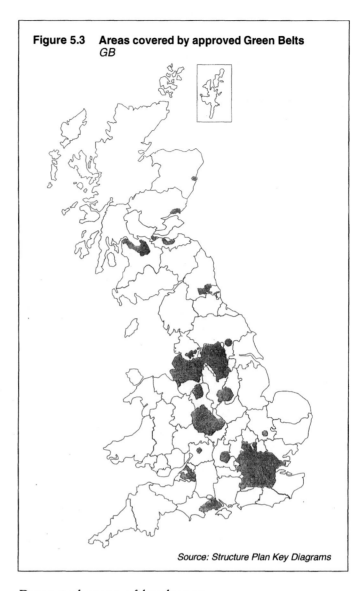

Figure 5.3   Areas covered by approved Green Belts
GB

Source: Structure Plan Key Diagrams

Protected areas of landscape

5.16 The planning system exercises tighter controls in areas designated for their landscape importance. In England and Wales, development control decisions under the Town and Country Planning Act 1990 in these areas are informed by a Planning Policy Guidance note on the Countryside and the Rural Economy.

5.17 In England, the Countryside Commission works with local authorities and others to implement landscape protection and improvement policies, to safeguard existing opportunities for public access and recreation, and provide new ones. English Nature, which took over the former Nature Conservancy Council (NCC) responsibilities for England, is the separate body which is mainly responsible for nature conservation and for advising the Government and others on matters affecting nature conservation.

5.18 In Wales, the Countryside Council for Wales (CCW) took over the functions of the Countryside Commission and the Nature Conservancy Council, in April 1991.

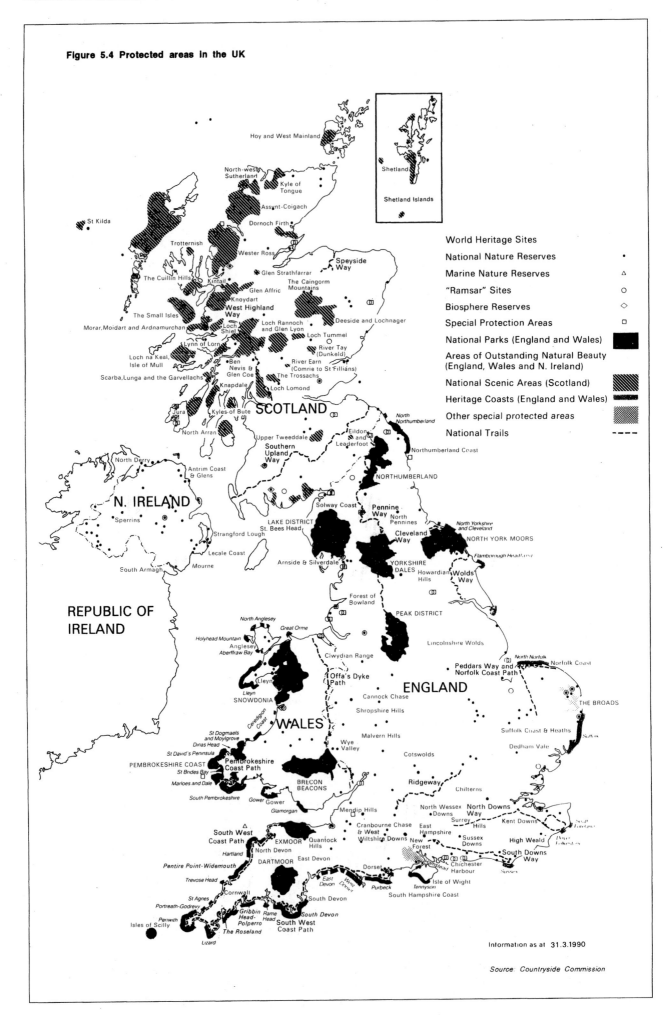

Figure 5.4 Protected areas in the UK

World Heritage Sites

National Nature Reserves •

Marine Nature Reserves △

"Ramsar" Sites ○

Biosphere Reserves ◇

Special Protection Areas □

National Parks (England and Wales) ■

Areas of Outstanding Natural Beauty (England, Wales and N. Ireland)

National Scenic Areas (Scotland)

Heritage Coasts (England and Wales)

Other special protected areas

National Trails – – –

Information as at 31.3.1990

Source: Countryside Commission

Scottish Natural Heritage (SNH) came into being in April 1992 to take over the functions of the Countryside Commission for Scotland and the Nature Conservancy Council for Scotland. The Joint Nature Conservation Committee (JNCC) was established in April 1991, through which the agencies in the three countries act together on UK and international nature conservation issues. There is no similar umbrella body for countryside responsibilities. The Environment Service of the Department of the Environment for Northern Ireland (DOE(NI)) is responsible for countryside protection in Northern Ireland.

5.19 National Parks in England and Wales, AONBs in England, Wales and Northern Ireland, and National Scenic Areas (NSAs) in Scotland, are the major national designations to protect areas of finest landscape, and together they account for nearly 20 per cent of the total land area of the UK. In addition, in England and Wales, the finest stretches of undeveloped coast are defined as Heritage Coasts to afford them protection; and the public has a right of way over the 225,000 km network of footpaths, bridleways and byways. A series of long-distance routes, or "national trails", has also been designated. Figure 5.4 shows the main protected areas and national trails. The background to these designated areas is described in more detail below.

5.20 The ten National Parks in England and Wales were designated in the 1950s under the National Parks and Access to the Countryside Act 1949, to protect some of the most beautiful parts of the countryside and provide opportunities for open air recreation. The ten Parks cover some 1.4 million hectares, 9 per cent of the land area of England and Wales. National Park designation does not imply any change in the ownership of the land, which remains largely in private hands.

5.21 Two other notable areas with unique features have special protection although they are not formally designated as National Parks. The Broads in East Anglia is an important area in terms of landscape, nature conservation and recreation and was given a status equivalent to a National Park in 1989. The Broads Authority brings together in one body responsibility for planning, development control, recreation and amenity provision, and navigation over most of the Broads river system. The New Forest in Hampshire is a generally unenclosed area of woodland, heath and bog, once a Royal Forest but now managed by the Forestry Commission.

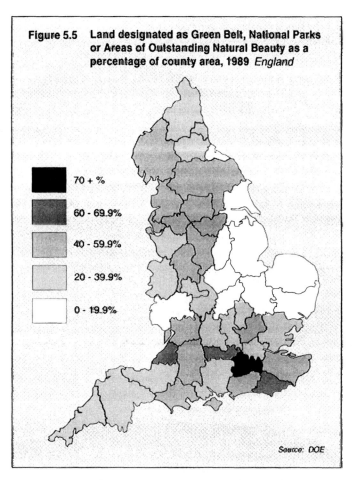

Figure 5.5 Land designated as Green Belt, National Parks or Areas of Outstanding Natural Beauty as a percentage of county area, 1989 *England*

70 + %

60 - 69.9%

40 - 59.9%

20 - 39.9%

0 - 19.9%

Source: DOE

5.22 There are thirty nine AONBs in England and Wales covering 2 million hectares, 13 per cent of the land area. In England and Wales, these areas may contain landscapes of equal merit to those in the National Parks, but have fewer opportunities for open air recreation. The current programme of AONB designation is nearing completion with the designation of areas in the Nidderdale moors in Yorkshire, the Tamar valley in south west England, and consideration of the Berwyn mountains in Wales.

5.23 In National Parks, the Broads and AONBs, the permitted development rights (ie development that may be carried out without local authority planning permission) are more restricted than elsewhere. Figure 5.5 shows the percentage of land in English counties designated as Green Belt, National Parks and AONBs.

5.24 In Northern Ireland, AONBs have been redefined in legislation passed in 1985. There are nine AONBs, four of which have been reviewed and redesignated since 1985 and now have a different legal status. The Environment Service of DOE(NI) is currently consulting on proposals for an additional two AONBs in County Fermanagh.

5.25 The forty NSAs in Scotland were identified by the Countryside Commission for Scotland in 1978 as areas of outstanding

---

Areas protected under international nature conservation obligations

Box 5.2

Special Protection Areas (SPAs) are established under the EC Directive 79/409/EEC on the Conservation of Wild Birds, to conserve certain rare and migratory species of birds. The Directive requires member states to take special measures to conserve the habitat of these species in order to ensure their survival and reproduction, and to safeguard designated areas from damage and disturbance. SPAs are also SSSIs.

"Ramsar" Wetland Sites are designated in accordance with the Ramsar Convention on Wetlands of International Importance (signed at Ramsar in Iran in 1971) as Waterfowl Habitat. Many Ramsar sites are also SPAs and all have SSSI status.

The EC Habitats Directive, agreed at the Environment Council in December 1991 was adopted in May 1992. It will give rise to the designation of **Special Areas of Conservation (SACs)**. These areas will be protected for the purpose of conserving Europe's rarest flora and fauna species and habitat types. Together with SPAs designated under the EC Birds Directive, SACs will contribute towards the creation of a Community-wide network of protected sites. Most SACs are likely to be drawn from the existing SSSI network.

Biosphere Reserves were devised by UNESCO and are of particular value as standards for the measurement of long term changes in the biosphere as a whole. They are protected areas of land and coastal environment representing significant examples of habitat throughout the world.

---

beauty and value meriting special protection. They total over one million hectares, or 13 per cent of the land area. The planning authorities are responsible for development control in these areas in conjunction with SNH, and for incorporating suitable landscape policies into structure and local plans.

5.26 The new designation of Natural Heritage Areas was introduced in Scotland in 1991 under the Natural Heritage (Scotland) Act 1991. It provides for an integrated management approach to landscape and nature conservation for large areas with special natural heritage significance.

5.27 In England and Wales, there are forty four Heritage Coasts, amounting to nearly 1,500 km or 33 per cent of the coastline.

The majority of Heritage Coasts get further protection through their inclusion within National Parks or AONBs. Like most protected areas, the majority of Heritage Coasts are in private ownership. The Heritage Coast programme has been closely associated with the National Trust's Enterprise Neptune project, which has taken 770 km of coastline into protective ownership, of which 540 km are in Heritage Coasts.

5.28 Over 7,500 km of the Scottish mainland and island coastline is designated as Preferred Coastal Conservation Zones through the National Planning Guidelines. Structure and local plans are expected to include policies which reflect the distinction between the conservation zones and the coastal development zones.

Protected areas of scientific and environmental value

5.29 A framework of statutory measures to safeguard wildlife habitats and natural features of the environment has existed since 1949. Some of these, such as National Nature Reserves (NNRs) and Sites of Special Scientific Interest (SSSIs) are established under UK legislation. Others are the result of UK involvement with international initiatives (see Box 5.2). Table 5.3 shows the numbers and total area covered for each type of statutorily protected area in 1991. Sites may be subject to more than one designation, for example, NNRs are also SSSIs, and many of the sites listed under the designations described in the following paragraphs are within National Parks or other designated areas (see also

Table 5.3   Statutory protected areas, 1991 *UK*

Status [1]	Number	Area (000's ha)
National Nature Reserves	286	172.5
Local Nature Reserves [2]	241	17.1
Sites of Special Scientific Interest (SSSIs) [2]	5,671	1,778.5
notified under 1981 Act	5,576	1,721.5
subject to S15 management agreements	(a)	98.5
Areas of Scientific Interest [3]	46	63.4
Areas of Special Scientific Interest [3]	26	6.9
Special Protection Areas	40	134.4
Biosphere Reserves	13	44.3
"Ramsar" Wetland Sites	44	133.7
Environmentally Sensitive Areas	19	785.6

(a) = 2032 agreements

1. Some areas may be included in more than one category. For example, in Great Britain NNRs, SPAs, Biosphere Reserves and Ramsar sites are all SSSIs (see also Figure 5.4).
2. Great Britain only.
3. Northern Ireland only.

Source: NCC, DOE (NI), MAFF

Figure 5.4). Environmentally Sensitive Areas (ESAs) are described later in the section on Agriculture.

5.30 NNRs are areas of national and sometimes international importance which are owned or leased by the appropriate statutory conservation body or bodies approved by them, or are managed in accordance with nature reserve agreements with landowners and occupiers. In GB, there were 242 NNRs covering 168,100 hectares in 1991. In Northern Ireland, forty four NNRs (4,400 hectares) have been designated by DOE(NI).

5.31 SSSIs are sites that are important to Britain's natural heritage because of flora, fauna, geological or physiographical features or a combination of these. The statutory bodies have a duty to identify and notify the landowners and occupiers, the local planning authority and the appropriate Secretary of State, of all those sites which have a special scientific interest.

5.32 Planning authorities must consult the appropriate statutory conservation body before granting permission for development likely to affect a SSSI. Since the Environmental Protection Act 1990, planning authorities also have to consult about planning applications in any consultation area around a SSSI defined by the statutory conservation body. In addition, a system operates to protect SSSIs from operations outside the scope of planning controls. In some cases owners and occupiers enter into formal management agreements with the statutory body. These may include payments to enhance the conservation interest of the SSSI, or payments to compensate the owner or occupier on the basis of net profit forgone.

5.33 The amended SSSI system, introduced in the Wildlife and Countryside Act 1981, was designed primarily to deal with the adverse effects on wildlife of intensive agriculture and forestry. Significant damage from which a site is not expected to recover has now reduced to around 1 per cent a

Table 5.4	Sites of Special Scientific Interest reported as damaged, 1990 - 91 *GB*			
Damage [1] caused by:	Short-term damage [2]	Long-term damage [3]	Partial or full loss [4]	Total
Agricultural activities	102	6	3	111
Forestry activities	3	2	-	5
Activities given planning permission	1	9	2	12
Activities of statutory undertakers and other public bodies not included above	13	5	1	19
Recreational activities	38	5	1	44
Miscellaneous activities [5]	60	6	3	69
Insufficient management	16	-	-	16

1. Some cases of damage are caused by more than one activity.
2. Damage from which the special interest could recover.
3. Damage causing a lasting reduction in the special interest.
4. Damage which will result in denotification of the special interest.
5. Including pollution, unauthorised tipping and burning.

Source: EN, CCW, SNH

year. Much of it results from planning decisions, where the nature conservation interests have been balanced against other legitimate interests. Short term damage also occurs and is often caused by accident or unthinking behaviour from which the majority of the sites will recover. Table 5.4 sets out reports of damage in 1990-91.

5.34 In Northern Ireland, the Areas of Scientific Interest are being replaced by Areas of Special Scientific Interest (ASSIs) and are protected by legislation similar to that for SSSIs in GB. Twenty six ASSIs covering about 6,900 hectares had been designated by 31 March 1991.

5.35 Local authorities have powers to establish Local Nature Reserves (LNRs) after consulting the appropriate statutory conservation bodies.

5.36 There are also two Marine Nature Reserves (MNRs) (not shown in Table 5.3), the Islands of Lundy and Skomer, which were designated in 1986 and 1990 respectively. Activities within MNRs are regulated by a combination of bylaws and a voluntary code of conduct. Five other sites identified as possible MNRs are the Menai Straits, Bardsey Island and the Lleyn Peninsula, Loch Sween, St Abbs Head and

Hill Livestock Compensatory Allowances   Box 5.3

Hill Livestock Compensatory Allowances (HCLAs) are paid under EC rules to farmers in the Less Favoured Areas (LFAs) to compensate for the permanent natural handicaps affecting farming in these areas. Payments in the UK are made on breeding cattle and sheep. The objectives of the scheme are to ensure the continuation of livestock farming in the hills and uplands, thereby helping to maintain a viable population in the LFAs and to conserve the countryside. HCLAs have made a major contribution to the maintenance of the variety of landscape in the hills and uplands. Nevertheless, increased stocking of grazing land has contributed to the decline of valuable habitats in some areas. The Government has announced a number of measures designed to address this issue.

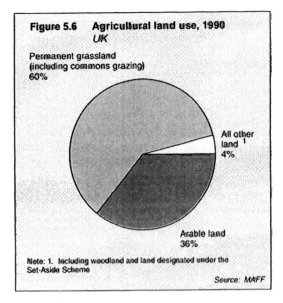

Figure 5.6 Agricultural land use, 1990 UK

Permanent grassland (including commons grazing) 60%

All other land [1] 4%

Arable land 36%

Note: 1. Including woodland and land designated under the Set-Aside Scheme

Source: MAFF

the Isles of Scilly. The Menai Straits and Loch Sween sites are being progressed.

International legislation

5.37 Table 5.3 also lists those areas protected under the UK's international nature conservation obligations which are described in Box 5.2.

Other protected areas

5.38 In addition to the above designated areas a variety of other protection and conservation arrangements exist in the UK, many of which are non statutory. Many

protected areas are owned or managed by voluntary bodies like the Royal Society for the Protection of Birds, the Royal Society for Nature Conservation and local Nature Conservation Trusts, the Scottish Wildlife Trust, the Wildfowl and Wetlands Trust, the Woodland Trust, the Field Studies Council and the National Trust. Some of these areas are SSSIs, or include MNRs. The later section in this chapter on Forestry gives information about forest parks and other conservation and recreational areas managed by the Forestry Commission.

Some individual uses and features

Agriculture

5.39 The total area of agricultural land in the UK in 1990 was 18.5 million hectares (about three quarters of the total land area) compared with 19 million hectares in 1980. Sixty per cent is permanent grassland and the rest is mostly arable (see Figure 5.6).

5.40 Agriculture is discussed in more detail in Chapter 15 on pressures on the environment. However, the following paragraphs describe three types of area where special measures apply to farming activities, which help to protect the environment.

Less Favoured Areas

5.41 Less Favoured Areas (LFAs) are designated under EC law and are eligible for subsidy under EC legislation depending on the nature of farming carried out on the land. Numerous schemes apply to LFAs and an example of one is shown in Box 5.3. The first LFAs were designated for the UK in the mid 1970s and were based on the old hill areas. They were updated and expanded in 1984 and 1990. About 53 per cent of farming land is designated as LFA.

Environmentally Sensitive Areas

5.42 ESAs were first designated in 1987-88 in the UK because of the major influence which agriculture can have on the conservation and enhancement of landscape, wildlife and historical features. The areas chosen for designation are recommended by the Countryside Commission and English Nature (CCW in Wales and SNH in Scotland) on the basis of their national importance. The scheme is voluntary and farmers within an ESA are encouraged to maintain or adopt farming practices that will conserve characteristic environmental features. The initial agreements were signed for five years. Since their inception, ESAs have been subject to monitoring in order to assess their environmental and economic effects. The

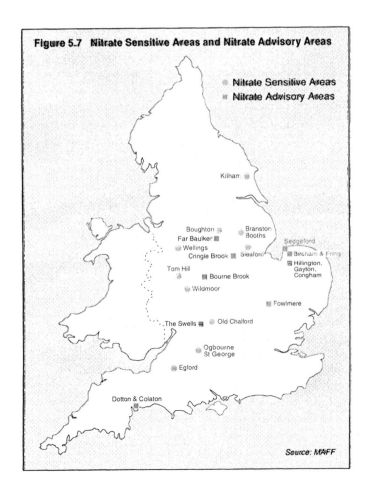

Figure 5.7 Nitrate Sensitive Areas and Nitrate Advisory Areas

● Nitrate Sensitive Areas
■ Nitrate Advisory Areas

Kilham

Boughton
Far Baulker
Wellings
Cringle Brook
Tom Hill
Bourne Brook
Wildmoor
The Swells
Ogbourne St George
Egford
Dotton & Colaton

Branston Booths
Sedgeford
Sleaford
Bircham & Fring
Hillington, Gayton, Congham
Fowlmere
Old Chalford

Source: MAFF

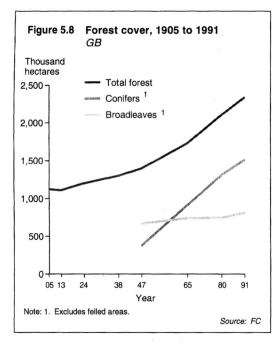

Figure 5.8   Forest cover, 1905 to 1991
GB

Thousand hectares

— Total forest
Conifers [1]
Broadleaves [1]

Year

Note: 1. Excludes felled areas.

Source: FC

encouraging first results have led to the progressive upgrading of existing schemes and proposals for the designation of further areas. The existing nineteen ESAs cover 785,600 hectares. On completion of the current programme the area involved could be approximately trebled.

Nitrate Sensitive Areas

5.43 Nitrate used in agriculture can find its way into water courses and create a risk of pollution of public water supplies (see also Chapter 7 on inland water quality). A pilot Nitrate Scheme began in 1990 to explore ways of controlling nitrate leaching from soil. Under the scheme ten Nitrate Sensitive Areas (NSAs) have been designated to test the effectiveness of agricultural measures in reducing nitrate leaching and to assist future policy development.

5.44 In the NSAs, farmers may qualify for annual payments for undertaking to restrict practices such as fertiliser use or switching from cereal production to low intensity grassland cultivation. Nearly 80 per cent of eligible farmers have put forward over 9,000 hectares (from a total of about 11,000 hectares overall) for the scheme. Figure 5.7 shows the locations of the NSAs and also the nine Nitrate Advisory Areas which cover about a further 20,000 hectares. In these areas farmers get free advice on ways to reduce the risk of nitrate leaching into water and are encouraged to follow practices designed to reduce leaching.

Forestry

5.45 Forests and woodland occupy some 10 per cent of the land area in the UK. The Forestry Commission (FC) own or manage

close to 40 per cent of these forests and woodlands and the remainder in GB are privately owned. The balance of ownership is different in Northern Ireland where most forestry is owned by the DANI Forestry Service. Table 5.5 shows woodland area by type for each UK country. Scotland accounts for 60 per cent of the UK conifer forestry, while more than 75 per cent of all broadleaved woodland is in England.

5.46 Since 1947, the area of forest and woodland cover in GB has increased from 1.4 million hectares to over 2.3 million hectares in 1991. This expansion was largely confined to marginal agricultural land, mainly on poor and remote sites in the uplands. Establishment of forests on these exposed sites was achieved mainly by planting new, highly productive species of conifers introduced from north west America. Coniferous forest cover increased from around 380,000 hectares to over 1.5 million hectares over this period (see Figure 5.8). Growing forests act as reservoirs for free carbon from carbon dioxide (CO_2) in the atmosphere, thereby helping to reduce the contribution of CO_2 emissions to global warming. However, the planting of new conifer forests has led to concern about the impact on the landscape and on the ecology of semi-natural areas. It has led to declines in populations of some comparatively uncommon birds, and often destroys existing semi-natural vegetation (see also Chapter 10 on wildlife). Conifers also increase run-off and may increase water acidity.

5.47 The areas of new planting for both conifers and broadleaved forests since 1979-80 are shown in Figure 5.9. About 23,500 hectares of conifers were planted in 1979-80, rising to a peak of almost 26,200 hectares in 1988-89, then falling sharply in 1989-90 to 14,100 hectares and 10,900 in 1991-92. This reduction has been due to woodland owners adjusting to the effects of the removal of forestry tax reliefs in the March 1988 Budget. New plantings of broadleaved forests have increased from just

Table 5.5   Woodland area by type of woodland and country, 1991 [1] *UK*

Thousand hectares

	High forest Conifers	High forest Broad-leaves	Coppice	Other Wood-land
England	385	434	38	102
Wales	172	61	2	12
Scotland	962	87	0	82
Northern Ireland	70	1	0	4
UK	1,589	583	40	200

Note: 1 .Data for NI refer to 1990

Source: FC and DANI

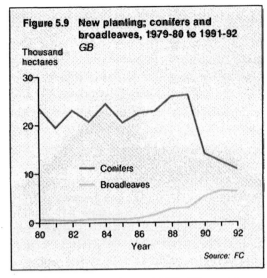

Figure 5.9 New planting; conifers and broadleaves, 1979-80 to 1991-92 GB

Thousand hectares

Conifers
Broadleaves

Year

Source: FC

over 600 hectares in 1979-80 to almost 6,300 hectares in 1991-92. The marked increase in new plantings since 1985 follows the introduction of the Broadleaved Woodland Grant Scheme and its successor

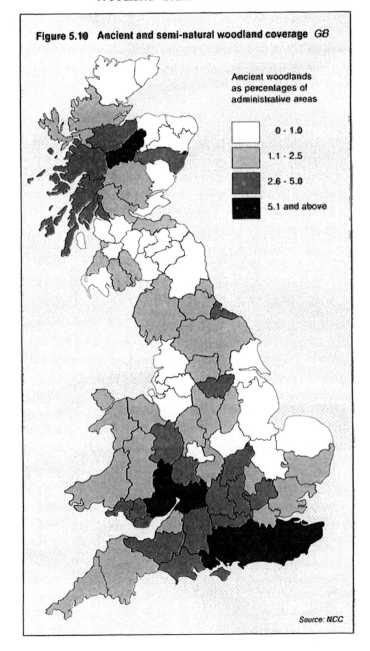

Figure 5.10 Ancient and semi-natural woodland coverage GB

Ancient woodlands as percentages of administrative areas

0 - 1.0

1.1 - 2.5

2.6 - 5.0

5.1 and above

Source: NCC

the Woodland Grant Scheme, introduced in 1988. The Farm Woodland Scheme, introduced in 1988 to encourage a productive alternative use for farmland, has also contributed to the increase in the new planting of broadleaves. This scheme was replaced in 1992 by the Farm Woodland Premium Scheme.

5.48 Most tree felling in GB is controlled by the FC under the Forestry Act 1967 and the Forestry (Exceptions from Restrictions of Felling) Regulations 1979. With certain minor exceptions, it is an offence to fell trees without having an agreed plan of operations or a felling licence. Most licences carry a requirement to restock with suitable species.

Recreation and conservation

5.49 Table 5.6 shows recreation facilities on FC land. Forest Parks (FPs) providing public access and recreational facilities, now cover a quarter of all FC land. Fourteen FPs have been designated totalling some 296,000 hectares. The Commission has also designated forty six Forest Nature Reserves (FNRs), wildlife sites within FC forests, which have been selected to represent and conserve rare species and habitats. FNRs designated so far cover some 21,000 hectares in GB. More recently the Commission have introduced Woodland Parks (WPs) in FC forests primarily aimed at providing places for local people to enjoy. At the end of 1991 five WPs had been declared - near Perth, in South Wales, the Chilterns, and in west Scotland.

5.50 Community Forests are also proposed to bring about landscape improvements and increase recreational opportunities for urban populations. None have yet been established but work is in progress in twelve locations in England. In Scotland, a Countryside Trust has been set up to manage and improve the degraded area of central Scotland, through woodland planting, environmental

Table 5.6	Forestry Commission public recreational facilities, 1992 GB			
		Number		
	England	Wales	Scotland	GB
Camping and caravan sites	22	1	9	32
Picnic places	357	94	193	644
Forest walks and nature trails	296	77	274	647
Visitor centres	10	5	9	24
Arboreta and forest gardens	13	6	3	22
Forest drives	6	1	3	10
Forest cabins and holiday homes	107	0	67	174

Source: FC

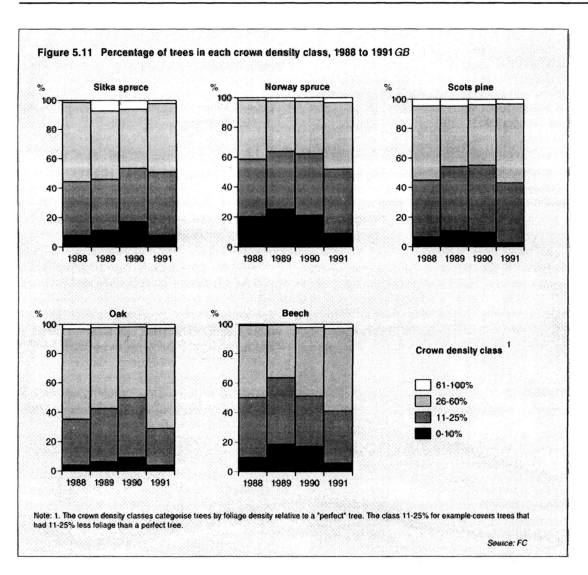

Figure 5.11 Percentage of trees in each crown density class, 1988 to 1991 *GB*

Crown density class[1]

- ☐ 61-100%
- ▨ 26-60%
- ▨ 11-25%
- ■ 0-10%

Note: 1. The crown density classes categorise trees by foliage density relative to a "perfect" tree. The class 11-25% for example covers trees that had 11-25% less foliage than a perfect tree.

Source: FC

improvements, and conservation and recreational initiatives. This initiative, known as Central Scotland Woodland, involves an area of about 100,000 hectares. Proposals are being developed for a new National Forest covering 50,000 hectares on the Leicestershire/Derbyshire border.

5.51 In Northern Ireland forests, there are fifty one areas designated as nature reserves. Fifteen of these reserves (some 304 hectares), are classed as NNRs of prime conservation importance. The remaining thirty six are FNRs (some 1,760 hectares), which are of slightly less importance but have a major role in education.

Ancient semi-natural woodlands

5.52 Ancient semi-natural woodlands are sites which have been continuously wooded for at least 400 years. They still have tree and shrub layers composed of species native to the site including oak, ash, hazel, birch, beech and Scots pine, which have regenerated naturally from seed. The environmental significance of ancient semi-natural woodlands is that they tend to be richer in plants and animals than other woodland areas, and

contain many rare and vulnerable species. Ancient semi-natural woodlands cover 2.6 per cent of England, 2.7 per cent of Wales and 1.9 per cent of Scotland, a fraction of the original natural woodland which once covered most of GB. As Figure 5.10 shows, the greatest concentrations now are in south east England, the southern Welsh borders and the central Scottish Highlands. In the last 50 years, 7 per cent of the remaining ancient semi-natural woodlands have been cleared mainly for agriculture and development, with the heaviest losses being in the Home Counties, the north Midlands and the Welsh borders. Over the same period, 38 per cent of ancient semi-natural woodlands have been replanted with both indigenous and introduced coniferous and broadleaved species, and the depletion of semi-natural woodlands has now largely ended. The statutory conservation bodies maintain registers of ancient woodlands in GB.

Forest health surveys

5.53 Forest health is characterised by crown density (the amount of light passing through the crown) and needle or leaf discolouration.

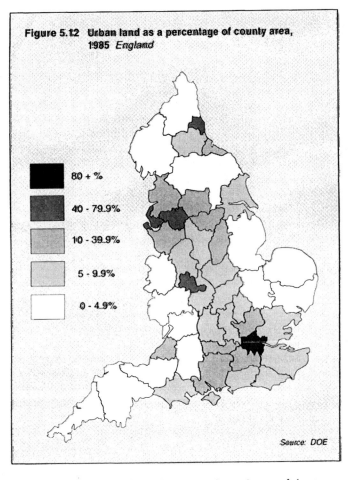

Figure 5.12 Urban land as a percentage of county area, 1985 England

80 + %

40 - 79.9%

10 - 39.9%

5 - 9.9%

0 - 4.9%

Source: DOE

Both reflect the general condition of the tree, but neither has an exclusive cause. Besides air pollution, other factors such as frost,

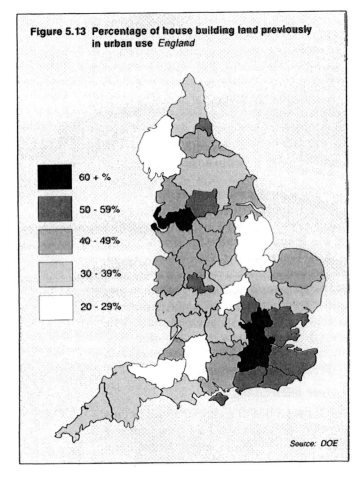

Figure 5.13 Percentage of house building land previously in urban use England

60 + %

50 - 59%

40 - 49%

30 - 39%

20 - 29%

Source: DOE

drought, wind and nutrient deficiencies can cause discolouration and reductions in crown density, as can a variety of fungi and insect pests.

5.54 The FC conducts an annual monitoring programme of forest health [7]. In the 1991 survey, 8,843 trees were assessed. Figure 5.11 compares tree crown density results between 1988 and 1991. The results relate to the percentage of trees in each crown density class (categorised by foliage density relative to a "perfect" tree, eg class 11-25 per cent covers trees that had 11-25 per cent less foliage than a perfect tree). The data show that the condition of individual trees can vary markedly from one year to the next. Trees with very high levels of defoliation are capable of recovery which may occur quite rapidly. The sharp increase in 1991 in the number of trees with more than 10 per cent foliage loss compared with 1990 can probably be attributed to recent very dry summers and severe weather conditions in the winter and late spring preceeding the survey. Beech trees severely affected by the 1989 and 1990 droughts continue to be in poor condition and some individual oak trees show a marked decline which may be related to drought stress. Apart from the changes in crown density, however, most trees showed far fewer drought-related problems in 1991.

5.55 Although a longer time series is required before any firm conclusions can be reached, air pollution, particularly in the form of acid rain, appears to be affecting some aspects of forest ecosystems in GB. Continued detailed monitoring is needed, in order to set targets for remedial action. See also Chapter 2 on air quality and pollution.

Storm damage

5.56 There have been five major storms in Britain within the past 50 years which damaged many trees. The storm of October 1987 caused extensive damage, especially in the south and east of England. Although not as severe, the storms of late January and February 1990 caused widespread damage to trees and woodlands in southern England, Wales, the Midlands and northern England. Around 3.9 million cubic metres of timber were brought down in 1987 and about 1.3 million cubic metres in 1990. An estimated 180,000 trees were planted in 1990-91 in urban, rural and historic landscapes under the Countryside Commission's "Task Force Trees", launched in partnership with local authorities, the private sector and voluntary bodies, to repair and improve landscapes following the 1987 storm, and extended after the 1990 storms.

Tree Preservation Orders

5.57 Important trees in urban areas, and woodlands of landscape value come under the control of Tree Preservation Orders (TPOs) which are designed to protect individual trees and woodland from wilful damage or destruction, uprooting, lopping, topping or felling without the consent of the local planning authority. Most TPOs apply to urban areas.

Urban areas

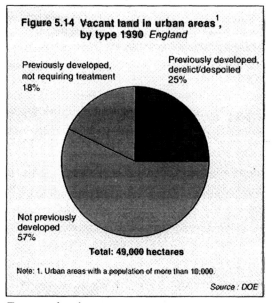

Figure 5.14 Vacant land in urban areas[1], by type 1990 England

Previously developed, not requiring treatment 18%

Previously developed, derelict/despoiled 25%

Not previously developed 57%

Total: 49,000 hectares

Note: 1. Urban areas with a population of more than 10,000.

Source : DOE

Rates of urbanisation

5.58 A report, Rates of Urbanisation in England 1981-2001, concluded that the growth of the urban area for the last twenty years of this century is likely to be relatively modest [8]. It estimated that land in urban use in England will increase from 10.2 per cent in 1981 to 11 per cent in 2001. In the South East where pressures for development are likely to be the greatest, the report forecast about 1.27 per cent of rural land transferring to urban use. In the South East outside London, land in urban use is expected to increase from 12.25 per cent to 13.5 per cent. Urban land as a percentage of county area in England in 1985 is shown in Figure 5.12. The recycling of urban land has already had a significant effect on urbanisation, particularly on the areas of greatest pressure. Thirty six per cent of the land currently developed for new housing is estimated to have been developed previously and a further 10 per cent was previously undeveloped land in built-up areas. Larger proportions of housebuilding land come from recycled land in areas already heavily urbanised, such as Greater London, Greater Manchester and Merseyside, West Yorkshire and Tyne and Wear. Figure 5.13 shows the percentage of housebuilding land in England previously in urban use.

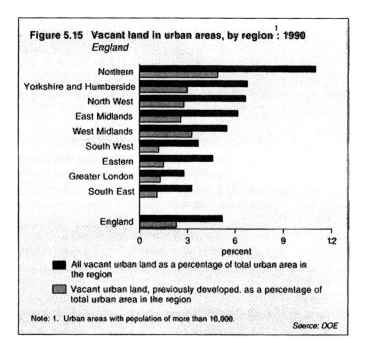

Figure 5.15 Vacant land in urban areas, by region[1]: 1990 England

percent

■ All vacant urban land as a percentage of total urban area in the region

▨ Vacant urban land, previously developed, as a percentage of total urban area in the region

Note: 1. Urban areas with population of more than 10,000.

Source: DOE

5.59 The need for new housing will continue well into the next century although probably with considerable regional variation. Changes in life style and life expectancy, leading to an increase in smaller households, are the main reasons rather than population growth or migration. In particular, factors such as the sharply increasing incidence of divorce and the reduced rates of remarriage, which emerged during the 1980s, are expected to lead to a significant increase in household formation. In the UK as a whole, the number of households will probably increase much faster than the growth in population and rise from 22.7 million now to 24.5 million by 2001.

5.60 Transport, both within and between towns and cities, and the demand for minerals extraction for fuel and building materials, also have important implications for land use. Chapter 15 on pressures on the environment

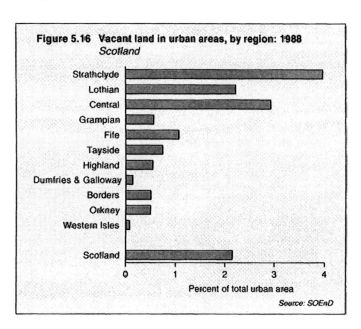

Figure 5.16 Vacant land in urban areas, by region: 1988 Scotland

Percent of total urban area

Source: SOEnD

discusses transport in more detail and minerals extraction is covered later in this chapter.

Vacant land

5.61 The first national sample survey of the stock of vacant land in urban areas of England was carried out in 1990[9]. The survey, which focused on urban settlements with a population of more than 10,000, estimated that there were 49,000 hectares of vacant land in such settlements (roughly 5 per cent of their total area) although not all of this land is available for development. Of this, some 21,000 hectares (43 per cent) of vacant land had been developed at some time in the past (see Figure 5.14). When the survey results were "grossed up" to include smaller urban settlements, the total national stock of urban vacant land was an estimated 60,000 hectares, of which some 25,000 hectares had previously been developed.

5.62 In proportional terms, as Figure 5.15 shows, vacant urban land was found to be most concentrated in the Northern region (11 per cent of the total urban area there), Yorkshire and Humberside (6.8 per cent), the North West (6.7 per cent) and the East Midlands (6.2 per cent). The Northern region also had the highest concentration of

---

Initiatives aimed at helping areas severely affected by dereliction and unemployment   Box 5.4

Urban Development Corporations (UDCs). Eleven have been set up since 1981; ten in England and one in Wales. A further UDC has recently been designated at Birmingham Heartlands. Table 5.7 shows the land reclaimed by the English UDCs as at June 1992.

In Wales, the Cardiff Bay Development Corporation has been established to revitalise the area of the city adjoining the waterfront. By the end of March 1991, just over 500 hectares of land had been reclaimed and 750 dwellings completed.

Enterprise Zones (EZs). Twenty seven EZs have been set up since 1981. Currently there are 21 EZs remaining in the UK. A number of zones have expired in 1991 and 1992 having reached the end of their ten year lifespans.

Garden Festivals. Garden Festivals have been held at five sites in the UK: Liverpool (1984), Stoke (1986), Glasgow (1988), Gateshead (1990) and Ebbw Vale (1992). The land reclaimed at the five sites totals 375 hectares.

The Urban Programme brings together Government, local authorities, the private and voluntary sectors and local communities to tackle the wide ranging problems of older urban areas. Grants are paid to cover 75 per cent of the costs of approved projects which form part of an annual programme to help to regenerate particular areas.

The Programme for the Valleys is a regeneration scheme for the economic and social development of the South Wales coal-mining region launched in 1988. By the end of 1991, almost 630 hectares of derelict land had been reclaimed (including the Ebbw Vale Garden Festival site) with a further 480 hectares scheduled for reclamation by March 1993.

City Grant supports private sector development projects in inner city areas. Since City Grant's inception in 1988, 276 projects have been approved and are proceeding.

City Challenge was launched in May 1991. It brings together seven existing programmes to tackle the problems of run down urban areas. Local authorities must form partnerships with the private sector, voluntary and community organisations, to put together regeneration schemes.

Simplified Planning Zones (SPZs) can help to promote the development or redevelopment of parts of an authority's area and are most likely to be useful in older urban areas. Any development started within ten years of making the scheme can bypass normal development control procedures. At present there are six adopted SPZ schemes in the UK and about ten more are being prepared.

Scottish Office led Urban Partnership initiatives in four public sector housing estates on the periphery of Glasgow, Edinburgh, Paisley and Dundee were established in 1988 to secure the economic, social and environmental regeneration of these severely deprived areas.

Local Enterprise Grants for Urban Projects (LEG-UP) is a scheme administered by Scottish Enterprise and its network of local enterprise companies. The scheme is aimed at encouraging private investment in projects which will benefit the economic development of urban areas with particular needs.

Northern Ireland: Urban Development Grant is paid to attract private investment.

Table 5.7	Land reclaimed by English Urban Development Corporations at March 1991 *England*
UDC	**Land reclaimed (ha)**
Black Country	146.0
Bristol	9.0
Central Manchester	3.5
Leeds	29.3
London docklands	582.6
Merseyside	312.0
Sheffield	28.5
Teesside	242.8
Trafford Park	58.0
Tyne and Wear	135.0
Total	**1,546.7**
	Source: DOE

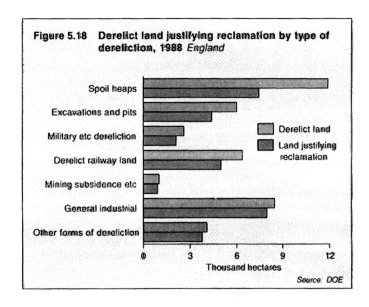

Figure 5.18 Derelict land justifying reclamation by type of dereliction, 1988 *England*

previously developed vacant land (4.9 per cent). The lowest concentrations of vacant land were found in Greater London (2.8 per cent), the South East (3.3 per cent) and the South West (3.7 per cent). These regions also had the lowest concentration (about 1 per cent) of previously developed vacant land.

5.63 In Scotland, the 1988 Pilot Scottish Vacant Land Survey [10] identified 5,060 hectares of vacant land within the built-up areas. This land was concentrated in Strathclyde region (67 per cent of the total area of vacant land), Lothian region (13 per cent) and Central region (8 per cent). The lowest amounts were found in the predominantly rural regions and the islands. See Figure 5.16. The survey showed that 2,247 hectares (44 per cent) had been vacant for at least ten years and that 2,490 hectares (49 per cent) were in private ownership.

Land registers

5.64 In 1989, arrangements were introduced for each public body to compile and maintain a register of all unused and underused land it owns in England at 31 March each year. The register is available for public inspection. Each local authority also provides annually, statistics on areas of land added, disposed of or brought into use.

Housing development

5.65 The estimated stock of permanent housing in the UK was 23.6 million dwellings at the end of December 1991. An estimated 183,000 new houses were completed in 1991. Figure 5.17 shows the trends over the last 20 years in the completion of new dwellings and numbers demolished. The annual publication, Housing and Construction Statistics, Great Britain, gives information on a wide range of housing matters over a ten year period [11].

5.66 DOE collects information from English local authorities about land with outstanding planning permission for private house building. At March 1990 there were permissions for 808,000 dwellings covering 32,000 hectares, compared with 726,000 dwellings and 30,000 hectares at March 1989 [12].

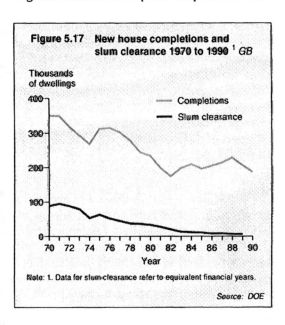

Figure 5.17 New house completions and slum clearance 1970 to 1990 [1] *GB*

Note: 1. Data for slum clearance refer to equivalent financial years.

Source: DOE

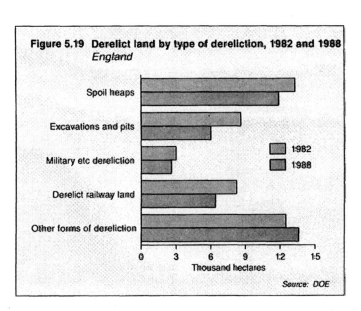

Figure 5.19 Derelict land by type of dereliction, 1982 and 1988 *England*

Source: DOE

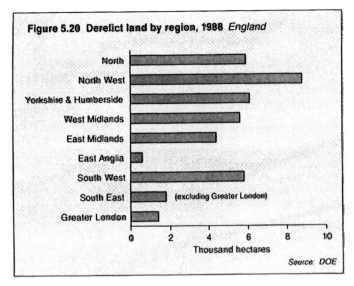

Figure 5.20 Derelict land by region, 1988 *England*

Thousand hectares

Source: DOE

Urban improvement

5.67 There are financial incentives and simplification of planning controls to encourage private companies to move to or to expand in particular locations affected by dereliction and unemployment. (See Box 5.4).

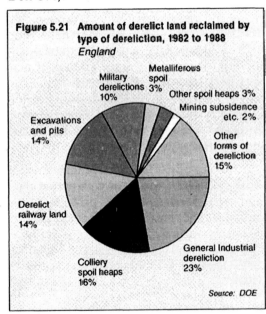

Figure 5.21 Amount of derelict land reclaimed by type of dereliction, 1982 to 1988 *England*

Military derelictions 10%
Metalliferous spoil 3%
Other spoil heaps 3%
Mining subsidence etc. 2%
Excavations and pits 14%
Other forms of dereliction 15%
Derelict railway land 14%
General industrial dereliction 23%
Colliery spoil heaps 16%

Source: DOE

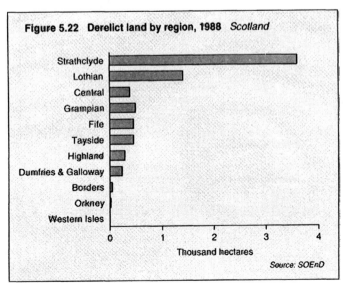

Figure 5.22 Derelict land by region, 1988 *Scotland*

Thousand hectares

Source: SOEnD

Derelict land

5.68 Derelict land describes land and buildings which have been so damaged by industrial and other development that they are incapable of beneficial use without treatment. Information about derelict land in England, Wales and Scotland is based on periodic surveys. In Northern Ireland data are held on derelict buildings on land owned by DOE(NI) which are regarded as contributing to urban dereliction.

England

5.69 The 1988 Survey of Derelict Land in England[13] recorded 40,500 hectares of derelict land (0.3 per cent of the area of England). Disused spoil heaps were the most common form of derelict land (11,900 hectares), but general industrial dereliction was also extensive (8,500 hectares). Approximately 31,600 hectares of the derelict land (78 per cent) were considered by local authorities to justify reclamation. Figure 5.18 shows derelict land justifying reclamation by type of dereliction in 1988.

5.70 The total area of derelict land had declined since 1982, in contrast to increases shown in the 1982 and 1974 surveys. Figure 5.19 compares the reductions in the different types of dereliction. There was an 11 per cent reduction in all derelict land, and an 8 per cent reduction in land justifying reclamation.

5.71 Figure 5.20 shows the amount of derelict land by region. The survey showed that most derelict land was found north of a line drawn from the Bristol Channel to the Wash. All regions except Yorkshire and Humberside had less derelict land than in 1982.

5.72 Between 1982 and 1988, 14,000 hectares of derelict land were reclaimed. Figure 5.21 shows derelict land reclaimed between 1982 and 1988 by type of dereliction. Of the land reclaimed, 63 per cent was subsequently used for sport and recreation, public open space, agriculture and forestry, while 27 per cent was used for industry, commerce and housing. The remaining 10 per cent was in miscellaneous use.

Wales

5.73 A derelict land survey was carried out in Wales in 1988 but the latest survey for which results are available, relates to 1971-72. This latter survey showed that 20,580 hectares of land in Wales were derelict. Since then, derelict land has been reclaimed at an average annual rate of nearly 330 hectares, with three quarters of the reclamation occurring in the industrial counties of south east Wales.

Scotland

5.74 Nearly 7,400 hectares of derelict land were recorded in Scotland in the 1988 pilot Scottish Vacant and Derelict Land Survey. The survey covered derelict land in urban and rural areas, but the major concentrations were in the urban areas. Strathclyde region with 49 per cent and Lothian region with 19 per cent predominated (see Figure 5.22). Almost 58 per cent of the total derelict land had lain derelict for ten years or more.

5.75 The private sector owned 59 per cent of the derelict land recorded in the pilot survey; 31 per cent was owned by the public sector (largely district councils, British Coal and British Rail); the remaining 10 per cent was either in mixed ownership or ownership was unknown[10].

5.76 During 1988 and 1989, over 300 hectares of derelict land in Scotland was reclaimed by the Scottish Development Agency.

Contaminated land

5.77 The major industrial conurbations of the North, the Midlands, South Wales and Scotland have large areas of land formerly used by industry and now derelict or reclaimed for other uses. Much of this land may be contaminated. However, many other parts of the country are or were centres for industries or trades which may cause contamination of the soil, and the sites of former town gas works, scrap yards and waste disposal facilities, for example, are found throughout the country.

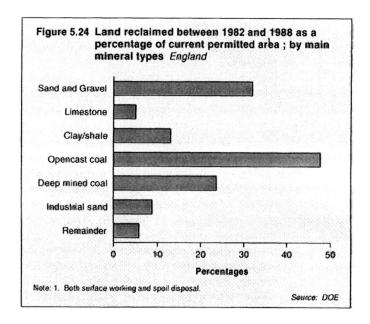

Figure 5.24 **Land reclaimed between 1982 and 1988 as a percentage of current permitted area ; by main mineral types** *England*

Note: 1. Both surface working and spoil disposal.

Source: DOE

5.78 Neither the total area of contaminated land nor the number of sites in the UK can easily be quantified at present. Current independent estimates suggest that there may be 50-100,000 hectares of contaminated land, but this should be regarded as no more than a broad estimate.

5.79 In 1982, the Welsh Office and the Welsh Development Agency commissioned a study of contaminated land in Wales. An updated version was published in 1988 which identified 749 sites in Wales. The Welsh studies have not covered sites in active use nor sites under 0.5 hectares[14].

5.80 The 1988 pilot Scottish Vacant Land Survey recorded 402 hectares of land known to be contaminated. Just over 800 hectares were suspected to be contaminated, some 4,000 hectares were thought not to be contaminated, and the state of the remaining 7,250 hectares covered by the survey was not known[10]. These estimates are approximate only.

Mineral workings

England

5.81 In England in 1988, 116,000 hectares of land had permission for surface mineral extraction or disposal of mineral wastes. The 1988 Survey of Land for Minerals Workings in England[15] looked at land affected by mineral activities, and described the nature and extent of mineral workings and spoil tips at 1 April 1988. The area of land affected by permissions for surface mineral workings was 96,100 hectares and this is shown by main mineral type in Figure 5.23. Permissions for sand and gravel extraction were the most extensive, accounting for 30 per cent of the total permitted area for all surface workings. Just over half of the total

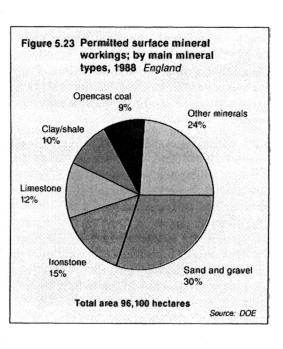

Figure 5.23 **Permitted surface mineral workings; by main mineral types, 1988** *England*

Opencast coal 9%

Other minerals 24%

Clay/shale 10%

Limestone 12%

Ironstone 15%

Sand and gravel 30%

Total area 96,100 hectares

Source: DOE

area permitted for surface workings had been worked but not reclaimed; the remainder had not yet been worked.

5.82 Information about underground mining is less certain. The survey points out that the 780,000 hectares recorded as being affected by permissions for underground mining understates the position. Details of the extent of rights under the General Development Order for underground coal mining are not available for some areas; they are thought to be extensive but unlikely ever to be worked to their full extent.

5.83 Between 1982 and 1988, 20,600 hectares of land were reclaimed from mineral workings. Figure 5.24 shows land reclaimed as a percentage of the current total permitted areas, both worked and not yet worked, for the main mineral types. Figure 5.25 shows the main uses following reclamation. Agriculture and amenity together account for almost 90 per cent of all uses. There are, however, wide variations between mineral types. For example, 80 per cent of the total area reclaimed to agriculture came from just three mineral types - sand, gravel and coal.

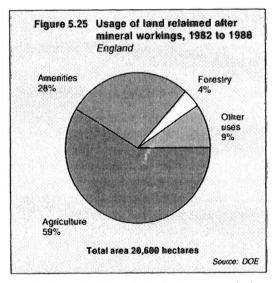

Figure 5.25   Usage of land reclaimed after mineral workings, 1982 to 1988
England

Amenities 28%
Forestry 4%
Other uses 9%
Agriculture 59%

Total area 20,600 hectares

Source: DOE

5.84 About half of all dereliction recorded in the DOE's 1988 Derelict Land Survey was a result of former mineral workings. The main source of mineral workings dereliction was sites worked under old planning permissions which had no provisions for reclamation, although some dereliction was also the result of provisions for reclamation being unfulfilled.

Wales

5.85 The 1988 Minerals Survey for Wales[16] estimated the area of land affected by permissions for surface minerals extraction or related disposal of mineral wastes at 14,100 hectares. The area affected by permissions for extraction alone was estimated at 10,800 hectares. Permissions for opencast

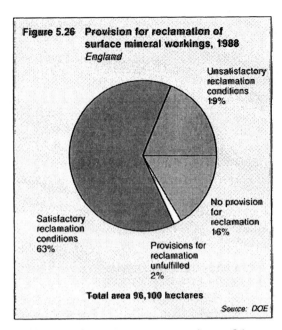

Figure 5.26   Provision for reclamation of surface mineral workings, 1988
England

Unsatisfactory reclamation conditions 19%
No provision for reclamation 16%
Provisions for reclamation unfulfilled 2%
Satisfactory reclamation conditions 63%

Total area 96,100 hectares

Source: DOE

coal were the most extensive (over 31 per cent of the total permitted area for all surface workings). Seventy eight per cent of the total area permitted for surface workings had been worked but not reclaimed; the remainder had not yet been worked.

5.86 As in England, the area recorded as being affected by permissions for underground mining (9,700 hectares) is thought to be understated.

5.87 Reclamation in Wales between 1982 and 1988 amounted to 2,400 hectares of land. Agriculture and amenity accounted for nearly 90 per cent and only 6 per cent was reclaimed to forestry. About two thirds of the total area reclaimed to agriculture came from opencast coal.

Regulation of mineral workings

5.88 In England and Wales, mineral planning authorities (mpas) are responsible for the control of mineral working. Mpas include county councils, the former metropolitan county council areas, London boroughs, the Peak Park Joint Planning Board and the Lake District Special Planning Board. They are responsible for undertaking periodic surveys of land affected by mineral workings for DOE.

5.89 Over 80 per cent of the land reclaimed between 1982 and 1988 was covered by conditions attached to planning permissions to ensure reclamation when mineral activities ceased. The rest was reclaimed by Derelict Land Grant or as a result of planning permissions for subsequent development on the land. Figure 5.26 shows the reclamation provisions attached to the planning permissions for areas of permitted surface mineral workings in England in 1988. Overall about 63 per cent had satisfactory conditions, nearly 19 per cent had unsatisfactory conditions, and some 18 per cent had no provisions for reclamation or the conditions were unfulfilled.

5.90 In Scotland, mineral workings planning is the responsibility of the district or Islands Councils, except in Borders, Dumfries and Galloway and Highland regions where the Regional Council is responsible. In Northern Ireland, mineral workings planning is the responsibility of DOE(NI).

Heritage

5.91 Most historically or aesthetically important buildings and land in the UK are owned privately, but responsibility for protecting this heritage rests with government departments and various statutory agencies. See Box 5.5 for details of those operating in England. A significant contribution is also made by the numerous national and local voluntary organisations including the National Trust. The approaches adopted by England, Wales, Scotland and Northern Ireland are discussed in the following sections.

England

5.92 The Department of National Heritage (DNH) is the lead Department for heritage matters in England. It relies on a number of bodies with expertise in particular aspects of the heritage, see Box 5.5.

Listing

5.93 Buildings of special architectural or historic interest are listed under the Planning (Listed Buildings and Conservation Areas) Act 1990. Buildings are listed in grades I, II* or II, according to their importance (a starred grade II being more important historically than grade II alone). Once listed, a building

---

Statutory agencies with heritage responsibilities **Box 5.5**

English Heritage (EH) was established in 1984 by the National Heritage Act 1983 to:

- manage and promote historic properties and sites in the Government's ownership and guardianship; EH manages some 400 castles, abbeys, historic houses and other sites and properties in England, most of which are open to the public;
- operate grant support schemes for archaeology and for the repair of outstanding historic buildings and conservation areas; and
- to advise the Government on the statutory protection of the heritage and on heritage matters generally.

The Royal Commission on the Historical Monuments of England (RCHME) was established by Royal Warrant in 1908 as the national body of survey and record for the historic environment. Its main functions are to:

- compile and maintain the National Monuments Record(NMR);
- survey archaeological sites and historic buildings in order to enhance and update the NMR;
- make the NMR available to the public and to promote its use by the public;
- oversee the system of local Sites and Monuments Records; and
- undertake the Survey of London which provides a detailed and comprehensive history of London's building fabric.

The National Heritage Memorial Fund (NHMF) was established under the National Heritage Act 1980 to give financial assistance towards the cost of acquiring, maintaining or preserving land, buildings, works of art and other objects of outstanding interest which are also of importance to the national heritage. It is funded partly by income from its investments and partly through grant-in-aid from the Department of National Heritage (DNH).

The Royal Fine Art Commission (RFAC), established by Royal Warrant in 1924, advises Government Departments, local planning authorities and other bodies in England, and also in Wales, on mainly architectural town planning and landscape matters, but also important matters of public aesthetic interest including sculpture, painting and engineering works.

The Royal Armouries cares for the national collection of arms and armour, to improve its presentation to the public, and to develop the Armouries museum as a centre of research and excellence.

The Redundant Churches Fund (RCF) was established in 1969 to preserve, in the interests of the nation and the Church of England, churches of outstanding historic or architectural importance which are no longer needed for regular worship (the RCF currently has some 270 churches vested in it). It is funded jointly by the DNH and the Church Commissioners.

Table 5.8	Consent granted for alterations and demolition of listed buildings; 1990-91 England			
Grade	Total number of buildings in grade	Alteration/ extension	Partial demolition	Total demolition
I	12,550	331	75	0
II*	25,800	567	171	4
II	461,650	19,998	634	105
				Source: DOE

cannot be altered, demolished or extended in any way which affects its character, without the consent of the local planning authority. At national level DOE remains responsible for policy on alterations to or demolition of listed buildings.

5.94 There are currently about 440,000 entries in the statutory lists representing about 500,000 individual buildings (some are listed in groups). Figure 5.27 shows the increase in the number of listed buildings over the last decade.

Alteration and demolition of listed buildings

5.95 In 1990-91 consent was given to the total demolition of 109 listed buildings (0.02 per cent of all listed buildings) and to the partial demolition of 880 listed buildings (0.18 per cent). Consent was also granted to the alteration or extension of almost 21,000 listed buildings (4.2 per cent). In each case the great majority of the buildings in question were listed grade II. See Table 5.8.

Buildings at risk

5.96 In 1988 English Heritage (EH) invited all local authorities to take part in a national survey of listed buildings at risk from neglect. The results of a sample survey, carried out during 1990-91, showed that most listed buildings are in reasonable condition and that only about 7 per cent of listed buildings are

at risk. But on the basis of the survey's findings 36,700 listed buildings in England may be at risk from neglect, with a further 72,850 being vulnerable[17].

5.97 As well as listed buildings, there are conservation areas. Local authorities designate them where they judge that the character or appearance of an entire area should be preserved or enhanced. Such designations introduce control over the demolition of most buildings, including unlisted buildings, in an area. Over 6,000 conservation areas have so far been designated.

Scheduled ancient monuments

5.98 Monuments of national importance are scheduled under the Ancient Monuments and Archaeological Areas Act 1979, and there are currently some 13,000 scheduled monuments in England. Once scheduled, consent is required before any works are carried out which would damage a monument.

Wales

5.99 In Wales, the lead is with the Welsh Historic Monuments Executive Agency (CADW). The arrangements for listing and dealing with applications for consent to carry out works to listed buildings are broadly similar to those in England. There are over 13,600 listed buildings, compared with 9,000 in 1981, and a current survey is expected to add 20,000 to the list. The Royal Commission on Ancient and Historical Monuments in Wales is charged with making an inventory of all historic monuments and constructions and specifying those most worthy of preservation. There are over 2,600 scheduled monuments in Wales.

Scotland

5.100 The responsibility for identifying and protecting the built heritage of Scotland lies with Historic Scotland.

5.101 Historic Scotland is also responsible for the scheduling and listing of buildings, monuments and sites, and for administering the consent systems which apply to such scheduled monuments and listed buildings. There are currently almost 5,000 scheduled monuments and some 37,000 listed buildings. A major survey is underway to review the list.

Northern Ireland

5.102 In Northern Ireland, the Environment Service of DOE(NI) is responsible for the built heritage, including the scheduling, listing and care of monuments and buildings; maintaining archives; licensing and carrying

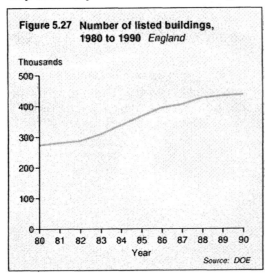

Figure 5.27 Number of listed buildings, 1980 to 1990 England

Thousands

Source: DOE

World heritage sites in the UK   **Box 5.6**

Hadrian's Wall Military Zone

Durham Castle and Cathedral

St Kilda

Fountain's Abbey and St Mary's, Studley Royal

The Castles and Town Walls of Edward I in Gwynedd

Ironbridge Gorge

Stonehenge, Avebury and Associated Sites

Giant's Causeway and Causeway Coast

Blenheim Palace

Palace of Westminster and Westminster Abbey

City of Bath

The Tower of London

Canterbury Cathedral, St Augustine's Abbey and St Martin's Church.

---

out archaeological excavations; and grant-aiding conservation of listed buildings. In 1991 there were 7,900 listed buildings, compared with 5,200 in 1981. The survey and listing of all pre-1960 buildings of architectural or historic merit will be completed by 1994. In addition, Conservation Areas increased from 13 in 1981 to 28 in 1991. Scheduled historic monuments rose from 437 in 1980 to 1,026 in 1991, and are expected to number 1,500 by 1995.

The Global Heritage

World Heritage Convention, UNESCO.

5.103 The World Heritage Convention (WHC) has two main purposes: to draw up a list of World Heritage sites (sites of outstanding universal value, which UN member states pledge to protect); and to operate the World Heritage Fund which gives practical support to conservation projects at threatened sites on the list. Thirteen sites in the UK have received recognition by the WHC. See Box 5.6.

References and further reading

1. Department of the Environment, (1992). Land Use Change in England Statistical Bulletins. No 7 Statistical Bulletin (92)4 Changes in Land Use in England in 1987 and No 6 Statistical Bulletin (92)3 Changes in Land Use in England in 1985 and 1986.

2. Department of the Environment and Countryside Commission, (1986). Monitoring landscape change. Hunting Surveys and Consultants Limited.

3. Institute of Terrestrial Ecology, (1991). Changes in Hedgerows in Britain between 1984 and 1990. ITE

4. Barr, C. J., et al. Landscape Changes in Britain. Institute of Terrestrial Ecology.

5. Countryside Commission, (1991). Landscape Change in the National Parks. Countryside Commission.

6. Statutory Instrument. The Town and Country Planning (Assessment of Environmental Effects) Regulations 1988 (SI 1988 No 1199). HMSO.

7. Innes, J. L., & Boswell, R. C., Forestry Commission, (1991). Forest Monitoring Programme 1991 Results. Research Information Note 209. HMSO.

8. Department of the Environment, (1990). Rates of Urbanisation in England 1981-2001. HMSO.

9. Department of the Environment (to be published). A National Sample Survey of Vacant Land in Urban Areas of England.

10. The Scottish Office/Scottish regional and district councils/Scottish Development Agency, (1990). Pilot Scottish Vacant and Derelict Land Survey 1988.

11. Department of the Environment/Scottish Office/Welsh Office. Housing and Construction Statistics Great Britain. HMSO.

12. Department of the Environment (to be published). Development Control Statistics, 1990-91.

13. Department of the Environment, (1991). Survey of Derelict Land in England 1988. HMSO.

14. The Welsh Office, (1988). Survey of Contaminated Land in Wales. HMSO.

15. Department of the Environment, (1991). Survey of Land for Minerals Workings in England 1988. HMSO.

16. Welsh Office, (1991). 1988 Minerals Survey for Wales. HMSO.

17. English Heritage, (1992). Buildings at Risk: a Sample Survey. English Heritage.

Ministry of Agriculture, Fisheries and Food, (1990). Agriculture in the UK 1990. HMSO.

Scottish Office Environment Department, (1992). The Scottish Environment Statistics No 3. Scottish Office.

Welsh Office, (1992). Environmental Digest for Wales, No. 6 1991. Welsh Office.

Countryside Commission, (1992). Protected Landscapes in the United Kingdom. Countryside Commission Publications.

Department of the Environment. Planning Policy Guidance Series. HMSO.

Department of the Environment. Minerals Planning Guidance Series. HMSO.

Scottish Office. National Planning Policy Guideline series.

Nature Conservancy Council, (1991). Seventeenth Report, 1 April 1990 - 31 March 1991. English Nature.

6 Inland water resources and abstraction

☐ Average annual replenishment of fresh water resources in the UK from normal rainfall is estimated to be around 120,000 gigalitres surface water resources and 9,800 gigalitres from main groundwater resources (6.12).

☐ Overall, relatively small proportions of fresh water resources are utilised - around a quarter of groundwater and less than 10 per cent of surface water resources. There are significant variations over time and regionally (6.12).

☐ Water abstractions increased by 4 per cent in England and Wales between 1980 and 1990 (6.16). Abstractions for piped mains water increased by 13 per cent, but were partly offset by a fall in abstractions by industry, because of a contraction in some industries and more efficient water usage and recycling (Figures 6.11 and 6.14).

☐ The National Rivers Authority have identified 40 rivers where low flow is a problem, partly because of over abstraction. Twenty are being investigated with a view to finding possible solutions (6.22).

☐ Of the 20,350 megalitres per day of piped mains water in the UK, 28 per cent is metered. Metered water is supplied mainly to non-domestic users (Figure 6.15).

☐ Average domestic consumption of water in 1990 in England and Wales was about 140 litres per person per day. Around 30 per cent of the average household use of water is for WC flushing (Figure 6.18).

☐ There have been three major dry periods in the last twenty years. 89 drought orders were made during 1989 and 61 in 1990 in England and Wales (Table 6.3).

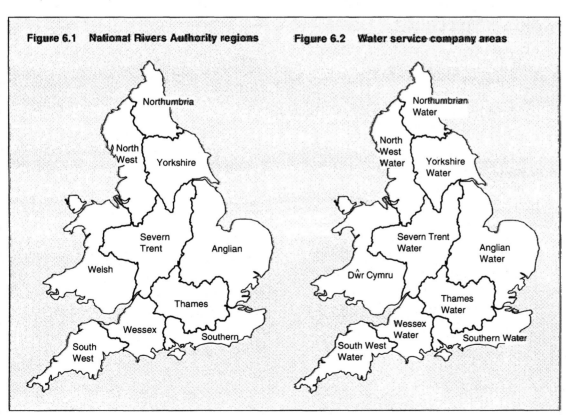

Figure 6.1 National Rivers Authority regions

Figure 6.2 Water service company areas

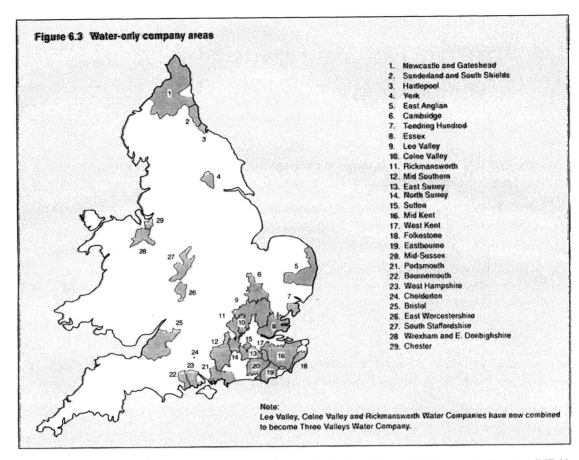

Figure 6.3 Water-only company areas

1. Newcastle and Gateshead
2. Sunderland and South Shields
3. Hartlepool
4. York
5. East Anglian
6. Cambridge
7. Tendring Hundred
8. Essex
9. Lee Valley
10. Colne Valley
11. Rickmansworth
12. Mid Southern
13. East Surrey
14. North Surrey
15. Sutton
16. Mid Kent
17. West Kent
18. Folkestone
19. Eastbourne
20. Mid-Sussex
21. Portsmouth
22. Bournemouth
23. West Hampshire
24. Cholderton
25. Bristol
26. East Worcestershire
27. South Staffordshire
28. Wrexham and E. Denbighshire
29. Chester

Note:
Lee Valley, Colne Valley and Rickmansworth Water Companies have now combined to become Three Valleys Water Company.

6.1 This chapter discusses surface and groundwater resources, the demand for water and abstractions, and low river flow. Later sections cover the public water supply and metering and the average household use of water. The final section looks at recent periods of very dry weather.

6.2 The National Rivers Authority (NRA) are responsible for managing water resources in England and Wales. Water service companies and water-only companies (collectively referred to as water undertakers) are responsible for public water supplies (see Box 6.1). In Scotland, public water supplies are provided by the Regional and Islands Councils; River Purification Authorities (RPAs) are responsible for maintaining and improving the water environment, and river gauging (taking measurements). In Northern Ireland, the Department of the Environment for Northern Ireland has responsibilities for water resources, supplies and sewerage services. Water resources must be managed effectively to take account of the needs of users (represented by the "abstractors", who include public water supply undertakers, agricultural and industrial concerns) and to take account of water quality and wider environmental and recreational concerns. Over-abstraction and over-usage can result in the drying up or depletion of lakes, rivers, streams and groundwater, thereby causing damage to wildlife habitats, deterioration in water quality and adverse impacts on recreation and amenity.

Water resources

6.3 Water resources fall into two categories; surface water resources such as lakes, reservoirs, rivers and streams; and

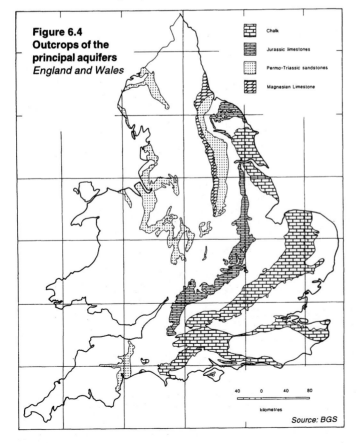

Figure 6.4
Outcrops of the principal aquifers
England and Wales

Chalk

Jurassic limestones

Permo-Triassic sandstones

Magnesian Limestone

40 0 40 80
kilometres

Source: BGS

groundwater resources held in water bearing rocks called aquifers. Successful development and management of water resources depends on the availability of adequate hydrological data. Detailed knowledge of the expected range of flows for a river, including the frequency and duration of low flows in droughts, is required for the estimation of reliable yields of water resource schemes, for flood control purposes and for environmental management such as the estimation of the dilution available for effluents. A national surface and groundwater archive is maintained by the Natural Environment Research Council (NERC) through the Institute of Hydrology (IH) and the British Geological Survey (BGS), incorporating data from over 1,200 gauging stations throughout the UK. A yearbook is published containing a broad selection of surface and groundwater data [1].

6.4 Figure 6.4 shows the outcrops of the principal aquifers in England and Wales. Major limestone and sandstone aquifers occur over much of England but are absent from most of the high ground in the west and Wales. The sites of the main public groundwater resources in Scotland are shown in Figure 6.5.

6.5 Surface waters are the predominant source of supply in the north and west of England and Wales because of the impermeable rocks in these areas and the relatively high rainfall. Aquifers are a particularly important source of water for public water supply in the south east of England and parts of the midlands. In Scotland abundant, easily developed, surface water sources are the main resource and the development of groundwater resources for public water supplies has been relatively limited.

6.6 Figure 6.6 shows the water cycle. The adequacy of water resources at any given time or place depends on the demand for water, the gross storage capacity, and the net volume of water in storage (which in turn depends on *inputs* ie rainfall, perhaps over several seasons, and *outputs* ie evaporation).

Inputs and outputs

6.7 Average annual rainfall varies substantially from region to region (see also the section on precipitation in Chapter 1). Three measures are discussed below; the *annual average rainfall*; *average effective rainfall* which is defined as average water resources from rainfall less evaporation (assuming no "runoff" and ignoring re-use of abstracted water); and a similar measure but

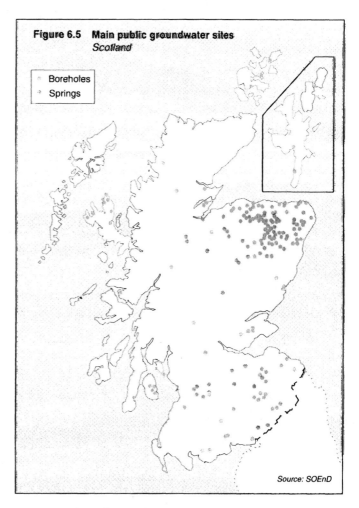

Figure 6.5   Main public groundwater sites
Scotland

□ Boreholes
∘ Springs

Source: SOEnD

assuming drought conditions, termed *average effective rainfall under drought conditions* (ie conditions which might be expected to occur once every fifty years, and generally used as a benchmark for assessing the

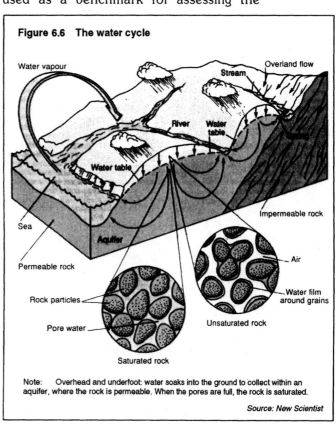

Figure 6.6   The water cycle

Water vapour
Overland flow
Stream
River
Water table
Water table
Impermeable rock
Sea
Aquifer
Permeable rock
Air
Water film around grains
Rock particles
Unsaturated rock
Pore water
Saturated rock

Note: Overhead and underfoot: water soaks into the ground to collect within an aquifer, where the rock is permeable. When the pores are full, the rock is saturated.

Source: New Scientist

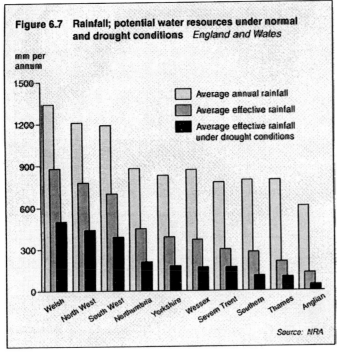

Figure 6.7 Rainfall; potential water resources under normal and drought conditions *England and Wales*

mm per annum

Average annual rainfall
Average effective rainfall
Average effective rainfall under drought conditions

Source: NRA

availability of water supplies). Figure 6.7 shows these three measures for each NRA region. Wales, the North West and South West England potentially have higher water resources than the South East. Adequacy of resources however also depends on demand and reservoir storage capacity.

Demand

6.8 The demand for water (measured in terms of actual abstractions), compared with potential water resources (effective rainfall under drought conditions), in each NRA region is shown in Figure 6.8. The Thames region meets a demand in excess of available resources through re-use of water.

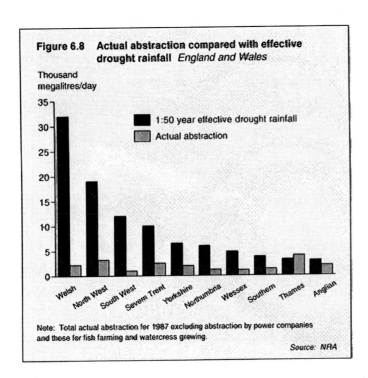

Figure 6.8 Actual abstraction compared with effective drought rainfall *England and Wales*

Thousand megalitres/day

1:50 year effective drought rainfall
Actual abstraction

Note: Total actual abstraction for 1987 excluding abstraction by power companies and those for fish farming and watercress growing.

Source: NRA

6.9 Figure 6.9 shows the balance between current demand and available resources for public water supply for regions in England and Wales. The ratios, showing surplus of resources over average demand, give only a broad indication of the availability of resources and may vary widely within regions. In Scotland, in 1990, there was a 52 per cent surplus of resources over average demand.

6.10 Many water undertakers in England and Wales are planning to increase their water storage capacity and major options for meeting additional demand are set out in a recent NRA report[2]. The report also gives projections which show for each region how actual and prospective demands for piped mains water (public water supplies, excluding industrial and agricultural use) vary in relation to actual resources, taking into account possible water resource developments. Figure 6.10 shows the average demand projections to 2021 by region. These are preliminary estimates only. In Scotland, the total available yield of surface water sources developed for public water supply is not expected to change substantially in the foreseeable future. The Department of the Environment for Northern Ireland is planning the development of a new source to supply an additional 130 megalitres of water per day to the Greater Belfast area by the end of the century (either by abstraction from Lough Neagh or the construction of an upland reservoir at Glenwhirry, County Antrim).

6.11 The Department of the Environment (DOE) issued a consultation paper[3] on water conservation in July 1992, focusing on the options for reducing wastage and using water more efficiently. Options include better use of water and ways of reducing consumption including cutting leakage, installing meters and recycling.

Water volume

6.12 Total annual water replenishment in the UK has been estimated as 120,000 gigalitres per year in the form of surface water resources and 9,800 gigalitres per year from main groundwater resources. Relatively small proportions of these resources are used; the percentage utilisation of groundwater resources is about a quarter and of surface water less than 10 per cent[4]. There are however, significant variations over time and regionally.

6.13 Water resources can also be measured by estimating the reliable yield of piped mains water available to water undertakers. Average reliable yields relate to the continuous supply that could be obtained during drought conditions (assuming that there is sufficient

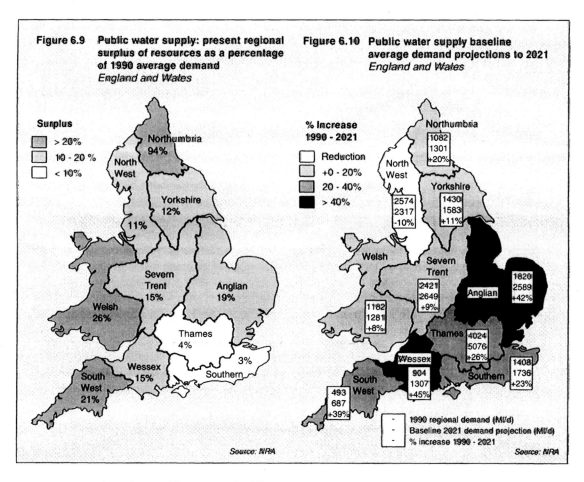

Figure 6.9 Public water supply: present regional surplus of resources as a percentage of 1990 average demand *England and Wales*

Surplus

- > 20%
- 10 - 20 %
- < 10%

Northumbria 94%

North West 11%

Yorkshire 12%

Severn Trent 15%

Welsh 26%

Anglian 19%

Thames 4%

3% Southern

South West 21%

Wessex 15%

Source: NRA

Figure 6.10 Public water supply baseline average demand projections to 2021 *England and Wales*

% Increase 1990 - 2021

- Reduction
- +0 - 20%
- 20 - 40%
- > 40%

Northumbria 1082 1301 +20%

North West 2574 2317 -10%

Yorkshire 1430 1583 +11%

Welsh

Severn Trent 2421 2649 +9%

1182 1281 +8%

Anglian 1820 2589 +42%

Thames 4024 5076 +26%

Wessex 904 1307 +45%

Southern 1408 1736 +23%

South West 493 687 +39%

- 1990 regional demand (Ml/d)
- Baseline 2021 demand projection (Ml/d)
- % increase 1990 - 2021

Source: NRA

capacity to utilise the yield potential). The total reliable yield for England and Wales is 20,200 megalitres per day. In Scotland, the total available yield of surface water sources so far developed for public water supply is about 3,500 megalitres per day. The estimated reliable yield in Northern Ireland is 500 megalitres per day.

Water abstractions

6.14 Abstractions of surface water and groundwater are subject to a licensing system in England and Wales. Licence holders include water undertakers (supplying piped mains water), industry (particularly electricity

generating companies), and agricultural concerns.

6.15 Water abstraction in Scotland is subject to separate controls for most public supply purposes and for agricultural and horticultural irrigation.

6.16 Estimated total abstractions (in England and Wales) since 1980 are shown in Figure 6.11. Abstractions fell from 33,900 megalitres per day in 1980 to 31,500 megalitres per day in 1985 (a reduction of 7 per cent), increasing to 35,200 megalitres per day in 1990 (an increase of 12 per cent). Over the decade abstractions have increased by 4 per cent. Total abstraction figures include a substantial amount of re-use of water.

Abstractions by region

6.17 Estimated abstractions by region in England and Wales are given in Figure 6.12 (showing separately abstractions from surface water and groundwater). The two main abstracting regions are Welsh and Severn Trent where considerable amounts are abstracted for electricity generation. Groundwater abstractions account for 20 per cent of all abstractions on average, over England and Wales but there are considerable regional variations, with abstractions ranging from 51 per cent in Wessex to 1 per cent in Wales.

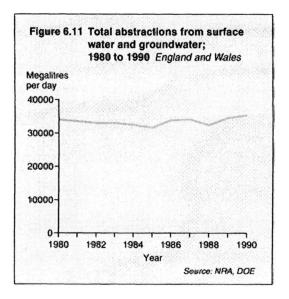

Figure 6.11 Total abstractions from surface water and groundwater; 1980 to 1990 *England and Wales*

Megalitres per day

40000

30000

20000

10000

0

1980 1982 1984 1986 1988 1990

Year

Source: NRA, DOE

> ## The National Rivers Authority, water service companies and water-only companies
> ### Box 6.1
>
> In 1989 the structure and organisation of the water industry in England and Wales was changed following the Water Act 1989; ten new private sector companies were created to replace the water authorities and also two new regulatory bodies, the National Rivers Authority (NRA) and the Office of the Director General of Water Services. The Drinking Water Inspectorate was also established under the Water Act 1989.
>
> The Water Act 1989 and other associated legislation were consolidated on 1 December 1991 into five separate Acts. The legislation relating to the NRA is now generally in the Water Resources Act 1991.
>
> The **National Rivers Authority's** responsibilities for the management of water resources include water resource planning, authorising and controlling abstractions and operational involvement in augmenting river flows to support abstractions, and protecting the environment through river regulation and water transfer schemes. (These latter schemes refer to the transfer of water from one river basin or catchment, across watersheds, to another to provide adequate supplies in these areas; and to regulate river flows by releasing water from storage during times of low river flow and transferring water to storage during high flows). The NRA maintains management agreements with water companies, taking account of the need to release water from reservoirs for the benefit of the downstream river environment, and other uses such as recreation.
>
> The NRA also has responsibilities for controlling pollution in inland, estuarial and coastal waters. Pollution control is carried out through discharge consents (any discharge by industry, farms, sewage treatment works etc into controlled waters has to be authorised by the NRA, having regard to the type and concentration of pollutants, volume of discharge etc), monitoring water quality and the achievement of water quality standards (see also Chapter 7 on inland water quality and pollution).
>
> In addition the NRA has responsibilities for flood defence and land drainage, salmon and freshwater fisheries, navigation in some areas, nature conservation and recreation in inland waters and associated land.
>
> Figure 6.1 shows the NRA administrative regions, which are delimited in terms of river catchments.
>
> The ten new private sector companies normally referred to as **water service companies** are responsible for the provision of water and sewerage services in the areas previously served by water authorities. Figure 6.2 shows the water service company areas.
>
> Statutory water companies normally referred to as **water-only companies** which were already in the private sector continue with their existing responsibility of serving some 25 per cent of the population in England and Wales with water supply. Figure 6.3 shows the water-only company areas.
>
> Water service companies and water-only companies are collectively referred to as water undertakers.

6.18 Percentage changes in amounts of water abstracted over the periods 1980 to 1985, and 1980 to 1990, are given in Table 6.1, by region. There have been significant reductions in the North West, Yorkshire and

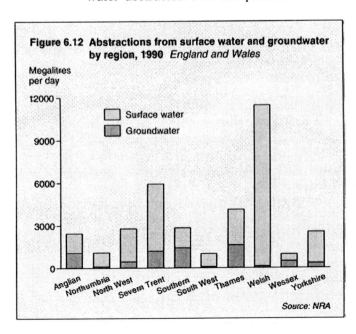

Figure 6.12 Abstractions from surface water and groundwater by region, 1990 *England and Wales*

Megalitres per day

Surface water
Groundwater

Source: NRA

Table 6.1 Percentage change in abstractions, by region, 1980 to 1990 *England and Wales*

NRA region	1980 to 1985	1980 to 1990
	%	%
Yorkshire	-38.0	-37.5
North West	-14.5	-88.0
Northumbria	3.1	4.0
Severn Trent	-5.3	-11.4
Anglian	-3.3	23.2
Thames	2.2	-1.4
Wessex	4.1	20.9
Welsh	14.2	55.1
Southern	18.8	124.5
South West	93.4	123.3

Source: NRA, DOE

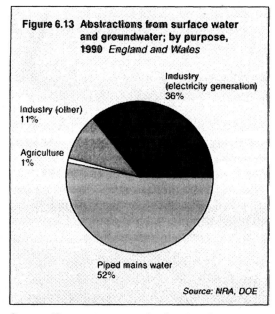

Figure 6.13 Abstractions from surface water and groundwater; by purpose, 1990 *England and Wales*

Industry (electricity generation) 36%

Industry (other) 11%

Agriculture 1%

Piped mains water 52%

Source: NRA, DOE

Severn Trent regions and substantial increases in Southern and South West regions.

Abstractions by purpose

6.19 Abstractions by purpose are shown in Figure 6.13. Abstraction for public water supply accounts for 52 per cent of all abstractions; electricity generating companies account for a further 36 per cent (but water abstracted for electricity generation is returned to source; the total volume abstracted is therefore of less importance environmentally).

6.20 Trends in abstractions by purpose since 1980 are shown in Figure 6.14. Abstractions for piped mains water supply increased by 13 per cent. Abstractions by industry (other than electricity supply) declined, reflecting in part a contraction in some industries but also more efficient water usage, including

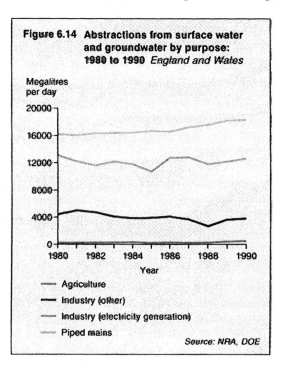

Figure 6.14 Abstractions from surface water and groundwater by purpose: 1980 to 1990 *England and Wales*

Megalitres per day

— Agriculture
— Industry (other)
— Industry (electricity generation)
— Piped mains

Source: NRA, DOE

recycling. Agricultural abstractions increased by 125 per cent since 1980 but they account for only 1 per cent of total abstractions. About half of agricultural abstractions are for spray irrigation and amounts abstracted fluctuate considerably, depending upon the weather.

6.21 Abstractions by region and purpose are shown in Table 6.2. Abstraction for piped mains water is greatest in the Thames, Severn Trent and Welsh regions. Abstractions for electricity generation are important in the Severn Trent and Welsh regions. Abstractions for agriculture, although small, are relatively important in the Anglian region.

Table 6.2 Abstraction from surface water and groundwater by purpose and region, 1990 *England and Wales*					
				Megalitres per day	
NRA Region	Piped mains water	Agriculture	Industry		Total
			Electricity generating companies	Other	
Anglian	1,928	231	2	295	2,456
Northumbria	1,060	1	-	38	1,098
North West	1,883	8	161	734	2,787
Severn Trent	2,421	77	2,991	451	5,940
Southern	1,621	38	-	1,184	2,843
South West	630	35	210	128	1,003
Thames	3,827	26	111	167	4,131
Welsh	2,671	18	8,475	310	11,474
Wessex	798	32	-	137	967
Yorkshire	1,498	39	662	351	2,551
England & Wales	18,336	507	12,612	3,795	35,249

Source: NRA

Low river flow

6.22 In some areas there are concerns that the high level of authorised abstractions has led to reductions in river flows and drying up of some rivers with a consequential loss of amenity and wildlife habitats. In England and Wales the NRA has undertaken a review which identified 40 rivers where reduced river flow is a problem. They are currently investigating 20 rivers with a view to finding possible solutions such as the introduction of additional water, river bed lining or the revocation of existing abstraction licences.

The water table

6.23 Abstractions from groundwater tend to lower the natural underground water level (the water table or groundwater level), which in turn reduces the flow of water in rivers which are supported by springs and groundwater seepage. In coastal areas the depletion of groundwater levels increases the risk of saline intrusion (the drawing of salt water into an aquifer). Mean groundwater levels have been declining in some areas but

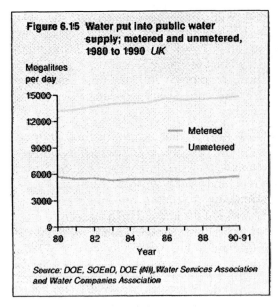

Figure 6.15 Water put into public water supply; metered and unmetered, 1980 to 1990 *UK*

Megalitres per day

Source: DOE, SOEmD, DOE (NI), Water Services Association and Water Companies Association

not generally. Groundwater levels have been rising, for example, in some cities and industrial areas, (including London, Birmingham and Liverpool) resulting from reduced groundwater abstraction by industry. In many aquifers there are strong seasonal variations in groundwater levels since recharge occurs mainly during the winter months.

6.24 Increasing groundwater levels may result in the flooding of tunnels and basements, and chemical attack on structural materials. Sewers may also receive more groundwater which reduces their effective capacity and increases the probability of overflows during storms with consequential surface water pollution by untreated sewage.

Public water supply and metering

6.25 Of the 20,350 megalitres per day of piped mains water put into supply in the UK, 28 per cent is metered, supplied mainly

to industry, and 72 per cent is unmetered, supplied mainly to domestic and commercial users; about a third of unmetered water however, is "not delivered" - see paragraph 6.29. Figure 6.15 shows recent trends in metered and unmetered public water supply. The use of unmetered water has increased steadily from 13,150 megalitres per day in 1980 to over 14,700 in 1990-91. The use of metered water fell from 5,680 megalitres per day in 1980 to about 5,260 in 1983 but since then the metered supply has increased to 5,630 megalitres per day. The number of metered commercial and domestic users has been increasing in recent years.

6.26 Figure 6.16 shows the supply of piped mains water (metered and unmetered) in 1990-91 by water service company areas (including supplies by water service companies and water-only companies - see Box 6.1) in England and Wales, and for Scotland and Northern Ireland. The proportion of metered to unmetered supplies depends on the charging policy of each company and the number of large industrial users of water within their area. Although water companies in the south east of England have the greatest water resource problems, they supply the lowest proportion of metered water because they have a higher proportion of domestic users.

6.27 In England and Wales, unmetered water is generally charged on the basis of rateable value but this method of charging will cease in April 2000. Future charging methods are currently being considered by the water industry. These include metering, a single flat rate charge per domestic property and the banding of property by type with a fixed charge for each band. National water metering trials for domestic premises were started in 1989: metering trial sites are shown in Figure 6.17.

6.28 In Scotland the cost of public water supplies is met through the personal community water charge, the non-domestic water rate, and direct (metered) charges. At present, the option of having a metered public water supply is available only to non-domestic users, the net annual value of whose premises exceeds a value determined annually by the authority. The option for all users to receive public water through a meter is provided for in the Local Government Finance Act 1992.

6.29 There are significant differences between the quantity of water output from treatment works and the quantity supplied ("water not delivered"). The difference is accounted for by water leakage from mains

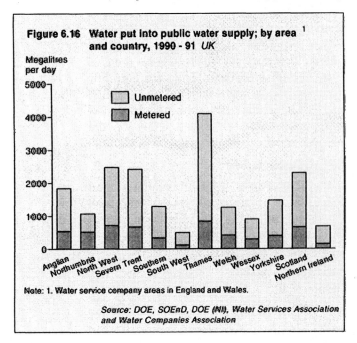

Figure 6.16 Water put into public water supply; by area [1] and country, 1990 - 91 *UK*

Megalitres per day

Note: 1. Water service company areas in England and Wales.

Source: DOE, SOEmD, DOE (NI), Water Services Association and Water Companies Association

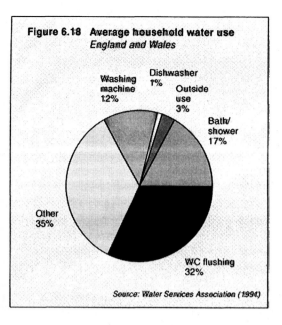

Figure 6.18 Average household water use
England and Wales

- Washing machine 12%
- Dishwasher 1%
- Outside use 3%
- Bath/shower 17%
- WC flushing 32%
- Other 35%

Source: Water Services Association (1991)

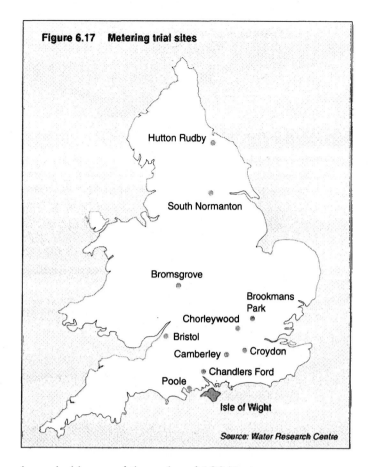

Figure 6.17 Metering trial sites

Hutton Rudby
South Normanton
Bromsgrove
Brookmans Park
Chorleywood
Bristol
Camberley
Croydon
Chandlers Ford
Poole
Isle of Wight

Source: Water Research Centre

and also usage such as water for mains flushing, sewer cleaning, hydrant testing and fire-fighting. Mains leakage however is thought to account for most of the difference. In England and Wales water not delivered accounts for around a quarter of water supplied.

Average household use of water

6.30 Of the 140 litres per person per day consumed in households in England and Wales, over 30 per cent is used in WC flushing and 35 per cent in a variety of miscellaneous usages. Figure 6.18 shows the relative proportions of average household water use. Box 6.2 shows the average water consumption for some domestic appliances and, for comparison, the usage of water to make a selection of products.

6.31 The results of a Scottish Development Department study in 1984 showed that domestic consumption of water in Scottish households was of the order of 120 litres per person per day (representing the median value; there were variations from area to area). The results of a new study of domestic consumption in Scotland are expected in 1992.

Droughts

6.32 There have been three major dry periods in the last twenty years; 1971 to 1976 (with 5 dry years out of six), 1984 (February to August) and from late 1988. In England and Wales, 89 drought orders, including renewals of expired orders, were

Average water consumption;   *Box 6.2*
selected domestic appliances and products

Average consumptions of various domestic appliances: litres of water used by appliances each time they are used

Automatic washing machine	100
Dishwasher	50
WC	10
Bath	80
Shower	30

And litres of water used to make the following products

Average car	30,000
One ton of steel	4,546
One ton of ready mixed concrete	454
A pint of beer	4.5

Table 6.3 Number of drought orders by region [1], 1974 to 1990 *England and Wales*

NRA Region	1974	75	76	77	78	79	80	81	82	83	84	85	86	87	88	89	90	Number Total
Anglian	-	1	15	-	-	-	-	-	-	-	-	-	-	-	-	1	3	20
Northumbria	-	-	2	-	-	-	-	-	-	-	-	-	-	-	-	-	-	2
North West	-	-	-	-	2	-	2	-	6	-	31	-	-	-	-	21	-	62
Severn Trent	-	1	13	-	-	-	-	-	-	-	6	-	-	-	-	5	-	25
Southern	1	1	4	-	-	-	-	-	-	-	-	-	-	-	-	19	25	50
South West	-	1	39	-	12	-	-	-	7	5	45	-	-	-	-	21	10	140
Thames	-	-	8	-	-	-	-	-	-	-	-	-	-	-	-	-	2	10
Welsh	-	-	20	-	2	-	2	-	2	1	22	-	-	-	-	13	1	63
Wessex	-	3	19	-	3	-	-	-	-	-	-	-	-	-	-	-	3	28
Yorkshire	3	2	16	-	-	-	-	-	-	-	-	-	-	-	-	9	17	47
Total	4	9	136	-	19	-	4	-	15	6	104	-	-	-	-	89	61	447

Note: 1. Water Authority region before 1989. Includes water supply companies.

Source: DOE

made during 1989 and 61 in 1990. These are made when a serious deficiency of supplies of water in any area exists or is threatened by an exceptional shortage of rain. Table 6.3 gives the number of drought orders which have been made in recent years in England and Wales by region. In Scotland, something over 30 drought orders have been made during the past 30 years; in that time the most severe drought occurred in the 1970s, with a further notable one in the spring and summer of 1984.

6.33 Drought orders do not always restrict water usage; apart from emergency orders, which are very rare, orders cover fixed periods of a maximum of 6 months (which can be extended once by further orders) and give water undertakers the power to relax their normal conditions of operation. For example, the orders may permit companies to abstract additional water, or to reduce compensating flows from reservoirs.

6.34 Hosepipe bans do not require the authority of a drought order; water companies can impose bans on the use of domestic hosepipes and sprinklers for watering gardens and washing cars (under section 76 of the Water Industry Act 1991) when this is considered necessary to conserve supplies.

6.35 The degree to which any area is affected by drought depends on the length of the drought, the total volume of water stored, the storage capacity and the demand on the resource system. Smaller surface water storages are particularly affected by prolonged summer droughts that start early in the year and continue through to the autumn months. Larger reservoirs and the major aquifers are generally less affected by one dry summer, but can be severely depleted by a succession of dry summers and winters such as those which have occurred in parts of eastern and southern England in 1988-89 and 1991-92. In 1990 and 1991 many of the areas most affected by the dry weather were those dependent on groundwater supplies.

6.36 When droughts mainly cover the summer months, areas which abstract large proportions of groundwater (mainly in the east and south east of England) are generally less affected than areas with a large proportion of abstractions from surface water. However, many of the areas affected by the drought in 1990 were those largely dependent on groundwater because the low rainfall in the winters of 1988-89 and 1991-92 did not adequately recharge aquifers.

References and further reading

1. Institute of Hydrology, British Geological Survey, (1990). Hydrological Data UK, 1990 Yearbook (also available for years 1985 to 1989). IH.

2. National Rivers Authority, (1991). Demands and Resources of Water Undertakers in England and Wales. NRA.

3. Department of the Environment, (1992). Using Water Wisely. DOE.

4. UN Economic Commission for Europe, (1989). Water Use and Water Pollution Control; Trends, Policies, Prospects. UN.

Monkhouse, R.A., & Richards, H.J., (1982). Groundwater Resources of the United Kingdom. European Economic Community, Atlas of Groundwater Resources. Th. Schaeffer Drukerei GmbH, Hannover.

Marsh, T. & Lees, M., (1985). Hydrological Data UK; The 1984 Drought. Institute of Hydrology.

National Rivers Authority, (1992). Water Resources Development Strategy; a Discussion Document. NRA.

Price, M., (1991). Water from the Ground. New Scientist No. 42, 16 February 1991.

Rodda, J.C., & Monkhouse, R.A., (1985). The National Archive of River Flows and Groundwater Levels for the UK. Journal of The Institution of Water Engineers and Scientists. Vol.39 No.4.

Robins, N.S., (1990). The Hydrogeology of Scotland. British Geological Survey, Keyworth.

Water Services Association, (1991). Waterfacts 1991. WSA.

7 Inland water quality and pollution

☐ In England and Wales in 1990, 89 per cent of rivers and 90 per cent of canals were of good or fair quality; in Scotland 99 per cent of freshwater rivers and canals were of good or fair quality; and in Northern Ireland 95 per cent of non-tidal rivers were of good or fair quality (Figures 7.1 and 7.3 and Table 7.1).

☐ Since 1980 there has been a slight net deterioration in water quality for both rivers and canals in England and Wales. The 1990 Survey results were affected to some extent by changes in survey methods, as well as changes in water quality due both to discharges and the effects of two hot, dry summers (7.6, 7.8).

☐ In 1989 the NRA surveyed a number of lakes and reservoirs in England and Wales for occurrences of algal blooms. Of the 915 waters surveyed, 594 had blue-green algae present and 169 had sufficiently high densities to warrant alerts to owners and to Environmental Health Officers (Table 7.2).

☐ In 1991 there were over 29,000 pollution incidents reported to the National Rivers Authority of which over 22,000 were substantiated; 386 of these were classified as major (Figure 7.7). Over 3,000 water pollution incidents were reported in Scotland in 1990-91 and 2,000 water pollution incidents were reported in Northern Ireland in 1991 (7.35).

☐ Sewage accounted for 28 per cent and oil 24 per cent of substantiated pollution incidents reported in 1991 in England and Wales. Farms and industrial sources each accounted for 13 per cent of incidents (Figure 7.6).

☐ Relatively few pollution incidents result in prosecutions but there have been significant increases in numbers of prosecutions since the establishment of the NRA. In respect of incidents which occurred in 1991, 348 prosecutions had been brought in England and Wales by the end of March 1992 and a further 185 were pending (7.41). In Scotland, reports of 65 pollution incidents were made to the Procurator Fiscal in 1990 (7.43). In Northern Ireland, 194 reported pollution incidents resulted in prosecution in 1991 (7.44).

☐ About 95 per cent of the polluting load of sewage is removed by treatment before being discharged into inland waters and 50 per cent of discharges into tidal waters are treated (7.46).

☐ The percentage of sewage treatment works in England and Wales in breach of compliance with their discharge consents declined from 23 per cent in 1986 to 8 per cent in 1990 (Table 7.3).

☐ Freshwater acidification estimates indicate that 200 squares (10 km by 10 km) of the UK will exceed critical loads after the year 2005 (7.67).

☐ 57 parameters have been defined to assess the quality of drinking water. 16 key parameters have been selected to give an overall indication of the quality of drinking water in England and Wales in 1991. Over 98 per cent of all determinations were within prescribed concentration limits (Table 7.4).

7.1 There are many pressures on inland watercourses: they are used for drinking water supply, and for industrial or agricultural abstraction; they may be used for waste disposal (mainly treated domestic sewage), industrial wastes and farm effluents; they are also increasingly in demand for leisure activities. Inland waters should be clean and litter-free, both to achieve an acceptable standard from a general amenity point of view and so that they are able to support fish, birds, mammals and other aquatic life.

7.2 Water quality is subject to regular monitoring and there are various controls to deal with water pollution. This chapter describes water quality, monitoring and control procedures. It also looks at some of the important factors influencing water quality and the measures taken to deal with problems. The final section looks at public drinking water supply.

Monitoring water quality

Rivers

7.3 Regular monitoring is carried out to determine both the overall state of the water environment and the factors which affect overall quality. Reliable estimates of the volume of flow at a water quality monitoring site are also required, for a variety of reasons, including the calculation of the dilution effect on effluent discharges for licensing or other purposes. Apart from ongoing, routine operational monitoring, a series of surveys has been carried out since 1958 in England and Wales specifically to assess the overall quality of rivers, canals and estuaries. Since 1970, surveys have been carried out at five year intervals. These categorise watercourses into four quality classes. Similar surveys have been carried out in Scotland since 1968, with a

separate biological survey also being carried out in 1980 and 1990. River quality monitoring on a routine basis has been carried out in Northern Ireland since 1973.

England and Wales

7.4 Water quality is currently assessed according to a classification introduced by the National Water Council (NWC) in the late 1970s which has a defined range of numerical quality standards. The classification is currently being reviewed as part of the development of the new scheme of statutory water quality objectives provided for by the Water Act 1989, and consolidated by the Water Resources Act 1991. The NRA's proposals for a possible new scheme have been published for public consultation [1].

7.5 The first results of the 1990 River Quality Survey [2] were published in December 1991, using the same NWC classification as used for the 1980 and 1985 Surveys. This ranks water in four quality classes, ie good (classes 1a and 1b), fair (class 2), poor (class 3) and bad (class 4). As with earlier surveys, rivers with a summer flow of 0.05 m^3/sec or greater were covered. Over 42,000 km of freshwater rivers and canals were covered by the 1990 survey.

7.6 The 1990 Survey of water quality showed that 89 per cent of rivers and 90 per cent of canals in England and Wales were of good or fair quality. Figure 7.1 provides a comparison of the proportions of rivers and canals in each water quality class for the last three surveys. Although there was an overall improvement in water quality between 1958 and 1980, since 1980 there has been a slight net deterioration in water quality for both rivers and canals, with the proportion in the best class (class 1a) falling

---

Water quality control responsibilities   *Box 7.1*

England and Wales: The National Rivers Authority (NRA) is responsible for maintaining and improving water quality and for pollution control, water resources, flood defence and fisheries, navigation, conservation and recreation. (More details are given in Chapter 6 on inland water resources and abstraction)

Scotland: The River Purification Authorities (RPAs), comprising the seven River Purification Boards and the three Islands Councils are responsible for maintaining and improving the quality of the water environment and for water pollution control.

Northern Ireland: The Department of the Environment for Northern Ireland is responsible for water quality and pollution control, water resources and conservation. A separate Directorate in the Department has responsibility for water supplies and sewerage services.

Her Majesty's Inspectorate of Pollution (HMIP) in England and Wales, and Her Majesty's Industrial Pollution Inspectorate (HMIPI) and the RPAs in Scotland are responsible for the new system of Integrated Pollution Control (IPC) introduced by the Environmental Protection Act 1990. HMIP liaises with the NRA in connection with discharges to water from industry in England and Wales.

from 34 per cent to 29 per cent while the proportion in the worst class (class 4) remained unchanged. A total of 6,267 km of rivers and canals were assigned a lower quality class in 1990 than in 1985, compared with 4,623 km which were assigned a higher quality class.

increased monitoring in some regions, provided more accurate information than was available in 1985. Second, the effects of two hot, dry summers will have had implications for water quality in rivers. Third, discharges from sewage works, industry and farms will also have had an effect on water quality. The NRA estimated that the three effects on water quality contributed in approximately equal proportion to the reported net deterioration.

7.9 The majority of the downgrading was accounted for by reported changes in the Thames and South West regions. In both regions, an important part of the changes may have been due to changes in survey methods - particularly, increased monitoring in the case of the Thames region, and difficulties in replicating earlier survey methods in the case of the South West - rather than to real changes in water quality. More generally, the differences in the way that the classification scheme had been applied

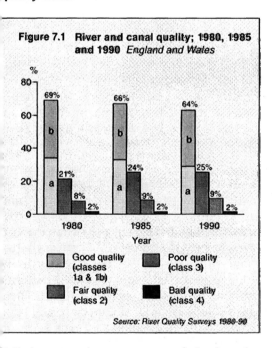

Figure 7.1 River and canal quality; 1980, 1985 and 1990 *England and Wales*

Source: River Quality Surveys 1980-90

7.7 A regional presentation of the lengths of rivers in England and Wales in each of the quality classes for 1990 is given in Figure 7.2. The Northumbria region recorded the lowest proportion of polluted rivers, with less than 3 per cent classified as poor or bad quality (class 3 or 4). The highest proportions of polluted rivers were recorded in the North West and South West regions, which both had 19 per cent of their river and canal lengths classified as of poor or bad quality.

7.8 The net overall downgrading since 1985 is mostly explained by three factors. First, changes in methodology, particularly

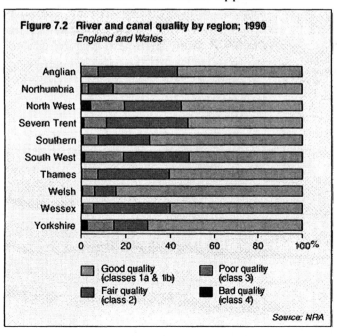

Figure 7.2 River and canal quality by region; 1990 *England and Wales*

Source: NRA

Table 7.1 River, loch and canal quality; 1980, 1985 and 1990 *Scotland*			
			Length in km
	1980	1985	1990 [1]
Class 1	45,352	45,695	49,452
Class 2	2,035	1,723	1,202
Class 3	260	272	238
Class 4	163	132	71
Total	47,810	47,822	50,963

Note: 1. 1990 survey includes rivers on islands which were not included in earlier surveys.

Source: SOEnD

throughout the country made accurate comparisons unreliable.

Scotland

7.10 There are over 50,000 km of rivers (including tidal rivers), lochs and canals in Scotland. In the 1990 Water Quality Survey of Scotland [3], 99 per cent were of good or fair quality. Table 7.1 shows the amounts in each of the 4 classes (which correspond broadly with the classes in England and Wales) and changes since 1980. The length of waters that show some degree of pollution (classes 2 to 4) has reduced by 38 per cent since 1980, and the length of class 4 rivers reduced by 56 per cent.

Northern Ireland

7.11 The 1990 survey of river water quality in Northern Ireland covered 1,435 km of river. Ninety-five per cent of non-tidal rivers were classified (according to the NWC scheme used in England and Wales) as being of good or fair quality. The proportions of rivers in each of the quality classes are shown in Figure 7.3.

Lakes and reservoirs

7.12 Although there is no specific classification scheme for lakes and reservoirs in the UK, many lakes have been monitored

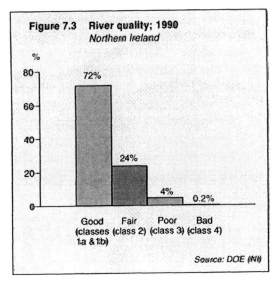

Figure 7.3 River quality; 1990
Northern Ireland

Source: DOE (NI)

and classified as part of associated river systems. The results form part of overall river quality survey reports. Monitoring of reservoirs and other lakes has been carried out on a largely ad hoc basis according to local needs and circumstances. A review of the eutrophic state of lakes and reservoirs has recently been undertaken as part of the process of implementing the EC's Directives on Nitrate and Urban Waste Water Treatment.

Algal Blooms

7.13 In the past decade there has been an increase in awareness of algal blooms in inland waters, and in particular the formation of the large floating masses caused by blue-green algae. Blue-green algae are found most abundantly in the summer in lakes, reservoirs and other still waters. Algal surface scums are created when changes in weather conditions break down a stable algal bloom. Surface scums tend to be carried by the wind to the shores of the lake or reservoir and may accumulate on the shore-line.

7.14 Certain species of blue-green algae can produce toxins which are potentially poisonous to fish and mammals. Illness, including rashes, eye irritation, vomiting, diarrhoea, fever and pains in muscles and joints, has occurred in some users of lakes and reservoirs after swallowing or swimming through algal scum. However, there is no evidence of adverse effects in the UK in treated drinking water associated with algal toxins.

7.15 Algal blooms are considered to be caused by a combination of long periods of calm, warm weather, sunlight and a sufficient supply of nutrients. Phosphorus is regarded as the nutrient limiting eutrophication in most cases in the UK (see Box 7.3). Phosphorus inputs to surface waters arise mainly from point sources (sewage treatment works, industrial discharges and pollution by farm wastes); as a result of soil erosion; and to a lesser degree from diffuse sources such as agriculture. The overall contribution from sewage treatment works accounts for about 75 per cent of the total and that from agricultural sources is estimated at 20 per cent, although this can vary greatly between catchments.

7.16 In September 1989 a bloom of toxic blue-green algae at Rutland Water in the Anglian region was thought to be responsible for the deaths of a number of sheep and dogs which ingested scum from the reservoir. As a result of this and other algal blooms, the NRA conducted a survey in 1989 of 915 waters in England and Wales [4]. 594 waters

Eutrophication

Box 7.3

There are several groups of algae which can be identified by colour. They provide essential food for animal plankton which form part of the food chain, eventually providing food for fish and then animals such as the otter. Certain species of blue-green algae can produce toxins which are potentially poisonous to mammals and fish.

In freshwater, phosphorus is regarded as the most important limiting nutrient, although nitrogen and silica also play major roles in controlling the growth of algae. The growth rate of algae is limited by the amount of nutrient available, even when the physical conditions such as light, temperature and stability are optimal.

The structure of the algal population in a lake or reservoir is dependent upon many factors since the optimum physical conditions for growth and the optimum balance of nutrients required vary with the species of algae concerned. In simplified terms, provided physical conditions for growth are favourable, when the concentration of nitrogen compared to phosphorus is high, green algae tend to dominate, and as the ratio of nitrogen compared to phosphorus becomes smaller, blue-green algae tend to dominate.

Waters in which the level of nutrients has increased such that the growth of algae is no longer limited can be described as eutrophic (meaning "well feeding"). Such waters can support a large population of algae. Eutrophication of a lake or reservoir can arise naturally or as a result of activities such as discharges from sewage treatment works, industry and agriculture.

Problems associated with excessive algal growth in lakes and reservoirs include the de-oxygenation of still waters and the reduction in light levels on the bed, both of which can result in the long-term disruption of the ecosystem. Excessive growths of blue-green algae can lead to surface scums and produce toxins which are potentially poisonous to fish and mammals.

were found to have blue-green algae present and 169 of these had sufficiently high densities to warrant alerts to owners and to Environmental Health Officers. Table 7.2 shows the number of waters with algae problems by region. When samples were taken for analysis, 68 per cent were found to contain toxins. A further survey in 1990 of 1,668 waters showed that 632 had blue-green algae present.

Table 7.2 Waters affected by toxic algae, by region, 1989 UK	
	Number of sites considered to have a problem
Anglian	53
Northumbria	0
North West	50
Severn Trent	18
Southern	1
South West	27
Thames	0
Welsh	1
Wessex	13
Yorkshire	6
Scotland	17
Northern Ireland	3
Total	**189**
Source: NRA, water authorities in Scotland, DOE (NI)	

7.17 In Scotland, corresponding surveys were carried out by the River Purification Boards and regional councils (as water authorities). Some occurrences of toxic algae were noted, but these were not extensive. See also Chapter 14 on environment and health.

Groundwater

7.18 Groundwater monitoring has been carried out largely in connection with its principal use, drinking water supply. A groundwater monitoring programme, due to be completed by mid 1993, is currently being carried out for the purposes of the EC's Nitrate Directive, which is concerned with the protection of water from pollution by nitrate from agricultural sources. More generally, a network of monitoring points has been set up but results are not yet available for publication.

Other water quality monitoring

7.19 In addition to overall water quality assessments, a considerable amount of specific information is also collected, relating to individual stretches of water. For a range of key rivers, assessments are made for specific features of water quality, including the amounts of certain "determinands" (ie attributes such as acidity, concentrations of pollutants including nitrates, suspended solids etc). From 1974 onwards data have been held centrally which record the level of many determinands in water samples taken at 230 sampling points in GB. These sampling points, situated mainly at tidal limits or at the

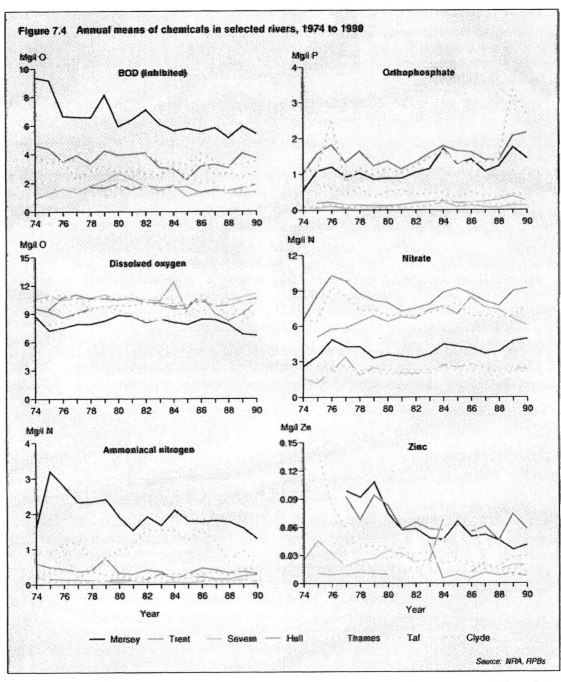

Figure 7.4 Annual means of chemicals in selected rivers, 1974 to 1990

Source: NRA, RPBs

point of confluence of a tributary, form a national network for water quality sampling, known as the Harmonised Monitoring Scheme. The data are used to investigate long term trends in the main determinands. Data on concentrations at tidal limits are used, with data on the rate of flow at the same points, to estimate the loads (in tonnes) of nutrients etc to the sea. These load estimates are used by the North Sea Task Force in its work for the Paris and Oslo Conventions.

7.20 Variations in sampling regimes, which may be required to respond to particular problems at individual sampling points, can affect the comparisons which can be obtained from the data. The data which are sampled most frequently and consistently relate to

oxygen regime and nitrogen and phosphate content. Figures 7.4 and 7.5 show data for five of these determinands: biological oxygen demand (BOD); dissolved oxygen; ammoniacal nitrogen; nitrate and orthophosphate. Data for other pollutants, including metals, are sampled less regularly and concentrations are often less than the limit of detection of the analytical techniques used. Charts for zinc, which is less affected by these considerations than other determinands, have been included in Figures 7.4 and 7.5.

7.21 Annual means are used to indicate trends at a sampling point or to make approximate comparisons of the state of the water at different points. Figure 7.4 shows annual mean concentrations from 1974 to

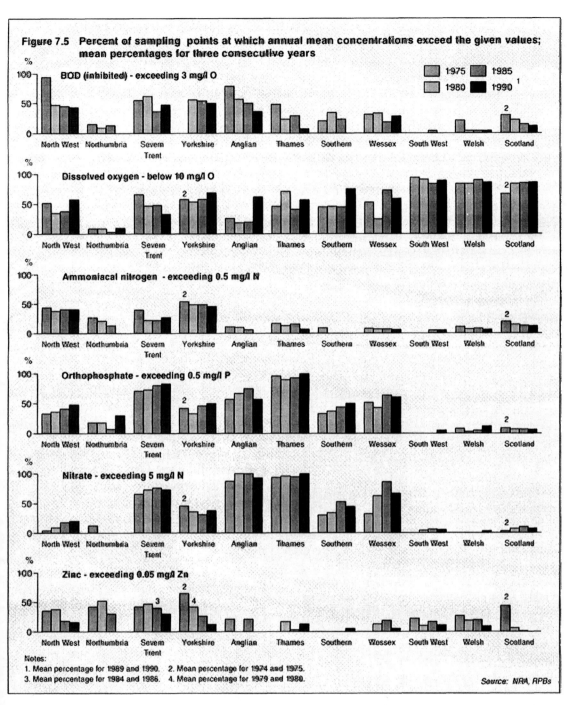

Figure 7.5   Percent of sampling points at which annual mean concentrations exceed the given values; mean percentages for three consecutive years

Notes:
1. Mean percentage for 1989 and 1990.   2. Mean percentage for 1974 and 1975.
3. Mean percentage for 1984 and 1986.   4. Mean percentage for 1979 and 1980.

Source: NRA, RPBs

1990 at seven points on major rivers. For each determinand shown, most sampling points do not show a trend over the period shown. The Mersey at Flixton shows an improving trend in BOD but a slightly deteriorating trend in dissolved oxygen. Similarly, the points with the highest concentrations of ammoniacal nitrogen (those on the Mersey and the Clyde) show generally improving trends in recent years. However, for nitrate and orthophosphate the highest concentrations tend to be associated with worsening trends. The apparent early improvements in zinc concentrations are largely due to analytical technique, but those of later years are more reliable.

7.22 Regional comparisons of data (shown in Figure 7.5) can be made in terms of the

percentage of sampling points in a region which exceed a particular concentration of a determinand. The concentration values used for these comparisons have no regulatory or health-related significance but are chosen to show trends and regional differences as clearly as possible. As some regions have only a few sampling points, three-year averages are used to smooth out trends. There are wide variations between regions: North West, Severn Trent, Anglian and Thames generally have the highest concentrations of most determinands, and Northumbria, South West, Welsh and Scotland the lowest. The improving zinc trend is partly due to improvements in analytical techniques.

7.23 Further water quality monitoring in the UK is carried out in compliance with EC

Directives (see Box 7.4) and this information is contained in the public registers maintained in England and Wales by the NRA, and in Scotland by the RPAs.

The water cycle

7.24 Water is essential to life and human activity. Water is needed for drinking, washing and cooking; it is also used for agriculture and many industrial processes. These needs have to be supplied, though some of them may conflict, and appropriate ways provided for the disposal of waste products.

7.25 Substantial volumes of waste are generated; some 30 to 40 gallons of domestic sewage waste are produced per person per day. These large volumes of water must be returned to rivers and streams; in many cases, without the return of cleaned waste water, rivers and streams would dry up or river flows be substantially reduced. The return of waste has an important bearing on water quality; wastes need to be treated to a suitable standard, and appropriate precautions taken to prevent accidental pollution. The primary objective of the EC's Urban Waste Water Treatment Directive, which was adopted in May 1991, is to ensure that all substantial discharges of sewage are treated before they are discharged either to inland surface waters, estuaries or coastal waters.

7.26 Water quality is inextricably linked to water quantity. If water levels are low, there is less water available to dilute waste effluents, and this may cause a reduction in water quality. (See also sections on water resources and abstraction in Chapter 6). Flooding and discharges from storm overflows in the sewerage system, which will be considered during the implementation of the Urban Waste Water Treatment Directive, may have an impact on water quality (see later section on sewage).

Controlling water quality

7.27 The NRA has responsibility for controlling pollution in inland, estuarial and coastal waters in England and Wales. In Scotland responsibility for controlling water pollution rests with the seven River Purification Boards and the Islands Councils. In Northern Ireland, protection of river quality is the responsibility of the Department of the Environment for Northern Ireland (DOE(NI)) See Box 7.1 on responsibilities for controlling water quality.

7.28 Any discharges by industry, sewage treatment works, farms, etc into rivers, estuaries and coastal waters must be authorised by the relevant authority through the system of discharge consents. Regulatory authorities have wide powers to consider the nature of the contaminant, its concentration and the volume of the discharge, before deciding whether to issue a discharge consent. If a consent is granted conditions may be imposed so that the quality of water can be maintained. There are also powers for the review and revocation of discharge consents, and for compliance with their terms to be enforced. Discharges of sewage and similar industrial effluents will in future be required to comply with the standards set out in the Urban Waste Water Treatment Directive, where those standards are more stringent than those set out in existing discharge consents.

7.29 Public registers have to be maintained by the NRA and by the RPAs, detailing discharge consents together with the results of samples taken for both the effluents discharged and the waters protected by the legislation. These registers are open to inspection by the public. In Northern Ireland, a public register of discharge consents is maintained by DOE(NI).

EC Directives   Box 7.4

The European Community has adopted a number of Directives for the protection and improvement of water quality. The standards of these Directives are implemented through the activities of the National Rivers Authority and the water companies in England and Wales, the River Purification Boards in Scotland and the Department of the Environment for Northern Ireland.

Directives cover the quality of surface water for abstraction for drinking water (75/440/EEC); the quality of bathing waters (76/160/EEC); the quality of water needed for freshwater fish (78/659/EEC); for waters for shellfish (79/923/EEC); the protection of groundwater (80/68/EEC); urban waste water treatment (91/271/EEC); the protection of waters against pollution caused by nitrate from agricultural sources (91/676/EEC); and the regulation of discharges of dangerous substances into the aquatic environment (76/464/EEC and "daughter directives" which set standards for discharges of dangerous substances such as mercury, cadmium and persistent organic chemicals).

Other Directives refer specifically to contamination from particular pollutants such as titanium dioxide or asbestos.

Water quality objectives   Box 7.5

Under the Water Resources Act 1991, statutory water quality objectives (WQOs) are to be introduced in England and Wales to ensure that controlled waters (ie most inland and coastal waters and groundwaters) are of sufficiently high standard for all their uses. The objectives are to be set by the DOE and the Welsh Office in consultation with the NRA and will reflect the differing uses to which all controlled waters may be put. They will apply to individual stretches of water, and each objective will specify the use for that stretch, the water quality standard, and the date by which it must be achieved. In December 1991, the NRA published a public consultation document [1] setting out proposals for new system of classifying water quality, and for introducing water quality objectives.

The regulatory authorities in Scotland and Northern Ireland are considering what arrangements would be appropriate there.

7.30 At the end of December 1990, the NRA's Public Registers contained a total of 134,587 discharge consents, covering discharges made to inland waters, estuaries and coastal waters and groundwaters. These consented discharges range in size and effect from major industrial and sewage effluent discharges to minor surface water outlets. The NRA has undertaken a major review of registers to remove redundant entries and the number is expected to fall to around 100,000.

7.31 Discharges into rivers and estuaries are an important factor influencing the marine environment. The introduction of Integrated Pollution Control by Her Majesty's Inspectorate of Pollution (HMIP) under the Environment Protection Act 1990 will bring inputs of dangerous substances to all media (air, land and water) in England and Wales under unified control. The introduction has begun and HMIP is undertaking a phased programme to take over from the NRA the authorisation of existing discharges from "prescribed processes" to controlled waters. This programme will be completed over the next four years. The NRA will retain its responsibilities to ensure that the environment is not polluted by such discharges. Controls of discharges of most of the most dangerous substances are also enforced by the RPAs in Scotland.

7.32 In addition to controls over regular discharges at known points, increasing attention is being paid to avoiding pollution which might result from accidental spillages and from diffuse sources eg agriculture. Regulations introduced in 1991 [5] set minimum construction standards for new or substantially altered facilities for the storage of farm waste - slurry and silage effluent. Existing installations may also be required to comply with the Regulations where the NRA is satisfied that there is a significant risk of pollution. The Regulations also cover

agricultural fuel oil stores, and subject to the outcome of consultation, it is the Government's intention to introduce similar controls on industrial fuel oil stores.

7.33 Other controls include the Town and Country Planning Act 1990 which covers the development and use of land - for example in determining the location of potentially polluting development and the location of other development near pollution sources.

Pollution incidents and prosecutions

7.34 Every year there are thousands of water pollution incidents. These generally result from accidental spillages of pollutants, whether by industry, by water companies or from agriculture, and represent unlawful discharges either because the discharge is made without consent, or because the terms of a consent have been breached. Other pollution sources are leaching from waste tips, accidents, flytipping and vandalism. Mining and local authority maintenance practices also contribute to the problem.

7.35 In 1991, there were over 29,000 pollution incidents reported to the NRA of which over 22,000 were substantiated; 386 of these were classified as major incidents [6]. A total of over 3,000 water pollution incidents were reported in Scotland in 1990-91 and in Northern Ireland 2,000 water pollution incidents were reported in 1991. These statistics reflect only reported incidents and probably understate the actual level of pollution incidents, although major incidents are more likely to be reported. See also Chapter 15 on pressures on the environment.

7.36 All NRA regions have 24 hour cover for reporting pollution incidents. The number of reported incidents in 1991 showed an increase of 4 per cent compared with 1990, but on the basis of the information available, it is not possible to quantify changes resulting

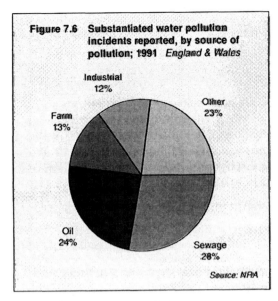

Figure 7.6 Substantiated water pollution incidents reported, by source of pollution; 1991 *England & Wales*

Industrial 12%

Farm 13%

Other 23%

Oil 24%

Sewage 28%

Source: NRA

from the provision of better reporting facilities. The RPBs in Scotland and DOE(NI) also provide a 24 hour pollution incident reporting service.

Incidents by source

England and Wales

7.37 Figure 7.6 shows the proportions of substantiated pollution incidents in 1991 by source of pollution. Sewage sources accounted for 28 per cent and oil sources 24 per cent of incidents. Farms and industrial sources each accounted for 13 per cent of substantiated incidents. However, it is not always possible to identify the source of pollution and 23 per cent of incidents were classified as other. One quarter of major incidents reported were farm-related, with roughly equal proportions attributable to sewage and to industry. Figure 7.7 shows changes in the number of reported pollution incidents since 1985.

7.38 The number of incidents varies appreciably from region to region and Figure 7.8 shows the number of substantiated

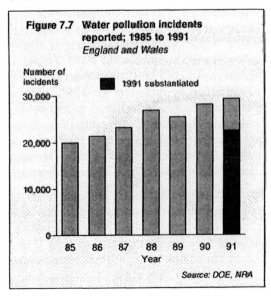

Figure 7.7 Water pollution incidents reported; 1985 to 1991 *England and Wales*

Number of incidents

■ 1991 substantiated

30,000
20,000
10,000
0

85 86 87 88 89 90 91
Year

Source: DOE, NRA

pollution incidents in each region in 1991. Regional variations are due mainly to differences in physical characteristics between areas (eg size, population and population density, lengths of rivers and coastline, industrial concentrations, etc). The Severn Trent region, for example, had the greatest number of pollution incidents but it is one of the largest regions, has areas of high population density, and contains 16 per cent of all river and canal lengths surveyed for water quality. The Wessex region on the other hand accounted for 5 per cent of substantiated pollution incidents, but the region is relatively small in terms of population, area and river length.

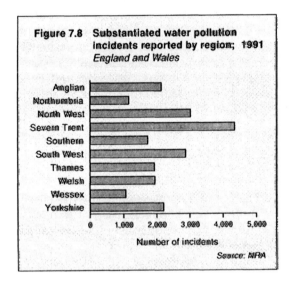

Figure 7.8 Substantiated water pollution incidents reported by region; 1991 *England and Wales*

Anglian
Northumbria
North West
Severn Trent
Southern
South West
Thames
Welsh
Wessex
Yorkshire

0   1,000   2,000   3,000   4,000   5,000

Number of incidents

Source: NRA

Scotland

7.39 Sewage and sewerage were responsible for 28 per cent of reported pollution incidents in 1990-91, with industry responsible for 24 per cent, and 18 per cent attributable to farms. The remaining 30 per cent were attributable to other causes. Figure 7.9 shows the proportions of pollution incidents by source of pollution.

Northern Ireland

7.40 Agricultural pollution accounted for some 50 per cent of reported pollution incidents in 1988, but this has decreased to 30 per cent in 1991. Roughly equal proportions are attributable to industry (19 per cent) and sewage (20 per cent).

Prosecutions and enforcement

7.41 Relatively few of the thousands of incidents reported in England and Wales result in prosecutions, although overall, in its first two years, successful prosecutions brought by the NRA have increased significantly. 348 prosecutions for pollution incidents occurring in 1991 had been heard in court by the end of March 1992 and 340 of these had led to

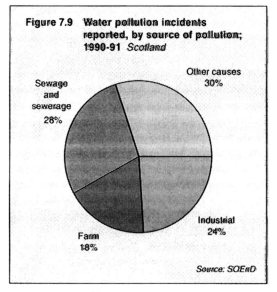

Figure 7.9 Water pollution incidents reported, by source of pollution; 1990-91 Scotland

Other causes 30%

Sewage and sewerage 28%

Industrial 24%

Farm 18%

Source: SOEnD

convictions; a further 185 cases had still to come to court. Of those convictions, 43 per cent related to farms, 27 per cent to industrial incidents, and around 11 per cent each to oil and sewage pollution.

7.42 The NRA also has powers under Section 161 of the Water Resources Act 1991 to carry out preventative or remedial works or operations to deal with pollution, and to recover the costs incurred from the persons responsible. These powers are widely used in connection with major pollution incidents, to safeguard fish populations at risk, or to carry out clean-up or restocking operations after a pollution incident has occurred. The costs involved can be significant.

7.43 In Scotland, 65 reports were made to the Procurator Fiscal in 1990. Of these, 36 related to industrial incidents, 21 to farms, 5 to sewage pollution and 3 to other causes. These proceedings represented between 4 and 5 per cent of the total number of farm-related and industrial incidents reported. The RPAs in Scotland have powers to require polluters to clean up the consequences of pollution incidents.

7.44 In Northern Ireland, 121 reported pollution incidents resulted in prosecution in 1990 and 194 in 1991. About 80 per cent are related to farms. DOE(NI) also has powers to deal with pollution affecting waterways and to recover the costs incurred from the persons responsible.

Major influences on water quality

7.45 Water quality can be influenced by a number of factors; directly by discharges from sewage works and agricultural and industrial installations, or from pollution incidents such as spillages; and indirectly from "runoff" (ie water entering watercourses from roads, industrial sites or land which may pick up contaminants on the way) and from leaching of contaminants from soil or rocks or from storage of hazardous chemicals. Water quality, particularly in lakes and reservoirs, can be influenced by surrounding vegetation, soils and deposition from pollutants in the air (and in particular "acid rain"). The following sections consider the main influences ie sewage and industrial waste, agriculture, air pollution and landfill sites.

Sewage

7.46 The discharge of sewage and sewage effluent is one of the most important influences on inland (and coastal) water quality. Waste is extracted as "sewage sludge" and the resultant treated effluent (essentially waste water), is discharged into watercourses. About 95 per cent of the polluting load of sewage is removed by treatment before being discharged into inland rivers and 50 per cent of discharges into tidal waters are treated. Sewage which is not treated by treatment works is either disposed of via septic tanks or by direct outfalls to streams, rivers and the sea. The provisions of the Urban Waste Water Treatment Directive will require that all significant discharges of sewage should be treated. Levels of treatment required by the Directive will vary according to the size of the discharge and the sensitivity of the receiving waters. The Directive includes a provision to end the dumping of sewage sludge at sea by the end of 1998. See also Chapter 8 on the marine environment and Chapter 11 on waste and recycling.

7.47 Sewage treatment and disposal is the responsibility of water service companies in England and Wales, the Regional and Islands Councils in Scotland, and in Northern Ireland, DOE(NI). Almost all the population of the

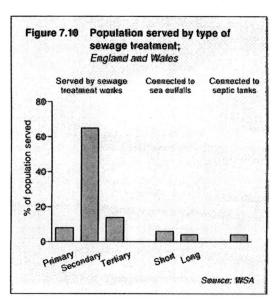

Figure 7.10 Population served by type of sewage treatment; England and Wales

Served by sewage treatment works

Connected to sea outfalls

Connected to septic tanks

% of population served

Primary Secondary Tertiary Short Long

Source: WSA

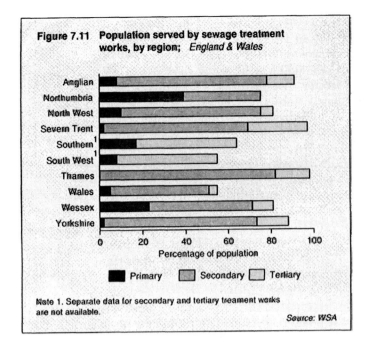

Figure 7.11 Population served by sewage treatment works, by region; *England & Wales*

Primary   Secondary   Tertiary

Note 1. Separate data for secondary and tertiary treament works are not available.

Source: WSA

UK (96 per cent) is served by sewers. The percentage varies regionally, from 88 per cent in the South West Water Service Company area to 98 per cent in the Thames region, Northumbria and the North West.

7.48 The polluting load of effluent, and as a result water quality in the area of discharge, will depend on the level of treatment provided to sewage. About three quarters of the population is served by sewage treatment works. The Urban Waste Water Treatment Directive specifies secondary treatment as the norm but provides for higher, tertiary standards of treatment for discharges to sensitive areas, and at least primary treatment for discharges to less sensitive areas. Primary treatment is likely to involve a physical and/or chemical process of settlement of suspended solids and secondary treatment generally involves biological treatment.

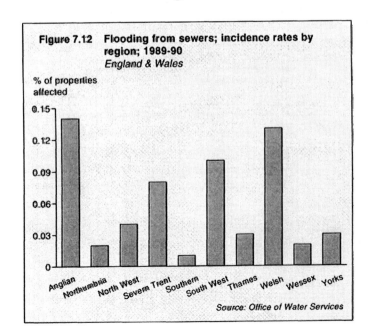

Figure 7.12 Flooding from sewers; incidence rates by region; 1989-90 *England & Wales*

Source: Office of Water Services

7.49 The proportions of population currently served by each of these treatment types in England and Wales is shown in Figure 7.10. Some 10 per cent of the population is served by sea outfalls and 4 per cent have septic tanks. There is however significant regional variation and this is shown in Figure 7.11. Relatively high proportions of populations in Southern, Northumbria and Wessex regions are served by primary treatment and correspondingly low proportions by secondary and tertiary treatment.

7.50 One way of assessing the efficiency or effectiveness of the sewerage system is by measuring the incidence of flooding of sewers, which may affect property, and by runoff, surface water quality. Figure 7.12 shows the incidence rates of flooding from sewers by region. The percentage of properties affected by flooding is extremely small (less than one tenth of one per cent in most regions).

7.51 A more direct impact on water quality can arise through the operation of storm overflows (these are necessary safety valves on the sewerage system) which are under increasing pressure as water usage has grown. Overflows often stem from old, overloaded sewers. They tend to be sporadic but can have a significant impact on water quality.

7.52 The 1985 River Quality Survey for England and Wales identified sewage works effluents as one of the main causes of local deterioration in water quality. From 1986 to 1988, water authorities in England and Wales were required to report works failing to meet the annual performance measures specified in discharge consents. Many of the works failing to comply did not have the capacity to deal with increased loads placed on them resulting from population increases; the majority of treatment works serve relatively small populations. From 1 September 1989, the monitoring of sewage treatment works has been the responsibility of the NRA in England and Wales, while in Scotland this work has for many years been carried out by the RPAs.

7.53 Table 7.3 summarises non-compliance with the terms of discharge consents for the years 1986 to 1990 for England and Wales (the information relates only to those works with numerical consent conditions; other, smaller works have descriptive consent conditions and compliance for these works is assessed qualitatively). The percentage of works in England and Wales in breach of compliance declined from 23 per cent in 1986 to 8 per cent in 1990.

Table 7.3 Sewage treatment works: non-compliance with discharge consents[1], 1986 to 1990 *England and Wales*					
	1986	1987	1988	1989	1990[2]
Number tested	4,354	4,230	4,271	4,315	4,145
Number in breach of discharge consents	1,002	887	742	551	333
% of works tested in breach of consent	23	21	17	13	8

Notes:
1. Works with numerical consents.
2. A number of works are covered by temporary consent variations associated with improvement schemes.

Source: NRA, water authorities

7.54 In 1989-90 a number of sewage works were granted temporary, time-limited consent variations associated with improvement schemes. The underlying rate of non-compliance with long term consent conditions in 1990 is estimated as around 12 per cent, compared with the overall 8 per cent non-compliance rate (which includes the effects of time-limited consent variations).

7.55 The percentage of works in breach of consents varies substantially from region to region, ranging from 1 per cent in Northumbria to 26 per cent in the South West (see Figure 7.13), which partly reflects historic differences in the structure and stringency of consents and partly regional differences in the water companies' improvement programmes.

7.56 Comparable statistics are not available in Scotland where different compliance standards are adopted. However, the Scottish RPAs' Annual Reports show an improvement in recent years in the rate of compliance with consents.

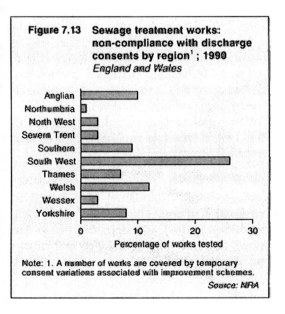

Figure 7.13 Sewage treatment works: non-compliance with discharge consents by region[1]; 1990 *England and Wales*

Percentage of works tested

Note: 1. A number of works are covered by temporary consent variations associated with improvement schemes.

Source: NRA

Agriculture

Farm waste and silage

7.57 Farm waste and silage effluent are normally spread on agricultural land but when accidents or carelessness occur, they can cause serious pollution of rivers and lakes [7]. Undiluted farm slurry (ie animal waste products) is up to one hundred times more polluting than raw sewage. Silage effluent (from the storage of grass fermented in the absence of air) for use as cattle feed is even more polluting - up to 200 times as much as untreated sewage. Pollution arises because bacteria use oxygen in the water to break down the high organic content, resulting in deoxygenation of the water which may cause severe damage to fish and other aquatic life. See also Chapter 11 on waste and recycling.

7.58 Capital grants for the handling of farm wastes have been available for many years. In 1989 the current Farm and Conservation Grant Scheme was introduced in order to help farmers to reduce pollution from farm waste. Grants are available for the provision, replacement or improvement of agricultural facilities for the storage, treatment and disposal of agricultural wastes and silage effluent.

7.59 Regulations setting minimum construction standards for farm waste installations came into full force in England and Wales in September 1991 [5]. The Regulations also enable the NRA to require existing installations to be brought up to an acceptable standard where it is satisfied that there is a significant risk of pollution. Similar regulations are in force in Scotland and the RPAs are responsible for their implementation; similar regulations will be made for Northern Ireland.

7.60 A number of other measures have been adopted to address the problem of water pollution from agricultural sources. These include the provision of free initial advice from the Agricultural Development and Advisory Service (ADAS), a free Code of Good Agricultural Practice for the Protection of Water giving advice to farmers on how to minimise the risk of water pollution, and a pilot study to see if pollution can be reduced by farmers drawing up "tailor-made" plans for the disposal of farm wastes on their holdings.

Nitrate

7.61 Nitrate in water is a pollutant above certain concentrations and can be a danger to human health. The main source of nitrate in water is agriculture although sewage discharges can also be an important factor.

The release of agriculture-derived nitrate from soil organic matter into surface and groundwater is a natural process but is increased by farming activities, especially the application of nitrogen as fertiliser, or organic manures in excess of crop requirements. The release of nitrate varies with crop types and climatic conditions. Controlling nitrate leaching is particularly difficult because the speed at which nitrate reaches underground water sources varies substantially, and present levels may reflect activity in the past, perhaps over several decades. Some of the areas worst affected by nitrate in water are in the Anglian and Severn Trent water regions. Nitrate concentrations in the public water supply are discussed later in this chapter.

7.62 Nitrate concentrations in water have received considerable attention over the last decade. In 1986 the DOE published Nitrate in Water [8] which assessed many of the complex issues of the nitrate problem. One of the main issues considered was whether controls on agricultural activity were a more viable option than water treatment, or whether both should be used in conjunction. In 1988 two reports were published; The Hatton Catchment Nitrate Study [9] and The Nitrate Issue [10], a study of the consequences of various local options for limiting nitrate concentrations in drinking water.

7.63 A recent initiative to reduce nitrate leaching into water has been the designation, on a pilot basis, of Nitrate Sensitive Areas (NSAs) in which the use of fertilisers and organic manures are regulated, and the introduction of Nitrate Advisory Areas (NAAs) in which there is an advisory campaign with the same objective (see also Chapter 5 on land use and land cover and Chapter 14 on health). Both the NSAs and the NAAs are aimed at evaluating ways of controlling nitrate leaching and will provide information necessary to implement the EC's Nitrate Directive by mid-December 1993. The Directive, which was adopted in December 1991, requires EC member states to identify waters affected by pollution from nitrate or waters which could be affected by pollution from nitrate if protective action is not taken. Lands draining into these waters and contributing to pollution will be designated as vulnerable zones within which action programmes will be required to reduce the pollution or prevent further pollution. The NRA and RPAs have begun a programme of monitoring in line with the requirements of the Directive.

Pesticides

7.64 In 1986, a statutory system of controls over pesticides was established which requires that before a pesticide is approved for use it must be considered by the independent Advisory Committee on Pesticides and Ministers in six Government Departments. The Advisory Committee considers key aspects of pesticide use, including risks from leaching into groundwater. Ministers must be satisfied that a pesticide can be used without risk to people and the environment. All approvals made before the statutory controls were introduced are being reviewed, and this UK programme is being integrated with the EC programmes now required under the Community's Directive on the approval of agricultural pesticides (91/414). A review by the Advisory Committee of the environmental behaviour of the two pesticides, atrazine and sumazine, most commonly found in water (mainly from non-agricultural sources) has been completed and approvals will be withdrawn at the end of August 1993 for non-agricultural uses. A statutory Code of Practice for the Safe Use of Pesticides on Farms and Holdings has been issued giving advice on the safe use and disposal of pesticides.

Industrial waste

7.65 Some industrial by-products, which are discharged into rivers and estuaries and eventually into the sea, are also potentially harmful to freshwater and marine ecosystems. Industrial discharges to water are generally covered by the same consent system as other discharges. Since 1 April 1991, all new or substantially changed major industrial processes require an authorisation from HMIP which takes into account releases to water as well as to air and land. Existing major industrial processes will require similar authorisations in accordance with a timetable which runs until 1996. Consents for direct discharges to surface waters are controlled by the NRA in England and Wales, by the RPAs in Scotland, and by DOE(NI), and particular controls are applied according to the nature of the effluent. The most dangerous discharges are the "Red List" substances (see also the references to Red List substances in Chapter 8 on the marine environment).

Air pollution

7.66 Pollutants deposited by air and in particular acid deposition - which can be dry deposition (gases), wet deposition (rain and snow) or occult deposition (fog and mist) - can also affect freshwater quality, especially lakes and reservoirs. Various natural features such as the type of soils, vegetation, sediments and climate all have an effect upon acidification of water. The activities of man have greatly increased the scale and rate at which waters have become acidified, and acid deposition arising from the burning of fossil fuels has been a major contributory factor. See also Chapter 2 on air quality and pollution.

7.67 The scale and extent of freshwater acidification in the UK has been mapped based on the "critical loads" concept (see later section in this chapter). The sensitivity of freshwaters (and soils) to acidification has been plotted on a 10 km by 10 km grid for the UK. By superimposing on these maps the dry, wet and occult deposition of acidic air pollutants, it is possible to determine those areas of exceedence where damage due to acidification will still occur. In addition, by modelling future emission and deposition levels and taking into account current and future emission reduction plans, maps of critical load exceedences can be produced for future years. It is estimated that over 200 squares (10 km by 10 km) in the UK will exceed critical loads and continue acidifying after the year 2005, based on current emission reduction plans.

7.68 In the UK, waters of moderately severe acidity are currently found in central and south west Scotland, the Pennines, parts of Cumbria and central and North Wales. Some lakes within these areas have become acidified to such an extent that they are now virtually devoid of fish.

Landfill sites

7.69 The leaching of pollutants from waste sites may also cause pollution particularly of groundwater. Controls have been in place for some time, originally through the planning system and more recently under the control of pollution legislation, to seek to prevent this happening and to ensure that proper monitoring arrangements are in place. See also Chapter 11 on waste and recycling.

Critical loads

7.70 Critical loads are defined as the levels below which significant harmful effects on specified sensitive elements of the environment do not occur. This concept is currently being developed as a way of assessing the impact of various pollutants on the environment (water, soils, vegetation, buildings and building materials). Critical load maps are a new development which provide information about the areas where critical loads are exceeded and identify areas at risk, thereby facilitating further analysis. See also Chapter 2 on air quality and pollution.

Public drinking water quality

7.71 A legal framework for drinking water quality in England and Wales was established by the 1989 Water Act. The Act has been consolidated into the Water Industry Act 1991. Regulations made under the Act [11] (which came into effect for the most part on 1 September 1989) set detailed quality standards for public water supplies. Similar regulations covering private water supplies came into force in January 1992 [12]. Regulations [13] have been made for Scotland

---

Drinking Water Inspectorates

Box 7.6

The Drinking Water Inspectorate (DWI) was established on 2 January 1990 under section 60 of the Water Act 1989. The Water Act 1989 and other associated legislation were consolidated on 1 December 1991. The legislation relating to the DWI is now generally in the Water Industry Act 1991.

The DWI is responsible in England and Wales for ensuring that water companies comply with their statutory duties. They do this by carrying out a "technical audit" to determine whether water companies are complying with regulation requirements; whether they are on schedule with improvement programmes; and whether they are following good operational practice.

The technical audit consists of three elements: annual assessment of water samples from treatment works, service reservoirs and water supply zones taken by the water companies, to check compliance with sampling and other requirements; inspection of individual companies (including a general check at the time of the inspection and an assessment of the quality of information collected by the company); and interim checks made on aspects of compliance-based information provided periodically by companies.

In Scotland, the monitoring of water authorities' progress with improvement programmes and of compliance with regulation requirements is carried out by the Scottish Office Environment Department.

It is planned to set up a Drinking Water Inspectorate in Northern Ireland in the near future.

under the Water (Scotland) Act 1990 which came into effect on 1 May 1990. Responsibility for enforcing the statutory standards lies with the relevant Secretary of State and local authorities.

7.72 Standards applied in England and Wales are implemented administratively in Scotland by the Scottish Office Environment Department and in Northern Ireland by the Department of the Environment for Northern Ireland. This function in Northern Ireland is to be transferred to a government owned company.

7.73 The EC Drinking Water Directive [14] sets standards, ie maximum admissible concentrations and minimum required concentrations, for more than 40 parameters. UK Water Regulations incorporate all these, plus standards for a further 11 parameters. In total, 57 parameters have been defined to assess the quality of drinking water. Water companies have a statutory responsibility for monitoring the quality of water supplies, subject to checks by local authorities and the Drinking Water Inspectorate (DWI). Monitoring of the quality of public supplies is carried out within designated water supply zones ie discrete areas, often served by a single source, in which not more than 50,000 people reside. In 1991, monitoring of water quality for public supplies was undertaken at

1,717 treatment works, 4,981 service reservoirs and within 2,577 water supply zones in England and Wales.

7.74 Most public supplies comply with national and European standards. A programme has been drawn up for improving those public supplies which may not yet comply with one or more of the statutory standards. By the end of 1995, most breaches of the national and European standards will have been remedied under the current programme of improvements.

7.75 The issues surrounding the supply and quality of drinking water are extremely complex. Different water sources produce water which varies widely in quality. Groundwater for example is generally of good quality, whereas upland surface water may be soft, coloured, and acidic. Lowland surface water may contain runoff and discharges from surrounding catchment areas. Water treatment varies from simple disinfection to multistage treatment.

7.76 Key parameters are discussed below to give an overall indication of the quality of drinking water. Table 7.4 provides a summary of the results of "determinations" (ie the measurement of the concentration or value of a single parameter within a single sample) for each of these key parameters [15] in England and Wales in 1991 and lists, for all supply zones:

- the total number of determinations carried out;
- the number of determinations which exceeded the prescribed concentration or value (PCV) (or a relaxation of the PCV authorised under regulations);
- the percentage of the total number of determinations carried out which exceeded the PCV (or a relaxation of the PCV authorised under regulations).

7.77 Table 7.5 gives similar information for Scotland, but only relates to the second half of 1990, the first period for which information in this form has been available. Some parameters have no particular significance for drinking water in Scotland. Chapter 14 on environment and health also discusses some of the parameters.

Key Parameters

Coliform standard

7.78 The coliform bacteria monitored are not usually themselves pathogenic (disease causing) but they serve as useful indicator organisms of faecal pollution and hence

Table 7.4	Drinking water quality; determinations and exceedences of permitted values; by parameter, 1991 England and Wales		
	Total number of determinations	Number exceeding PCV[1] or relaxed PCV[1]	Percent exceeding PCV[1] or relaxed PCV[1]
Coliforms	173,909	2,385	1.4
Faecal coliforms	172,376	324	0.2
Colour	87,127	49	<0.1
Turbidity	90,849	203	0.2
Odour	28,668	67	0.2
Taste	28,479	68	0.2
Hydrogen ion	99,759	584	0.6
Nitrate	41,430	1,170	2.8
Nitrite	46,058	2,228	4.8
Aluminium	75,305	552	0.7
Iron	89,185	2,515	2.8
Manganese	75,711	731	1.0
Lead	58,635	1,736	3.0
Polycyclic aromatic hydrocarbons	17,360	354	2.0
Trihalomethanes	20,541	761	3.7
Total pesticides	48,034	4,426	9.2
Individual pesticides	828,990	23,159	2.8
Other parameters	641,300	412	<0.1
Total of all parameters	**2,623,716**	**41,724**	**1.6**

Note: 1. Prescribed concentration of value (PCV): the numerical value assigned to water quality standards defining the maximum or minimum legal concentration or value of a parameter. A relaxation of the PCV may be granted subject to the completion of improvement works, in emergencies, as a result of exceptional meteorological conditions or by reason of the nature and structure of the ground in the area from which the supply emanates.

Source: DWI

possible contamination with pathogens. If water is free of coliform bacteria it is unlikely that pathogenic organisms are present. The coliform standard in water supply zones is breached if 5 per cent or more of determinations carried out over the preceding year on treated water samples contain coliforms (except when less than 50 samples have been taken in the year, in which case the criterion applies to the last 50 samples).

7.79 Coliforms were detected in 1,001 water supply zones in England and Wales in 1991. In all but 177 of these zones the 95 per cent compliance criteria (see above) were met. A further 200 zones in Scotland also did not meet the criteria in 1990. However, breaches of the standards for coliforms and faecal coliforms in samples taken at consumers' taps may be caused by the consumers' plumbing and may not necessarily reflect the quality of water supplied by companies.

Nitrate

7.80 In groundwater, the major source of nitrate is leaching from agricultural farmland. As far as surface waters are concerned nitrate is derived mainly through leaching from agricultural land but also through discharges from sewage treatment works. In England and Wales, a total of 94 water supply zones with nitrate concentrations above the standard were detected at some time during 1991. In Scotland there is virtually full compliance with the standard.

Aluminium

7.81 Aluminium is a natural constituent of many water sources, particularly upland surface waters, and aluminium compounds are widely used in water treatment as coagulants to remove impurities including pathogenic organisms. There has been concern about a possible link between aluminium and Alzheimer's disease, but medical advice is that the link is too tentative to justify changes in current practice in the use of aluminium compounds in water treatment. Aluminium concentrations above the standard were detected in 225 zones in England and Wales in 1991, and 87 zones in Scotland in 1990.

Lead

7.82 Lead concentrations complied with the standard in 1,916 water supply zones. However, samples taken from consumers' homes may contain lead derived from consumers' pipework and plumbing, and domestic tap water in these supply zones may not comply with the lead standard. The major source of lead in water is from consumers' pipes and there is considerable variation in lead concentrations in samples taken from consumers' homes. Lead concentrations above the standard were detected in 661 zones in England and Wales in 1991. In Scotland, an assessment of water supply zones for compliance with the lead standard has recently been completed.

Table 7.5 Drinking water quality; determinations and exceedences of permitted values, by parameter, 1990 Scotland			
	Total number of determinations	Number exceeding PCV[1] or relaxed PCV	Percent exceeding PCV[1] or relaxed PCV
Total coliforms	13,417	1,273	9.5
Faecal coliforms	13,417	539	4.0
Colour	5,972	125	2.1
Turbidity	4,639	111	2.4
Aluminium	5,878	534	9.1
Iron	4,891	209	4.3
Manganese	3,629	84	2.3
Lead	2,615	115	4.4
Trichalomethanes	703	105	14.9
All others	49,013	178	0.4
Total of all parameters	104,174	3,273	3.1

Note: 1. See footnote to Table 7.4.

Source: SOEnD

Trihalomethanes (THM)

7.83 This parameter consists of the sum of the concentrations of four individual substances: trichloromethane, bromo-dichloromethane, dibromochloromethane and tribromo- methane. THMs can be formed as a by-product of the disinfection of water by chlorine. Data on sampling in Scotland are incomplete, but those which are available are shown in Table 7.5.

Polycyclic aromatic hydrocarbons (PAH)

7.84 Most water mains made of cast or ductile iron laid before the mid 1970s were given an internal anti-corrosion coating of coal tar pitch which can contain up to 50 per cent of PAH. Water supplied through such lined mains has been found occasionally to contain PAH in solution and in suspension. PAH in solution arises mainly through leaching and PAH in suspension arises from shedding of coal tar pitch particles as the coating of tar pitch deteriorates. Although concentrations in water are very low, there is concern because some PAH are known to be carcinogenic in animals. In 1991, PAH concentrations above the standard were detected in 176 zones in England and Wales. No failures in the standard were recorded in Scotland.

Pesticides

7.85 The Regulations set standards of 0.1 µg/l for individual substances and 0.5 µg/l for the total of the detected concentration of individual substances and are based on the EC Drinking Water Directive.

7.86 There are approximately 450 individual substances used as pesticides in England and Wales. In 1991, 34 of these substances were found in water supplies at concentrations in excess of the standard, although the amounts were far smaller than those which are known to be harmful or likely to damage public health. Only 3 per cent of the 800,000 samples analysed exceeded the standard. In Scotland, no sample exceeded the standard for individual pesticides in 1990.

7.87 When the new Regulations [11,13] were introduced, some contraventions of the standards for key parameters were expected to occur. Many of the water supply zones in which contraventions have occurred have improvement programmes in hand in order to ensure future compliance.

Water fluoridation

7.88 Fluoride occurs naturally in water, but is also added to water at treatment works by certain water companies in England and Wales. Fluoridation of water supplies began in the early 1960s following the successful completion of trials which showed a substantial improvement in the condition of teeth of young children. Water companies are permitted to increase the fluoride content of water supplied within a specified area at the written request of a Health Authority. The Health Authority making such arrangements with a water company is required to ensure that the concentration of fluoride in the water supplied is, so far as is reasonably practicable, maintained at 1.0 mg/l.

7.89 Figure 7.14 shows the concentration of fluoride in public water supplies in England and Wales, distinguishing between areas supplied with water to which fluoride has been added and areas where it occurs naturally. The map shows areas supplied with water fluoridated at the optimum concentration of 0.9 to 1.1 mg/l, and areas where the operation of water distribution systems results in fluoridated water mixing with unfluoridated water to give fluoride concentrations below 0.9 mg/l.

7.90 At present, no Scottish public water supplies are treated to increase the naturally occurring level of fluoride. However, several Health Boards are currently investigating the costs of fluoridating water supplies in their areas. In Northern Ireland, two small water fluoridation schemes in County Down and in County Armagh have been in operation for many years and Health Authorities are considering new schemes. Like the water companies in England and Wales, DOE(NI) is permitted to increase the fluoride content of water supplied at the written request of a Health Authority.

Private water supplies

7.91 Local authorities have responsibility for monitoring private water supplies in England and Wales and district and Islands Councils are responsible in Scotland. Water quality regulations [12] making provision for the monitoring of private water supplies by local authorities were introduced in England and Wales on 1 January 1992. Equivalent legislation was introduced in Scotland in 1992 and is planned for Northern Ireland. Responsibility for monitoring in Northern Ireland will rest with DOE(NI).

Figure 7.14   Flouride concentrations in water supplies

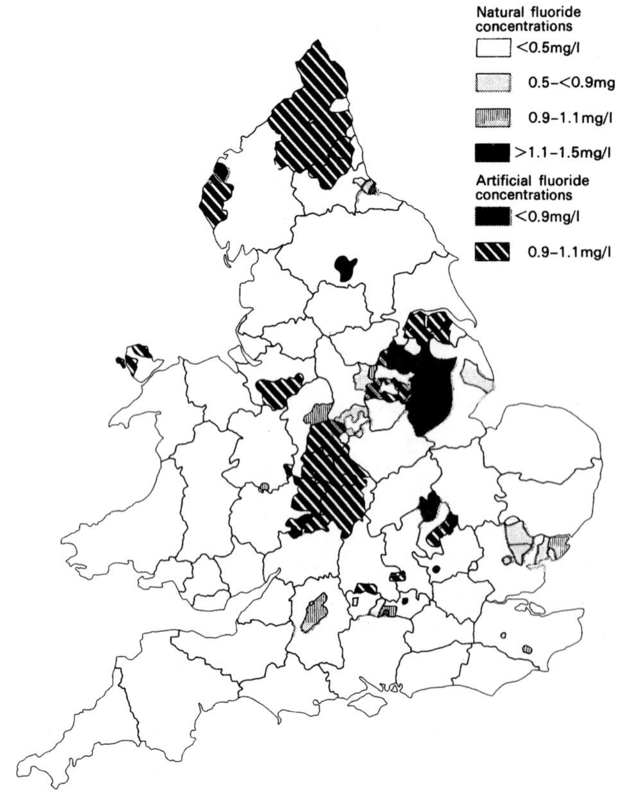

Natural fluoride concentrations
- <0.5mg/l
- 0.5–<0.9mg
- 0.9–1.1mg/l
- >1.1–1.5mg/l

Artificial fluoride concentrations
- <0.9mg/l
- 0.9–1.1mg/l

Note: Information as at 31.10.90

Source: DOE

References and further reading

1. National Rivers Authority, (1991). Proposals for Statutory Water Quality Objectives: Report of the National Rivers Authority, Water Quality Series No 5. NRA.

2. National Rivers Authority, (1991). The Quality of Rivers, Canals and Estuaries in England and Wales: Report of the 1990 Survey, Water Quality Series No 4. NRA.

3. Scottish Office Environment Department, (1992). Water Quality Survey of Scotland 1990. Scottish Office.

4. National Rivers Authority, (1991). Toxic Blue Green Algae, Water Quality Series No 2. NRA.

5. The Control of Pollution (Silage, Slurry and Agricultural Fuel Oil) Regulations, 1991.

6. National Rivers Authority, (1992). Water Pollution Incidents in England and Wales in 1991. Water Quality Series No 9. NRA.

7. National Rivers Authority, (1992). The Influence of Agriculture on the Quality of Natural Waters in England and Wales, Water Quality Series No 6. NRA.

8. Department of the Environment, (1986). Nitrate in Water: A Report by the Nitrate Coordination Group, Department of the Environment Pollution Paper No 26. DOE.

9. Ministry of Agriculture, Fisheries and Food, Severn Trent Water, Department of the Environment, (1988). The Hatton Catchment Nitrate Study: A Report of a Joint Investigation on the Control of Nitrate in Water Supplies. Severn Trent Water.

10. Department of the Environment, (1988). The Nitrate Issue: A Study of the Economic and other Consequences of Various Local Options for Limiting Nitrate Concentrations in Drinking Water. DOE.

11. Water Supply (Water Quality) (Amendment) Regulations 1991.

12. The Private Water Supplies Regulations 1991 SI No 2790.

13. Water Supply (Water Quality) (Scotland) Regulations 1990.

14. Drinking Water Directive (80/778/EEC).

15. Department of the Environment and Welsh Office, (1992). Drinking Water 1991: A Report by the Chief Inspector, Drinking Water Inspectorate. HMSO.

8 The marine environment

☐ Rivers in the UK (mainly the Forth, Humber and Thames) account for about 20 per cent of contaminants entering the North Sea by river. The Rhine transports discharges from the main industrialised areas of Germany, France, the Netherlands and Switzerland, accounting for 40 to 50 per cent of contaminants entering the North Sea by river (8.8).

☐ In 1990, around 9.3 million tonnes of sewage sludge (wet weight) were dumped in the sea around the UK, accounting for about 30 per cent of all UK sludge arisings. The quantity has remained fairly constant at around nine million tonnes (wet weight) annually since 1986. However, the metal content in sewage sludge has declined over the same period (Figure 8.4 and Table 8.2). The UK will phase out dumping of sewage sludge at sea by the end of 1998 (8.13).

☐ Liquid industrial waste dumped in seas around the UK by the UK declined from around 0.3 million tonnes (wet weight) in 1985 to just over 0.2 million tonnes in 1990. The measured content of metals in liquid industrial waste has declined over the period (Figure 8.6 and Table 8.3). The UK will cease sea dumping of liquid industrial waste by early 1993 (8.17).

☐ Solid industrial waste dumped in seas around the UK by the UK increased from 3.9 million tonnes (wet weight) in 1985 to 4.9 million tonnes in 1990. Amounts of chromium, nickel and cadmium dumped in solid industrial waste were higher in 1990 than in 1985 (Figure 8.6 and Table 8.3). The UK will cease dumping power station ash waste at sea by early 1993 and colliery waste by the end of 1997 (8.17).

☐ About 800 oil spill incidents around the UK were reported in 1990, 22 per cent of which were over 100 gallons. Forty five per cent of oil spill incidents in 1990 were related to offshore oil installations in the North Sea (Figure 8.8).

☐ In 1990, clean up of 136 oil spills was necessary. This was about 30 per cent of all spills (excluding offshore North Sea spills) (Figure 8.9).

☐ In England and Wales, 90 per cent of estuary length was classed as of "good" or "fair" quality in 1990. In Scotland, 96 per cent of estuary area was of good or fair quality in 1990. In Northern Ireland, 83 per cent of estuary length was of good quality in 1985 (Figures 8.10 - 8.12).

☐ In 1991, 343 out of 453 identified UK bathing waters complied with the European Community standards set for coliform bacteria, a compliance rate of 76 per cent overall (Figure 8.13).

☐ In 1992, 17 beaches in the UK were awarded the Blue Flag (Figure 8.14). Ninety three beaches won Seaside Awards (8.63).

☐ North Sea herring stocks recovered following closure of fisheries for five years, from 219,000 tonnes in 1977 to 4.4 million tonnes in 1986, but fell again to 3 million tonnes by 1990. Stocks of mackerel in west coast waters remained constant at just under 3 million tonnes throughout the 1980s. North Sea cod stocks have declined from 900,000 tonnes in 1980 to just over 300,000 tonnes in 1990 (Figure 8.15).

8.1 This chapter considers the natural processes and human activities which affect the quality of seas and estuaries of the UK. It also covers briefly the contamination of marine waters, sediments and marine life. Later sections give information about marine and estuarine water quality. The final sections cover quality of bathing waters and beaches, and fisheries.

8.2 The quality of marine and estuarine water is affected by many natural factors including the type of sea, underlying sediments, the pattern of tides, currents, prevailing winds, weather and climate. It can also be affected by human activities both on land and at sea, and particularly by the size and number of rivers carrying waste and contaminants to the sea. Discharges from pipelines and sea outfalls, the disposal of operational refuse from ships, waste dumping from ships, and deposition from the atmosphere all influence the quality of seas and estuaries.

8.3 There are two types of sea around the shores of the UK. The Irish Sea, North Sea, English Channel and Bristol Channel can be described as "semi-enclosed"; while the seas off the Atlantic coast of Scotland and Northern Ireland are "open" seas. Figure 8.1 shows in a simplistic way the general circulation of surface waters in the seas around the UK and Box 8.1 briefly explains the natural processes at work and the marine ecosystem in these areas.

Monitoring and management of the marine environment

8.4 The UK Fisheries Departments (Ministry of Agriculture, Fisheries and Food (MAFF) in England and Wales, Scottish Office Agriculture and Fisheries Department

(SOAFD) in Scotland and Department of the Environment in Northern Ireland (DOE (NI)) monitor the quality of the marine environment to meet national and international requirements. Descriptions of the various international organisations currently concerned with management of the marine environment, and recent initiatives are in Box 8.2.

8.5 The Marine Pollution Monitoring Management Group (MPMMG) is made up of experts from government and other organisations which have a responsibility for monitoring saline waters. In 1991, the MPMMG proposed a national monitoring programme for UK coastal waters to meet the UK's national and international commitments[1]. The MPMMG is currently formulating the detailed content of this monitoring programme.

Contaminant inputs

8.6 A contaminant is a substance which is present in higher concentrations than the natural background level. A substance becomes a pollutant if it is present at concentrations which cause harm. In the marine environment, dilution means that contaminants rarely become pollutants. The different kinds of inputs are usually compared by the amount of the input (the "load"), but it is other factors, such as the concentration of a substance which determines its effect on the environment.

8.7 Table 8.1 shows a summary of UK riverborne and direct inputs (sewage effluent and industrial effluent) in 1990. Data given for each contaminant show the upper limit value of discharges for each substance into the different seas around the UK during 1990. A number of these are on the Red List of dangerous substances (see Box 8.3). Estimates of the annual inputs of all Red List substances to saline waters of the UK are to be published in 1992.

Riverborne and direct discharges

8.8 The quality of the North Sea in particular is influenced by the major rivers which flow into it. From the continent these are the Rhine, Meuse, Elbe, Weser and Scheldt; and from the UK the Forth, Tyne, Tees, Humber and Thames. All these rivers carry drainage from land, outfalls from sewage works and effluent from industry. The Rhine accounts for between 40 and 50 per cent of the load of contaminant inputs entering the North Sea by river. Rivers in the UK account for about 20 per cent of the total.

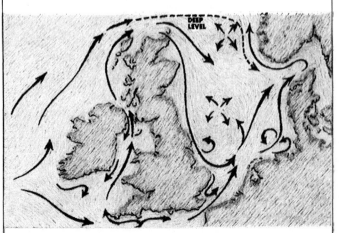

Figure 8.1   Surface current system in UK waters

DEEP
LEVEL

Note: Studies of water movements show that there is a long term anti-clockwise circulation.

Source: DOE

Natural processes   Box 8.1

Surface currents

The seas around the UK are all on the continental shelf and generally fairly shallow and for large parts of the year are fully mixed. Under such circumstances surface currents play a major role in the movement and dispersal of contaminants entering coastal waters. Figure 8.1 shows the surface current system in UK waters which illustrates the general circulation pattern of water around the UK. No attempt has been made in Figure 8.1 to represent fine detail or occasional eddy currents, and under the influence of this current system there is a clear flushing mechanism for the seas around the UK. The flushing times for different areas vary markedly; on average it takes three years for water in the German Bight to leave the North Sea, but only six months for water from the northern North Sea. However, there can be major alterations to this pattern under particular weather conditions. Day to day variations in flow due to the weather can be more significant than the long term circulation patterns for dispersion of contaminants from localised sources. Many contaminants adsorb strongly to particulates and may be deposited on the sea bed and therefore not flushed out of the area.

Food chain

The marine ecosystem is an extremely complex one and can be more accurately described as a food web than a single food chain. Few species are totally dependent on another single species for their survival. Nevertheless, in very simple terms, microscopic algae or phytoplankton are fed on by zooplankton which in turn are fed on by crustacea or small fish. These are in turn eaten by larger fish, birds or mammals, including seals and humans. This ecological pattern can lead to progressive accumulation of certain contaminants up the food chain.

Natural degradation and dispersion of contaminants

Most contaminants disperse naturally and subsequently degrade in the sea. Similarly, contaminants transported by rivers and discharges of contaminants from other coastal sources are not usually detected beyond a few miles of the coast or river mouth. In the case of direct disposal from ships, many contaminants cannot normally be detected above background levels within a few minutes of disposal.

8.9 Figure 8.2 illustrates the type of data available on riverine inputs from UK rivers in 1990. The substances shown in Figure 8.2 represent three types of contaminant input to the sea; nutrients (eg nitrogen and phosphorus), metals (eg lead, mercury, cadmium, copper) and organic compounds (eg lindane, PCBs and DDT). A number of these are Red List substances.

8.10 Direct inputs include all inputs from point sources (eg pipes) downstream of tidal limits in estuaries, or other inputs to coastal waters. Similar data to those shown in Figure 8.2 are collected for direct inputs of

---

Table 8.1   Summary of direct and riverine inputs from the UK to saline waters around the UK 1990

upper estimates

	Atlantic	North Sea East Coast	North Sea Channel	Celtic Sea	Irish sea	Total
Metals (tonnes)						
Cadmium	10	9.8	5.5	7.3	31	63.6
Mercury	1.1	4.3	0.2	1.0	5.2	11.8
Copper	220	330	91	88	120	849
Lead	193	190	48	140	99	670
Zinc	430	1,600	440	690	760	3,920
Nutrients (000 tonnes)						
Nitrate	26	91	11	29	34	191
Orthophosphates	3	18	4	5	8	38
Total Nitrogen	35	166	22	39	47	310
Total Phosphates	5	3	..	..	1	9
Organic compounds (kg)						
Lindane	202	340	46	127	63	778
PCBs	1,300	1,720	200	780	150	4,150
Other (000 tonnes)						
SPM [1]	433	1,180	132	425	690	2,860

Notes:
1. Suspended Particulate Matter   Source: DOE (for PARCOM)

Management of the marine environment:International organ-isations with management responsibilities, and the initiatives taken. *Box 8.2*

North Sea Conferences

A series of North Sea Conferences in recent years has agreed measures for the protection of the marine environment. The UK applies measures adopted at the conferences in all UK coastal waters and not just the North Sea.

The Second North Sea Conference. A comprehensive Quality Status Report on the North Sea was produced in 1987 which included estimates of the contaminant inputs, how they are dispersed and their effects. A number of measures were agreed to reduce pollution; these included the reduction of inputs of dangerous substances (Red List substances, see Box 8.3) into rivers by around 50 per cent between 1985 and 1995. A North Sea Task Force was subsequently established and this is preparing a new quality assessment of the North Sea for publication in 1993.

The Third North Sea Conference held in 1990 agreed further measures to improve the North Sea. For example it agreed to ensure that the use of all identifiable polychlorinated biphenyls (PCBs) is phased out and that they are destroyed or stored safely by the end of 1999 at the latest. Other initiatives agreed at the conference included new controls on operational discharges from ships and offshore platforms and new international measures on research, survey and protection of marine wildlife and especially small cetaceans (dolphins and porpoises). More details are given in a UK guidance note[2].

International Council for the Exploration of the Sea (ICES)

ICES is concerned with development of marine science generally, with a particular emphasis on investigations related to the sustainable exploitation of living and non-living marine resources. It acts in an advisory capacity to a number of Conventions (eg Oslo and Paris Conventions). It concentrates on developing strategies and scientific tools for monitoring and environmental assessment.

The Oslo and Paris Conventions

The Convention for the prevention of marine pollution by dumping from ships and aircraft, known as the Oslo Convention, entered into force in 1974. It aims to prevent marine pollution in the north east Atlantic caused by dumping of waste from ships and aircraft and from incineration of waste at sea. The Oslo Commission, which oversees implementation of the Convention, has agreed to the phasing out of most forms of dumping at sea. The Commission publishes statistics on all dumping and incineration in annual reports.

The Convention for the prevention of marine pollution in the north east Atlantic from land-based sources, known as the Paris Convention, entered into force in 1978. It aims to control all discharges into the sea from land-based sources including rivers and other coastal sources, oil platforms, and the atmosphere. Implementation of the Convention is overseen by the Paris Commission (PARCOM). It has Working Groups dealing with inputs to the sea, industrial sectors, diffuse sources of pollution, oil, nutrients and atmospheric inputs.

A Joint Monitoring Programme of the Oslo and Paris Commissions was established in 1978. Regular reports are made on the concentrations of contaminants in fish and shellfish, sea water and sediments and published in their annual reports.

International Maritime Organisation (IMO)

IMO is concerned with ensuring the safety at sea and protecting the marine environment by the control or prevention of pollution from ships. IMO is responsible for the development of conventions, codes and recommendations which are implemented by member governments. The International Convention for the Prevention of Pollution from Ships 1973, modified by a 1978 protocol (MARPOL 73/78) covers marine pollution. This Convention applies limits to operational discharges from ships of oil (Annex I), and discharges of noxious liquid substances (Annex II), and ship-generated refuse (Annex V). Stowage standards for harmful bulk liquid and packaged cargoes, designed to minimise the risk of inadvertent release of these substances into the sea, are also covered (Annex III). The Convention will also control the discharge of sewage from ships (when Annex IV of the Convention enters into force).

The London Dumping Convention

The Convention regulates dumping at sea on a global basis. It has reached agreement on the phasing out of incineration at sea and sea dumping of industrial wastes. It is also the international convention which controls sea dumping of radioactive waste.

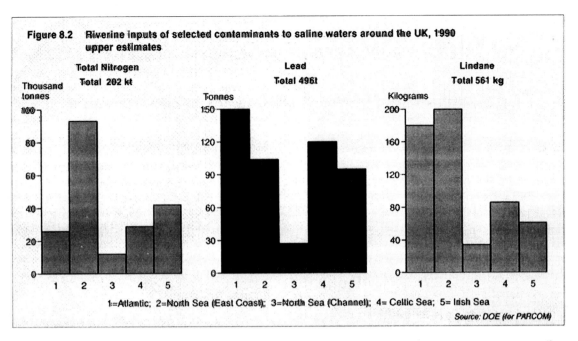

Figure 8.2 Riverine inputs of selected contaminants to saline waters around the UK, 1990 upper estimates

1=Atlantic; 2=North Sea (East Coast); 3=North Sea (Channel); 4= Celtic Sea; 5= Irish Sea

Source: DOE (for PARCOM)

Waste disposal at sea

8.11 The UK fisheries departments (MAFF, SOAFD, and DOE(NI)) control the disposal of wastes at sea by issuing licences under the Food and Environment Protection Act 1985. Conditions on the licence specify the nature of the waste, the disposal site and the method and volume of disposal including the total quantity of waste and specified contaminants in the waste allowed each year. Details of all disposal licences and of the monitoring conducted are sent to the Oslo Commission annually, and the latest data are to be published in 1992. The Ministry of Agriculture, Fisheries and Food (MAFF) publish reports annually covering all UK waters. These include information about waste disposal at sea[3].

contaminants, and the type of information available is illustrated in Figure 8.3.

8.12 The UK is phasing out sea disposal of both industrial waste and sewage sludge.

Sewage sludge

8.13 Sewage sludge is the residue left after treatment at sewage works. In 1990 around 9.3 million tonnes of sewage sludge (wet weight) were dumped in the sea around the UK, accounting for about 30 per cent of all UK sludge arisings. The quantity dumped has remained fairly constant at around nine million tonnes (wet weight) annually since 1986 (see Figure 8.4). However, the metal content in sewage sludge has declined over the same period (see Table 8.2), and this reflects tighter controls and better treatment of industrial effluent. The UK will phase out disposal of sewage sludge at sea by the end of 1998.

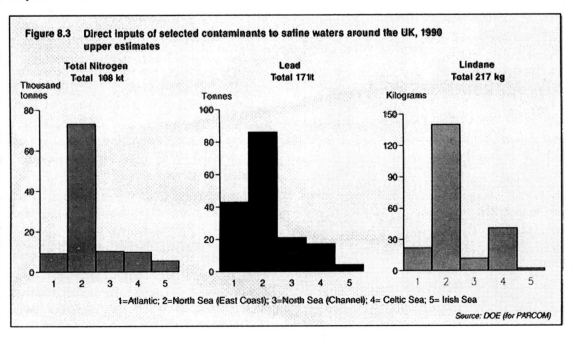

Figure 8.3 Direct inputs of selected contaminants to saline waters around the UK, 1990 upper estimates

1=Atlantic; 2=North Sea (East Coast); 3=North Sea (Channel); 4= Celtic Sea; 5= Irish Sea

Source: DOE (for PARCOM)

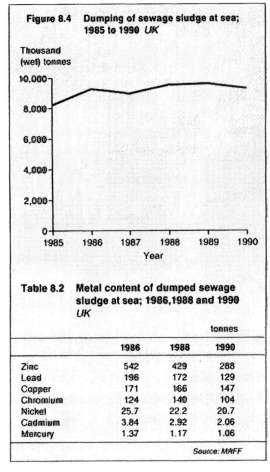

Figure 8.4 Dumping of sewage sludge at sea; 1985 to 1990 UK

Table 8.2 Metal content of dumped sewage sludge at sea; 1986, 1988 and 1990 UK

			tonnes
	1986	1988	1990
Zinc	542	429	288
Lead	196	172	129
Copper	171	166	147
Chromium	124	140	104
Nickel	25.7	22.2	20.7
Cadmium	3.84	2.92	2.06
Mercury	1.37	1.17	1.06

Source: MAFF

Sewage treatment

8.14 Figure 8.5 shows the UK population served by sewers discharging into coastal waters and estuaries in 1990, and the level of treatment (discharges serving more than 10,000 people). The sewage produced by 2.5 million people received no prior treatment, but this will have to change under the EC Urban Waste Water Treatment Directive which sets timetables between

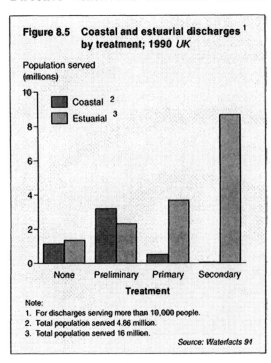

Figure 8.5 Coastal and estuarial discharges [1] by treatment; 1990 UK

Note:
1. For discharges serving more than 10,000 people.
2. Total population served 4.86 million.
3. Total population served 16 million.

Source: Waterfacts 91

1998-2005 for the treatment of coastal discharges according to the size of the discharge and the sensitivity of the receiving water. More information about sewage treatment and discharges is given in Chapter 7 on inland water quality.

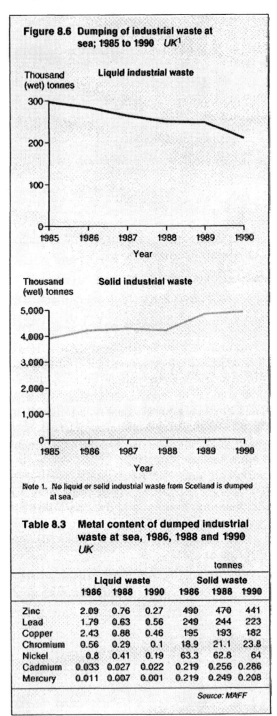

Figure 8.6 Dumping of industrial waste at sea; 1985 to 1990 UK[1]

Note 1. No liquid or solid industrial waste from Scotland is dumped at sea.

Table 8.3 Metal content of dumped industrial waste at sea, 1986, 1988 and 1990 UK

	Liquid waste			Solid waste		
	1986	1988	1990	1986	1988	1990
Zinc	2.09	0.76	0.27	490	470	441
Lead	1.79	0.63	0.56	249	244	223
Copper	2.43	0.88	0.46	195	193	182
Chromium	0.56	0.29	0.1	18.9	21.1	23.8
Nickel	0.8	0.41	0.19	63.3	62.8	64
Cadmium	0.033	0.027	0.022	0.219	0.256	0.286
Mercury	0.011	0.007	0.001	0.219	0.249	0.208

tonnes

Source: MAFF

Industrial waste

8.15 Liquid industrial waste dumped at sea consists mostly of liquid wastes which cannot be directly discharged to rivers or estuaries and which are difficult to treat, such as acidic effluents from chemical manufacturing. Solid industrial waste is largely colliery waste (minestone) and flyash from power stations. Information on liquid industrial and solid industrial waste dumped at sea is shown in

Figure 8.6 and Table 8.3. Scotland dumps no liquid or solid industrial waste at sea.

8.16 Amounts of liquid waste dumped declined from around 0.3 million tonnes in 1985 to just over 0.2 million tonnes (wet weight) in 1990. Dumping of solid industrial waste has however increased, from 3.9 million tonnes to 4.9 million tonnes (wet weight) over the same period.

8.17 Two licences for the disposal of liquid industrial wastes at sea currently remain in force. These will have expired by early 1993, and sea dumping of such wastes will then cease. The dumping of minestone on beaches will cease by the end of 1995 and at sea by the end of 1997 unless no practical land based methods are available. Flyash dumping at sea will cease by the end of 1992 or by early 1993 at the latest.

Dredged material

8.18 Dredging is required for maintenance of harbours, and much of the dredged material dumped in the sea is material which has previously been deposited by rivers. Figure 8.7 shows the amounts dumped in the sea by the UK between 1985 and 1990. Over the period, these amounts have varied between 36 million tonnes (wet weight) in 1990 and over 44 million tonnes in 1989. The content of metals in this dredged material is given in Table 8.4 for 1986, 1988 and 1990. The amounts of all measured metals in dredged spoil have fallen successively between 1986 and 1990, and are likely to continue to decline in the future as a result of PARCOM agreed measures being taken to reduce contaminant inputs to the sea transported by rivers.

Incineration

8.19 Incineration of wastes in the north east Atlantic ceased in early 1991.

Discharges of oil from ships and offshore installations

8.20 Oil enters the marine environment mainly as a result of spills or discharges from ships at sea, from offshore oil and gas platforms and from land based sources. The effects of oil pollution from slicks are most

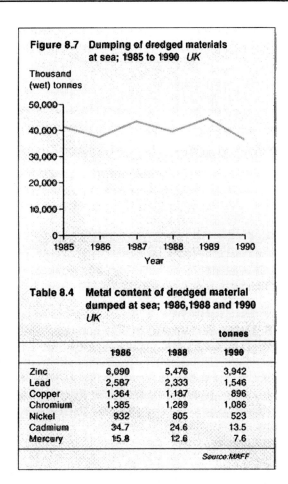

Figure 8.7   Dumping of dredged materials at sea; 1985 to 1990 UK

Table 8.4   Metal content of dredged material dumped at sea; 1986,1988 and 1990 UK

			tonnes
	1986	1988	1990
Zinc	6,090	5,476	3,942
Lead	2,587	2,333	1,546
Copper	1,364	1,187	896
Chromium	1,385	1,289	1,086
Nickel	932	805	523
Cadmium	34.7	24.6	13.5
Mercury	15.8	12.6	7.6

Source: MAFF

visible among sea birds and sea mammals. The impact of slicks on other marine species is generally not significant unless the spillage is very large, although this will depend on the local circumstances and the type of oil discharged. Permitted discharges of water containing oil from offshore oil installations, generally amounting to around three thousand tonnes of oil each year, mostly affect marine benthic animals in the vicinity of installations. These discharges, and any other spills from installations, invariably disperse before they reach the shore.

8.21 The Advisory Committee on Protection of the Sea (ACOPS) is a voluntary body which carries out annual surveys, and monitors the effects of any oil pollution around the coasts of the UK. Reports are published annually[4,5]. The following sections summarise information available from the surveys.

Red List substances   *Box 8.3*

The most recent Red List of dangerous substances was defined by the UK in April 1989. Substances on the list are those which are considered to pose the greatest potential threat or danger to the aquatic environment. Most (though not all) are on the Red List because they are persistent, toxic, and liable to accumulate in living tissues. They include heavy metals, such as mercury and cadmium; organic compounds such as pesticides and chlorinated industrial chemicals.

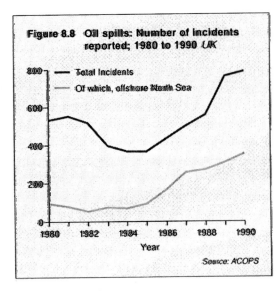

Figure 8.8   Oil spills: Number of incidents reported; 1980 to 1990 *UK*

Source: ACOPS

Oil spill incidents

8.22 Figure 8.8 shows the number of oil spills reported for waters around the UK between 1980 and 1990. There were nearly 800 such incidents reported in 1990, 45 per cent of which were from offshore oil installations in the North Sea. More comprehensive reporting using improved surveillance techniques, began in 1986.

8.23 The simple number of spills reported does not give a reliable indication of their effect on the environment, since many spills are small and will disperse naturally.

8.24 Around 70 per cent of the reported spills over 100 gallons in 1990 involved offshore installations in the North Sea. The number of these large spills from North Sea installations has increased from 32 in 1980 to 124 in 1990, while the total number in UK waters has increased less sharply, from 84 in 1980 to 174 in 1990.

8.25 Twenty six spills over two tonnes were reported in 1990 and six incidents involved

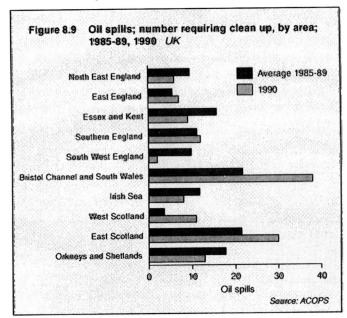

Figure 8.9   Oil spills; number requiring clean up, by area; 1985-89, 1990 *UK*

Source: ACOPS

spills of over 50 tonnes. The largest of these was the spillage of 1,100 tonnes of crude oil following a collision between the tanker "Rosebay" and a trawler off Start Point in Devon in May 1990.

Clean up of oil spills

8.26 Most oil spills are small, and clean up action is generally not necessary since the oil is dispersed and degraded by natural processes. It is spills of crude oil which are more likely to be washed ashore and cause more damage than the less persistent fuel oils released by ships. Most UK oil fields are distant from land, and so any spills from offshore installations are unlikely to threaten any seabird populations at sea or ashore, since most of these spills degrade naturally in the open sea and are not likely to reach the coast. Coastal oil pollution is therefore usually traceable to shipping incidents. There has been a steady decline in the number of beach pollution incidents reported over the last decade, falling to 31 during 1990 [5].

8.27 The responsibility for dealing with major spillages of oil or chemicals in UK waters rests with the Department of Transport. Any practical action required is undertaken by the Department's Marine Pollution Control Unit (MPCU).

8.28 Figure 8.9 shows the number of oil spills requiring clean up by UK coastal region for 1985-89 and 1990. The regional pattern broadly reflects the location of oil terminals, where most clean up operations took place. In 1990, there was a total of 136 clean up treatments, covering about 30 per cent of all spills in these coastal regions. Over a quarter of all clean up treatments took place in the Bristol Channel and South Wales coastal region and just over 20 per cent took place in the east Scotland coastal region. These data exclude offshore North Sea spills, for which information on clean up action is not available; in most cases it is likely that oil has been left to degrade naturally without treatment.

Prosecutions and convictions

8.29 It is often difficult to get enough evidence to prosecute oil spill offenders. In 1990 a total of 22 prosecutions were brought, each of which resulted in a conviction. However, this represented less than 3 per cent of all incidents (including offshore North Sea incidents).

Operational discharges from ships other than oil

8.30 Operational discharges from ships of noxious liquids carried in bulk by ships are controlled by Annex II of the MARPOL 73/38 regulations (see Box 8.2). Substances are identified in one of four categories (A,B,C,D) according to their potential pollution hazard. The most noxious of these substances (category A, for example, diphenyl ether and creosote) cannot be discharged at all. The least noxious (category D, for example, vegetable oils and ethylene glycol) are considered to pose only a minimal hazard to the marine environment and are practically non-toxic to aquatic life. No input estimates are available relating to this category[6].

8.31 The discharge of substances in the middle two categories of noxious liquid substances (category B, for example, trichloroethylene and phenol; and category C, for example, sulphuric acid and benzene) is controlled. Very approximate estimates have been made of the amounts of substances in these two categories, based on overall tonnages of substances shipped in bulk across the North Sea and average size of ships and cargo tanks. The estimated discharges in 1988 are 480 tonnes for category B and 2,300 tonnes for category C. Discharge provisions ensure that the substances enter the sea (in a highly diluted form) outside the 12 mile zone from the nearest land. In sample surveys carried out in the North Sea, virtually all samples taken confirmed that concentrations of these chemicals were below the limit of detection[6].

Marine water quality

8.32 This section concentrates mainly on information available about marine water quality in the North Sea and the Irish Sea. The two most recent reports covering the North Sea[6] and the Irish Sea[7] indicate that the condition of the marine environment is generally good and that environmental problems are confined only to specific areas. There are however many gaps in the information available.

8.33 A report published in 1990[6] provided an interim update to the 1987 report on the Quality Status of the North Sea[8], and confirmed the conclusion reached at the Second International Conference on the Protection of the North Sea in 1987. This was that damage from pollution is largely confined to specific areas of the North Sea such as the German Bight and Dutch Wadden Sea. Both these areas have shallow water, with several major rivers entering them carrying a heavy load of contaminants and a relatively low circulation of water. The North Sea Task Force established at the 1987 conference (see Box 8.2) is currently preparing the next Quality Status Report

(QSR) for the North Sea, which it will publish in 1993. This QSR will be based on an extended monitoring and research programme and will include a detailed study of North Sea sub regions.

8.34 The 1988 report on the environmental status of the Irish Sea published by the International Council for the Exploration of the Sea (ICES)[7] concluded that whilst environmental quality is generally good, there are a number of locations around the Irish Sea where the adverse effects of human activity are apparent. This is particularly the case near urban and industrial areas. In Liverpool Bay mercury levels in fish in the early 1980s were close to the EC environmental quality standard, but by 1988 these had been brought down to about two thirds of the set limit of 0.3 mg mercury per kilogramme (wet weight) in fish[3].

8.35 More monitoring information is needed however, to allow a better assessment of the cumulative effects of contaminants in localised areas, and so an Irish Sea Science Coordinator was appointed in 1992 to review existing studies and develop a research and monitoring programme for the Irish Sea.

Concentrations of contaminants in sea water

8.36 Work done before 1988 established the general distribution of the trace metals mercury, lead, cadmium and copper in sea water for coastal waters around England and Wales[9]. Offshore concentrations of all the metals were relatively low, but there were some higher concentrations in some estuaries, associated with inputs to estuarine waters. However, at that time, all values complied with UK environmental quality standards. The NRA in England and Wales and the RPAs in Scotland monitor estuarial and coastal waters near discharge points to check that EC or UK National Environmental Quality standards are being met. In 1990, quality standards for some EC List I substances were not met in the following areas:

Thames Estuary	Lindane
Irvine Bay	Mercury
Whitehaven	Cadmium
Tees Estuary	Chloroform

Corrective action is being taken in all of these cases.

8.37 MAFF have also estimated concentrations of lindane in sea waters outside estuaries (lindane is a Red List substance used as an insecticide). These were found to be low and within safe levels recommended by the EC (Directive 84/491/ EEC). Further data on the distribution of some of these contaminants in sea water are available following work carried out in 1988-89[10,11].

Concentrations of contaminants in sediments

8.38 MAFF also monitors the sediment concentrations of a range of heavy metals and organic compounds including some organochlorine pesticides. MAFF publish results from this monitoring programme in their Aquatic Environment Monitoring Reports[3]. Concentrations found in sediment samples have generally been low, and do not give cause for concern.

Concentrations of contaminants in marine life

8.39 Concentrations of contaminants, mainly of metals and organic compounds in

marine fish and shellfish, are monitored by MAFF Fisheries Laboratories, and the Scottish Office Agriculture and Fisheries Department (SOAFD). A number of standards and guidelines apply for a range of different contaminants in fish and shellfish, and some of those which apply to fish flesh, and the expected concentration of some contaminants are summarised in Box 8.4.

8.40 The results of MAFF monitoring surveys are presented in their Aquatic Environment Monitoring Reports[3]. These show that there is considerable variation in contaminant levels of mercury, PCBs and some organochlorine pesticides in fish species around the coast of England and Wales. There is little indication of serious contamination in the North Sea and Channel areas. Levels of contamination are greatest in the Irish Sea off the north west English coast and, in particular, relatively high levels of mercury, organochlorine pesticides and PCBs are still detectable in

---

Examples of standards, guidelines and expected concentration values of contaminants in fish Box 8.4

"Expected" values are those commonly found in areas not known to be severely contaminated, and are derived from past work undertaken by MAFF Directorate of Fisheries Research. Some of these are described here. A fuller description of the relevant guidelines and expected concentration values is given in the 1992 MAFF Aquatic Environment Monitoring Report[11].

Metals

Mercury European Community and Paris Commissions have adopted an Environmental Quality Standard (EQS) which requires mean concentrations of mercury in representative fish flesh to be less than 0.3 mg/kg wet weight.

The Joint Monitoring Programme (JMP) of the Oslo and Paris Commissions have adopted guidelines with a mid range of 0.1 - 0.3 mg/kg of mercury in fish flesh (wet weight).

Cadmium There are no standards or guidelines in England and Wales for concentrations in fish flesh. Expected values are less than 0.2 mg/kg wet weight.

Lead GB Lead in Food Regulations 1979 state concentrations in fish should not exceed 2.0 mg/kg wet weight [26]. Expected values are 0.2 - 0.3 mg/kg wet weight in fish.

Copper UK recommendations for limits for copper content of food are that levels should not exceed 20 mg/kg wet weight in fish[12]. Expected values in fish are up to 0.6 mg/kg wet weight, and greater than 1.0 mg/kg wet weight in fatty fish such as herring.

Zinc UK guidelines (1953) on levels of zinc in food are that levels should not exceed 50 mg/kg wet weight, except in some foods which naturally contain more than 50 mg/kg, such as herring[13]. Expected values commonly found are up to 6.0 mg/kg wet weight in most fish flesh, and in excess of 10 mg/kg in fatty fish.

Pesticides and PCBs

There are no set standards for concentrations of these substances in fish and shellfish from waters around England and Wales.

Dieldrin Expected values are 0.2 - 0.3 mg/kg wet weight in fish liver.

PCBs JMP guidelines have a mid range of 0.01 - 0.05 mg/kg wet weight in fish muscle, 2.0 - 5.0 mg/kg wet weight in cod liver, and 0.5 - 1.0 mg/kg wet weight in flounder liver. Guideline values for cod and flounder are used by MAFF as guidelines for concentrations in all roundfish and flatfish respectively.

fish from Liverpool and Morecambe bays. Levels are within relevant EC Environmental Quality Standards, however, and do not generally give cause for concern, although the implications for the health of certain coastal populations of dolphins and porpoises of PCB concentrations in fish from the Irish Sea are the subject of further study. Data on contaminants in marine mammals such as seals, dolphins and porpoises are also included in MAFF's Aquatic Environment Monitoring Reports.

8.41 Some results from the MAFF monitoring programme for metal contaminants in fish are given in Table 8.5. The table shows average concentrations of heavy metals in samples of cod and plaice taken in the Thames, Liverpool Bay and North Sea coastal regions for 1980 and 1986 to 1990. Some specific areas are known to receive more substantial inputs of contaminants than others, for example Liverpool Bay[3]. Concentrations in some non-commercial populations of fish and shellfish, especially in estuaries, have also been found to be relatively high, for example cadmium concentrations in mussels from the Bristol Channel and Whitehaven. The levels found in commercially exploited populations, however, pose no hazard to fish or human health. Measures are nevertheless being taken to reduce inputs of contaminants.

8.42 Shellfish are susceptible to contamination by raw or inadequately treated sewage discharged directly into the sea. The incidence of diseases most commonly associated with shellfish consumption is given in Chapter 14 on environment and health.

8.43 Tributyltin (TBT) compounds, which are used as antifouling agent in paints, have affected the health of a variety of UK marine species, in particular two types of shellfish, oysters and dogwhelks. Measures were introduced in 1986[14] and 1987[15] to control the sale and use of TBT, and as a result TBT can no longer be used in antifouling formulations on fish nets or on boats of less than 25m in length. Research by MAFF[10,16] and others[17,18] shows that concentrations of TBT in water and animal tissues are decreasing as are concentrations in sediments. In 1990, average levels of environmental concentrations of TBT in sea water derived from antifouling paints, were around a quarter of 1986 figures, and in some places less than a tenth. The Shellfish Association, which pressed for the 1987 ban on TBT use on small boats, report widespread recovery of oyster farming operations[19]. However, environmental recovery is not complete, and it is likely that the most sensitive species, for instance the dogwhelk, will take many years to recover.

8.44 The Department of the Environment has recently started a survey of the concentration and effects of TBT in the coastal waters of all North Sea countries, using the dogwhelk as the species under study. Results from this survey will be incorporated in the 1993 North Sea Quality Status Report.

Exceptional phytoplankton blooms

8.45 Phytoplankton is made up of a diverse range of free floating, mostly microscopic, aquatic plants. It constitutes a major source of food for fish and other species in the marine ecosystem. Under some conditions, some species of phytoplankton may form scums. Although these look and smell unpleasant they are generally harmless to marine life and people. Under some conditions, other species of phytoplankton naturally produce toxins, and these toxins

Table 8.5 Average concentrations of heavy metals in fish in selected areas; 1980, 1986 to 1990

mg/kg wet weight

	1980	1986	1987	1988	1989	1990
Thames						
Cod						
Mercury	0.15	..	0.07	0.11	..	0.09
Cadmium	0.1	..	..	..	..	..
Lead	0.2	..	..	..	..	..
Copper	0.2	..	0.3	0.2	..	0.1
Zinc	4.0	..	3.3	4.2	..	3.5
Plaice						
Mercury	0.09	0.04	..	0.06	..	0.06
Cadmium	0.1	..	..	..	..	..
Lead	0.2	..	..	..	..	..
Copper	0.4	0.3	..	0.2	..	0.2
Zinc	6.2	5.6	..	4.9	..	4.3
Liverpool Bay						
Cod						
Mercury	..	0.15	0.25	0.17	0.15	0.11
Cadmium	0.1	..	..	..	..	..
Lead	0.6	..	..	..	..	..
Copper	0.3	0.2	0.2	0.3	0.1	0.2
Zinc	3.7	3.5	3.2	3.2	3.2	3.1
Plaice						
Mercury	0.26	0.2	0.18	0.15	0.13	0.11
Cadmium	0.1	..	..	..	..	..
Lead	0.2	..	..	..	..	..
Copper	0.2	0.3	0.2	0.3	0.3	0.1
Zinc	6.3	5.1	4.4	4.4	4.0	3.9
North Sea (Southern Bight)						
Cod						
Mercury	0.04	0.08	0.07	0.08	0.1	0.07
Cadmium	0.1	..	..	..	..	..
Lead	0.2	..	..	..	..	..
Copper	0.3	0.2	0.2	0.2	0.2	0.2
Zinc	3.6	3.1	3.5	3.3	3.3	3.4
Plaice						
Mercury	0.04	0.04	0.05	0.06	0.05	0.06
Cadmium	0.1	..	..	..	..	..
Lead	0.2	..	..	..	..	..
Copper	0.2	0.2	0.2	0.3	0.3	0.3
Zinc	6.6	4.0	4.4	4.5	3.8	..

Source: MAFF

can be harmful to people and marine life. Some toxins can accumulate in shellfish, and if people then eat them paralytic shellfish poisoning (PSP) may result.

8.46 In GB, it has been well established over the last twenty years that PSP toxic algae occur seasonally between April and August along the north east coasts of England and Scotland. They occur infrequently and usually only affect small areas of the coastline. In recent years, it has become clear that PSP blooms may occur over much of the Scottish coastline including the west coast and the Northern Isles. A shellfish monitoring programme was established by MAFF in 1968, and currently a routine shellfish monitoring programme is carried out by MAFF and SOAFD on the affected coasts in England and Scotland. Public health authorities are alerted if toxin concentrations in shellfish approach levels which could affect human health. When toxins rise above such levels, the affected fisheries are closed. Annual results from the programme are published in the Aquatic Environment Monitoring Reports series [3]. See also Chapter 14 on environment and health.

Water quality in estuaries

8.47 Since 1958 there has been a series of surveys in England and Wales to assess the water quality of rivers, canals and estuaries; similar surveys have also been carried out in Scotland since 1968 and routine river quality monitoring has gone on in Northern Ireland since 1973. The results from the latest surveys covering estuaries are given in this section; information about freshwater rivers and canals is given in Chapter 7 on inland water quality.

8.48 Throughout the UK, estuaries have been categorised into four quality classes[20], from "good" quality (class A) to "bad" quality (class D). Overall assessments are based on the biological, aesthetic and chemical quality of the water (together with the minimum dissolved oxygen content). Data for England and Wales, Scotland, and Northern Ireland are given separately.

England and Wales

8.49 There are over 2,700 km of estuaries in England and Wales. The 1990 survey[20] showed that about 90 per cent of estuary length was classed as good or fair. Figure 8.10 shows the percentage of estuary length in each quality class for 1980, 1985 and 1990. Since the surveys began in 1958 there has been an overall improvement in the quality of estuaries in England and Wales, and in particular, an improvement in estuaries

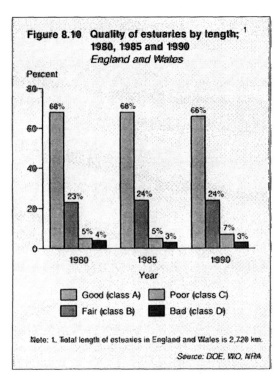

Figure 8.10 Quality of estuaries by length; [1] 1980, 1985 and 1990 England and Wales

Note: 1. Total length of estuaries in England and Wales is 2,720 km.

Source: DOE, WO, NRA

in the lowest quality class. Improvements are mainly due to investment in sewerage and sewage treatment works, the construction of new sea outfalls, and the diversion of unsatisfactory industrial discharges to sewers or to treatment works.

8.50 The 1990 survey showed a 2 percentage point reduction of estuaries in the top quality class (good) compared to the 1985 survey. Sixty six per cent of estuary length remained good.

Scotland

8.51 The Scottish Office has recently published survey results showing the quality of all estuaries in Scotland in 1990[21]. Results are given on the quality class of estuaries in Scotland, but by area (sq km) rather than by

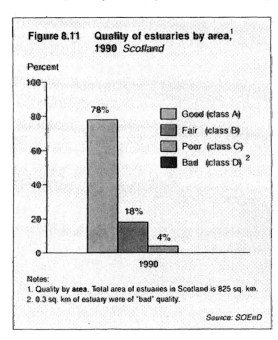

Figure 8.11 Quality of estuaries by area,[1] 1990 Scotland

Notes:
1. Quality by area. Total area of estuaries in Scotland is 825 sq. km.
2. 0.3 sq. km of estuary were of 'bad' quality.

Source: SOEnD

length (km), as in England and Wales and Northern Ireland. The results of this survey are shown in Figure 8.11, and show that 96 per cent of estuary area was of good or fair quality, the remaining 4 per cent was of poor quality, with just 0.3 sq km of estuary classified as bad quality.

Northern Ireland

8.52 Northern Ireland has just under 120 km of estuaries. The most recent data available are for 1985 and show that 83 per cent of estuary length was class A (good) with the remaining 17 per cent class C (poor). No estuaries were found to be in quality class B (fair) or D (bad). Estuary lengths of poor quality are confined to the tidal reaches of the Newry and Lagan rivers. Figure 8.12 shows the length of estuaries by class in Northern Ireland in 1985.

Bathing waters and beaches

8.53 The quality of beaches and bathing waters is of great public interest. People have been concerned for many years about risks to public health from beach pollution and poor quality bathing water and in 1975 an EC Directive (76/160/EEC) setting quality standards for bathing waters was adopted. The standards set are mainly microbiological.

8.54 A report by the House of Commons Select Committee on the Environment, published in July 1990[22], concluded that on the available evidence, the risk of the public contracting serious diseases from bathing in sea water contaminated by sewage is minimal. However, the Committee also concluded that bathing in polluted waters may be associated

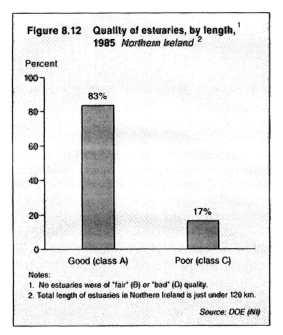

Figure 8.12   Quality of estuaries, by length,[1] 1985 Northern Ireland[2]

Notes:
1. No estuaries were of "fair" (B) or "bad" (D) quality.
2. Total length of estuaries in Northern Ireland is just under 120 km.

Source: DOE (NI)

with minor infections such as inflammation of the ear, nose and throat, gastroenteritis, and certain skin irritations.

Compliance with the European Community Bathing Water Directive

8.55 In 1991 samples were taken in 453 bathing waters (363 were in England, 51 in Wales, 23 in Scotland and 16 in Northern Ireland). 343 out of the 453 bathing waters complied with the EC coliform standards, a compliance rate of 76 per cent overall. The compliance rate varies in different regions, ranging from 30 per cent in the North West (33 bathing waters sampled) to 92 per cent in the Wessex region (39 sampled) to 100 per cent in Northern Ireland (16 sampled).

The European Community Bathing Water Directive   Box 8.5

Directive 76/160/EEC sets water quality standards for a range of parameters which must not be exceeded in bathing waters except in specified circumstances. The aim is to improve or maintain the quality of bathing water for amenity and to protect public health. Under the Directive, a bathing water is where "bathing is not prohibited and is traditionally practised by a large number of bathers". Additional bathing waters may be added each year. Further criteria used when considering identification of bathing waters include beach facilities like toilets, car parks, first aid and life guards.

The Directive itself lists 19 physical, chemical and microbiological parameters relevant to bathing water quality, some of which are mandatory and some of which are more stringent guideline values to be observed on a voluntary basis. The most relevant and most widely used parameters are total and faecal coliform bacteria, which are indicators of the presence of sewage pollution.

The Directive requires bathing waters to be sampled frequently and regularly throughout the bathing season. In the UK, a minimum of 20 samples are taken at each site through the season, which runs from mid May to the end of September in England and Wales, but is shorter in Scotland and Northern Ireland. In England and Wales, it is the NRA which carries out sampling and implements the Directive. In Scotland, it is the River Purification Authorities, and in Northern Ireland it is DOE (NI).

For the coliform parameters the Directive requires that at least 95 per cent of samples must have counts not exceeding the mandatory standards of 10,000/100 ml for total coliforms and 2,000/100 ml for faecal coliforms. The section on compliance with the EC Directive gives information on UK compliance with these EC standards set.

The standards are described in Box 8.5 and Figure 8.13 summarises the results for 1988 to 1991.

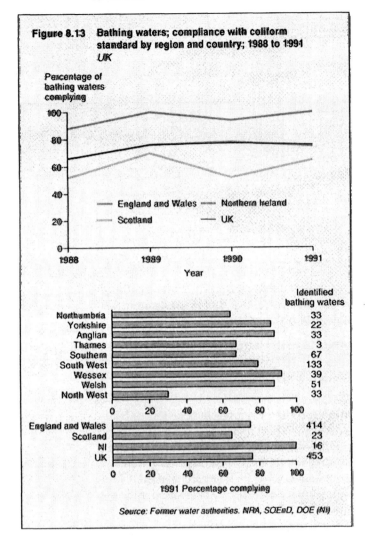

Figure 8.13 Bathing waters; compliance with coliform standard by region and country; 1988 to 1991

Source: Former water authorities. NRA, SOEnD, DOE (NI)

8.56 The overall compliance rate in the UK has risen from 66 per cent in 1988, to 77 per cent in 1990, and 76 per cent in 1991. Figure 8.13 shows the compliance rates separately for England and Wales, Scotland and Northern Ireland for the years 1988 to 1991. By country, compliance has varied between 52 per cent in Scotland in 1988 and 1990, and up to 100 per cent in Northern Ireland in 1989 and 1991. Northern Ireland has the best compliance rate generally over the period, averaging 96 per cent over 1988-1991, and Scotland has the lowest, averaging 60 per cent over 1988-1991.

8.57 Full results for sampling tests undertaken on bathing waters are available on public registers, which can be inspected at regional offices of the NRA in England and Wales, the River Purification Boards in Scotland, and DOE(NI) in Northern Ireland. The report to the EC Commission providing a detailed summary of UK monitoring results is placed in the libraries of both Houses of Parliament. This summary forms part of the annual report

produced by the EC Commission, which includes compliance tables and maps showing bathing water quality[23]. The NRA has published reports on the quality of bathing waters in England and Wales from 1987 to 1991[24,25].

8.58 In 1992, 455 coastal bathing waters were identified as being within the scope of the Directive (two, both in England, were additional to the 1991 identified bathing waters).

Public display of information on bathing water quality

8.59 In 1991 the UK launched a poster scheme for local authorities to display at beaches or other prominent sites, giving up to date and easily understandable information on bathing water quality. Virtually all coastal local authorities in England and Wales participated in this scheme in 1991, using water quality information provided by the NRA. A similar scheme was established in Scotland in 1991 using information provided by the River Purification Boards. The scheme is in operation in 1992 throughout the UK, including Northern Ireland where the information is provided by DOE(NI).

Blue Flag beaches

8.60 The EC Directive is concerned only with water quality. A much more comprehensive quality assessment of beaches, called the Blue Flag campaign, was launched in 1987 by the Foundation for Environmental Education in Europe. The scheme aims to improve the standard of beaches and marinas used by large numbers of holidaymakers. It is sponsored by the EC Commission and by a number of public and private bodies throughout the EC.

8.61 Blue Flags are awarded annually by a jury at the start of the bathing season and can be withdrawn in the course of the season if the bathing beach ceases to meet the required standards. To be eligible for a Blue Flag award, the water must comply with some of the microbiological standards specified in the EC Bathing Water Directive and in addition the beach has to satisfy a broad range of other criteria, including good management and safety, provision of basic facilities and environmental information to beach users.

8.62 In 1992, 17 UK beaches were awarded the Blue Flag. They are shown in Figure 8.14.

Seaside Awards

8.63 The Tidy Britain Group introduced new Seaside Awards in 1992. They complement the Blue Flag scheme. In 1992 there were 107 applications for the award and 93 winners. Thirty six beaches won Premier awards, achieving the guideline water quality standards required under the EC Bathing Water Directive (as with Blue Flags), and also meeting tougher standards for beach management and facilities at beaches than required under the Blue Flag scheme.

Fisheries

8.64 The state of fish stocks in the seas around the UK is determined primarily by the intensity with which they are exploited by fishing vessels and by natural factors, including the extent to which species prey on each other. The state of the stocks in UK waters is not generally affected by the relatively low level of contaminants found in these waters. Fish within UK waters are exploited by fishing vessels of several different nationalities, and mostly belong to stocks which are also exploited within other countries' fishery limits. Extensive data on the main fish stocks and regular scientific assessments are produced through internationally coordinated research and monitoring, mainly under the auspices of the International Council for the Exploration of the Sea (ICES, see Box 8.2).

8.65 The North Sea herring population was seriously affected by over fishing in the 1960s and the early 1970s. This fishery was closed for five years between 1978 and 1982. This led to a recovery of the stock from an estimated low of 219,000 tonnes in 1977 to just under 4.4 million tonnes in 1986, though in more recent years the stock has fallen again and was estimated at just over 3 million tonnes in 1990 (see Figure 8.15). While the North Sea herring fishery was closed, the stock of mackerel fished to the west and north of the UK showed a decline, but now seems to have stabilised, helped by the closure of a large area to the south west of England to directed fishing for mackerel.

8.66 At present the main stocks of roundfish (cod, haddock, whiting) and flatfish around the UK are fully exploited and in some cases heavily over-exploited. For example, cod stock in the North Sea was estimated at around 400,000 tonnes in 1963, rising to about 1.1 million tonnes in 1971. Thereafter, the stock declined to 525,000 tonnes in 1976 but recovered up to 1980 when the stock was estimated at just under 900,000 tonnes. Since then the stock has again

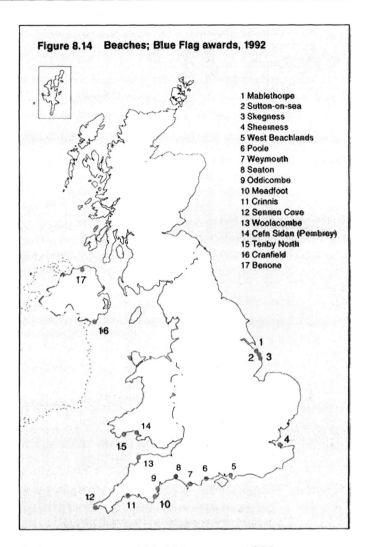

Figure 8.14 Beaches; Blue Flag awards, 1992

1 Mablethorpe
2 Sutton-on-sea
3 Skegness
4 Sheerness
5 West Beachlands
6 Poole
7 Weymouth
8 Seaton
9 Oddicombe
10 Meadfoot
11 Crinnis
12 Sennen Cove
13 Woolacombe
14 Cefn Sidan (Pembrey)
15 Tenby North
16 Cranfield
17 Benone

declined to just over 300,000 tonnes in 1990 (see Figure 8.15).

8.67 Under the Common Fisheries Policy, the conservation of fish stocks within the fishery limits of EC member states depends primarily on measures introduced by the EC

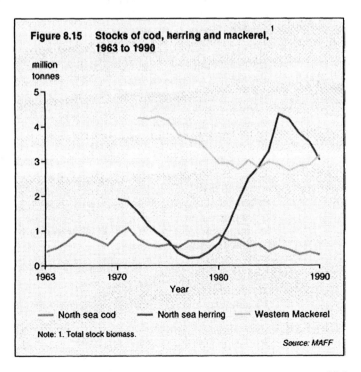

Figure 8.15 Stocks of cod, herring and mackerel,[1] 1963 to 1990

million tonnes

—— North sea cod   —— North sea herring   —— Western Mackerel

Note: 1. Total stock biomass.

Source: MAFF

121

and individual member states. Where necessary, the EC also works with non-EC countries to agree joint management arrangements. In addition to UK national measures, the EC has introduced a series of measures to improve conservation. Further action at EC and national level is being considered.

References and further reading

1. Department of the Environment, (1991). The Principles and Practice of Monitoring in UK Coastal Waters; A Report from the Marine Pollution Monitoring Management Group. DOE.

2. Department of the Environment, (1991). Third International Conference on the North Sea; UK Guidance Note on the Ministerial Declaration. DOE.

3. Ministry of Agriculture, Fisheries and Food, (annual). Aquatic Environment Monitoring Reports. MAFF Directorate of Fisheries Research, Lowestoft.

4. Advisory Committee on Protection of the Sea, (1980, 1981). Annual Reports.

5. Advisory Committee on Protection of the Sea, (1982 to 1991). Oil Pollution Surveys.

6. North Sea Conference, (1990). 1990 Interim Report on the Quality of the North Sea. The Hague, March 7 and 8, 1990.

7. Dickson, R.R. & Boelens, R.G.V., (Eds), (1988). Cooperative Research Report No 155. The Status of Current Knowledge on Anthropogenic Influences in the Irish Sea. International Council for the Exploration of the Sea.

8. Department of the Environment, (1987). Quality Status of the North Sea.

9. Ministry of Agriculture, Fisheries and Food, (1990). Monitoring and Surveillance of Non-radioactive Contaminants in the Aquatic Environment, 1984-1987; Aquatic Environment Monitoring Report No 22. MAFF Directorate of Fisheries Research, Lowestoft.

10. Ministry of Agriculture, Fisheries and Food, (1991). Monitoring and Surveillance of Non-radioactive Contaminants in the Aquatic Environment and Activities Regulating the Disposal of Wastes at Sea, 1988-89; Aquatic Environment Monitoring Report No 26. MAFF Directorate of Fisheries Research, Lowestoft.

11. Ministry of Agriculture, Fisheries and Food, (1992). Monitoring and Surveillance of Non-radioactive Contaminants in the Aquatic Environment and Activities Regulating the Disposal of Wastes at Sea, 1990; Aquatic Environment Monitoring Report No 30. MAFF Directorate of Fisheries Research, Lowestoft.

12. Ministry of Agriculture, Fisheries and Food, (1958). Circular FSH 13/58.

13. Ministry of Food, (1953). Food Standards Committee Report on Zinc.

14. The Control of Pesticides Regulations 1986, Statutory Instrument 1986 No 1510. HMSO.

15. The Control of Pollution (Anti-Fouling Paints and Treatment) Regulations 1987. Statutory Instrument 1987 No 783. HMSO.

16. Waldock, M.J. et al, (1990). Reductions in TBT Concentrations in UK Estuaries Following Legislation in 1987. Proceedings of the Third International Organotin Symposium, Monaco, April 1990.

17. Cleary, J.J., (1990). Organotins in Coastal Waters of Southwest England; An Assessment of Environmental Recovery. Proceedings of the Third International Organotin Symposium, Monaco, April 1990.

18. Abel, R., (1990). The Development of Legislation on TBT Antifoulants in the UK and Europe. J. Oil and Colour Chemists Association 73 (8) p332-333, 336.

19. Edwards, E., (1991). Polluted Areas will Recover from TBT. Fish Farming International. Nov 1991, p16-17.

20. National Rivers Authority, (1991). The Quality of Rivers, Canals and Estuaries in England and Wales: Report of the 1990 Survey, Water Quality Series No 4.

21. Scottish Office Environment Department, (1992). Water Quality Survey of Scotland, 1990.

22. House of Commons, (1990). Pollution of Beaches. A Report by the Select Committee on the Environment. HMSO.

23. Commission of the European Communities, (annual). Quality of Bathing Water.

24. National Rivers Authority, (1991). Bathing Water Quality in England and Wales 1990. Water Quality Series No 3. NRA.

25. National Rivers Authority, (1992). Bathing Water Quality in England and Wales 1991. Water Quality Series No 8. NRA.

26. The Lead in Food Regulations 1979, SI (1979) No 1254. HMSO.

Department of the Environment, (1991). The Principles and Practice of Monitoring in UK Coastal Waters; The Government Response to a Report from the Marine Pollution Monitoring Management Group. DOE.

North Sea Task Force, (1990). United Kingdom North Sea Action Plan 1985 - 1995. HMSO.

North Sea Report, (1990). The Marine Forum for Environmental Issues, London.

9 Coast erosion, flooding and sea level change

☐ The extent of erosion varies greatly around the coast of the UK. Rapid erosion occurs along parts of the coast of southern and eastern England (9.4).

☐ Special protection is given to outstanding scenic coasts under the Heritage Coast Protection scheme. There are 44 Heritage Coast covering one third of the coastline of England and Wales (Box 9.1).

☐ Nearly 80 per cent of Scotland's coast is designated as Preferred Coastal Conservation Zones (Box 9.1).

☐ A recent survey of National Rivers Authority (NRA) owned or maintained sea defences showed that 85 per cent of structures are in good condition, 13 per cent require moderate repair and 2 per cent are in need of significant works. Two per cent of local authority owned or maintained sea defences and 4 per cent of private or corporately owned defences require significant works (9.16).

☐ Over the past century the mean sea level has risen by around 10 to 15 cm. Estimates of the effects of global warming suggest rises in the global mean sea level of perhaps 20 cm by 2030, with the likely range being 10 to 30 cm (9.18).

9.1 This chapter considers coast erosion and flooding in the UK, and in the medium to longer term the implications of the sea level rise which will result from projected global temperature increases (see also Chapter 3 on the global atmosphere).

9.2 A number of factors influence flooding and coast erosion: local geology and geography (softer rocks are more rapidly eroded and low lying areas may be subject to flooding); weather and tides (flooding may occur when storm surges coincide with high tides); and in the long term, land movements

---

Flood defence and coast protection; responsibilities and control   Box 9.1

In England and Wales the National Rivers Authority (NRA) have operational responsibility for local flood warning systems and some flood defence works; they also have a general responsibility for all matters relating to flood defence, both inland and coastal. (See Chapter 6 for more details of NRA responsibilities generally). Internal Drainage Boards and local authorities are also empowered to undertake flood defence works; coast protection is the responsibility of maritime district councils. Departmental responsibility for flood defence and coast protection policy rests with the Ministry of Agriculture, Fisheries and Food (MAFF) in England, and the Welsh Office in Wales.

In Scotland, Regional and Islands Councils have powers as coast protection authorities to carry out works to protect land from erosion and encroachment by the sea and also to prevent flooding of non agricultural land. Departmental responsibility lies with the Scottish Office Environment Department.

In Northern Ireland, the Departments of Economic Development, Environment, and Agriculture, have responsibilities for coastal protection. The Department of Agriculture has responsibility for certain sea defences which are "designated".

In England and Wales, protection is also given to scenic coasts under the Heritage Coast protection scheme. There are 44 Heritage Coasts covering one third of the coastline and one third of the total length of Heritage Coast is owned by the National Trust. Nearly 80 per cent of Scotland's coast is designated as Preferred Coastal Conservation Zones through the National Planning Guidelines.

Legislation relating to the carrying out of flood and coast protection works is included in the Water Resources Act 1991, the Land Drainage Act 1991, the Coast Protection Act 1949 and the Flood Prevention (Scotland) Act 1961.

and the effect of climate change on sea levels. These may act separately or in combination to produce both coast erosion and flooding.

9.3 Responsibilities for coast protection and flood defence (including sea defence) are given in Box 9.1. Coast protection concerns the protection of land from permanent destruction by erosion and encroachment by the sea; flood defence concerns the alleviation of flooding of land by rivers or the sea.

Coast erosion

9.4 Coast erosion rates vary greatly around the UK and depend on the geology of the coast and its exposure to wind, wave and current action. Rapid erosion is found along parts of the coast of southern and eastern England where relatively soft geological formations are being eroded. At the same time however, parts of the same coast may be moving seawards as sediments build up.

9.5 Although there is a lot of information about rates of coastal erosion in the UK, much is held locally and is not available from a central source. Some illustrative information on the rates of coast recession in selected localities in the UK is given in Table 9.1.

9.6 Coast erosion is also a cause of landslides and rock fall. Of about 8,500 records which exist of landslides in GB, about 15 per cent were in the coastal zone [1]. These latter include many of the largest landslide complexes such as The Undercliff on the Isle of Wight, Folkestone Warren and Black Ven in Dorset.

9.7 The main concern of coast protection is to protect coastal areas by halting or reducing coast erosion and thereby saving homes, farmland, coastal paths etc. However, other considerations need to be taken into account such as the possible adverse effects of remedial measures on the natural environment. Coast protection measures in one area may also have detrimental "knock

Table 9.1	Rates of coast recession in selected areas	
Locality	Geology	Recession rate (metres per year)
Humberside	glacial drift	0.3 to 3.3
Norfolk	glacial drift	0.2 to 5.7
Suffolk	glacial drift	0.6 to 5.1
Kent	London clay	0.7 to 3.4
South coast	chalk	0.05 to 1.0
Wales	Lias shale	0.008 to 0.1
N.Ireland	glacial drift	0.2 to 0.8

Source: DOE, EC

on" effects elsewhere; the protection of a particular section of coastline for example, may result in erosion further down the coast. Conversely, removal of natural barriers to erosion such as beach deposits, may cause rapid recession of the coastline; natural deposition by wave action and longshore drift is a very effective form of coast protection. Box 9.2 lists various coast protection and land use planning options.

9.8 Coast protection schemes must be technically sound, environmentally acceptable and economically worthwhile with benefits greater than costs. In practice, schemes on undeveloped coasts or those protecting small areas of agricultural land, are unlikely to be financially viable and therefore much less likely to be carried out.

Flooding

9.9 Storm surges present the major flooding threat to low lying coastal areas. They are caused by a combination of low atmospheric pressure and wind stress on the sea surface. A storm surge gets stronger in shallowing water and converging coastlines, for example the North Sea from the north and the Bristol Channel from the south west. The risk of flooding is greater in winter, with the most dangerous period about the fortnightly high waters ("spring" tides). In estuaries and tidal rivers the problem is worse after prolonged

---

Coast protection and land use planning options

Box 9.2

There are a number of coast protection options such as:

- Land drainage to reduce risk of landsliding.
- Creating offshore breakwaters or sills to reduce the strength of waves hitting the shore.
- Direct protection by building sea walls or adding extra beach material.
- Strengthening the cliffs and building strong points at intervals.
- Allowing the natural process of erosion to go unchecked. This may mean extra protection elsewhere by sediments being deposited at other places along the coast.

There are also planning options, ie directing development to areas which are least at risk, and some local authorities have adopted policies to achieve this. Ventnor on the Isle of Wight for example (which lies in a landslide complex called The Undercliff) has been extensively surveyed and maps have been produced so that land instability can be taken into account in matters relating to land use planning and development [2].

The Storm Tide Warning System (STWS)   *Box 9.3*

The STWS provides early warning of storm surges that may cause flooding in low lying parts of GB. The Class A network of tide gauges provides continuous records of sea levels at key points around the coast and these are transmitted direct to the STWS based at the Meteorological Office at Bracknell. The data from the gauges are used to validate STWS forecasts, which are based on computer models developed by the NERC Proudman Oceanographic Laboratory. The tide gauge information is also used in national and international studies, which are becoming increasingly important in the context of potential global warming and the need to establish its effects on sea levels. See also Chapter 3 on the global atmosphere.

heavy rain. Although storm surges present the major threat, coastal defences may be overtopped or breached by wave action resulting in flooding, and recent notable examples are Portland (1979) and Towyn (1989) where this was the primary cause.

9.10 Major surges in the North Sea result from depressions tracking north eastwards across northern Scotland. These are exacerbated when the depression reaches the northern North Sea when the winds become northerly, helping the surge on its way, whilst the Coriolis force (the force generated by the earth's rotation) confines the surge to the east coast of Britain. Such conditions were responsible for the east coast floods of 1953, affecting areas from the Humber estuary to the Thames. Figure 9.1 shows the areas flooded. Following these floods, a national network of tide gauges and the Storm Tide Warning System (STWS) were established to get tidal information and to warn of possible recurrence (see Box 9.3).

Table 9.2	Major flooding incidents (fluvial [1] and coastal) by region, 1989-90 and 1990-91 *England and Wales*	
NRA region	**1989-90**	**1990-91**
Northumbria	1	2
Yorkshire	2	2
North West	12	9
Welsh	67	0
Severn Trent	5	0
Anglian	1	1
Thames	8	0
Southern	1	0
Wessex	14	0
South West	26	0
Total	**137**	**14**

Note: 1. Relating to rivers

Source: NRA

9.11 In the longer term, the projected change in sea level which will result from global warming would be expected to increase the incidence of flooding in low lying areas but, in general, the heights of surges are greater than the predicted rises in sea level and these will continue to be the incidents which may cause major damage. If however the sea level does rise, more areas will be at risk from surges, and more frequently.

9.12 Planning Policy Guidance on coastal planning, providing advice on development control in areas at risk from coastal flooding and erosion has been issued by government.

9.13 The number of major river and coastal flooding incidents in England and Wales is given in Table 9.2 by NRA region. There were 137 such incidents in 1989-90 (half of which were in the Welsh region) and 14 incidents in 1990-91.

9.14 Major tidal flooding is relatively rare in Scotland but in January 1991 a combination of spring tide, tidal surge and heavy wave activity resulted in significant flooding in the Firth of Clyde and in some parts of the west of Scotland. The sea level at its maximum was the highest experienced in the area for about 100 years. Only one major coastal

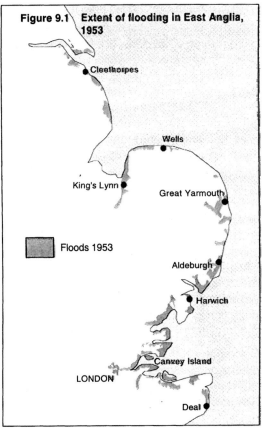

Figure 9.1 Extent of flooding in East Anglia, 1953

Cleethorpes

Wells

King's Lynn

Great Yarmouth

Floods 1953

Aldeburgh

Harwich

Canvey Island

LONDON

Deal

The Thames Barrier
Box 9.4

The Thames Barrier was completed in 1982 to protect London from flooding which might otherwise result from exceptionally high surges and tides. High levels in central London have been rising for a number of reasons; a rise in the sea level, the gradual subsidence in the South East due to downwarping of the margins of the North Sea basin, and local subsidence which may be due to water abstraction and clay shrinkage. The threat of flooding may be compounded by peak river flows. An early warning system operates which can predict exceptional levels from surge, tide and river flows several hours in advance, giving time for the barrier to be closed.

---

flooding incident occurred in Northern Ireland in recent years, in 1989-90.

9.15 The NRA have responsibility for around 2,700 km of tidal and estuarial defences (including defences on both banks) and over 700 km of sea defences. Table 9.3 shows NRA tidal/estuarial and sea defences by region. Estuarial and sea defence and coast protection works are also maintained by local authorities and privately.

9.16 The NRA have recently completed a major review of sea defences around the coastline of England and Wales. The survey was conducted in three phases. Phase 1, covering NRA owned or maintained sea defences showed that 85 per cent of structures are in good condition, 13 per cent require moderate repair and 2 per cent are in need of significant works. Phase 2, covering local authority owned or maintained defences (242 km), showed that 93 per cent of structures are in good condition, 5 per cent require moderate works and 2 per cent require significant works. Phase 3, covering private or corporately owned defences (around 200 km), found 71 per cent of structures are in good condition, 25 per cent require moderate repair and 4 per cent require significant works.

The survey has been extended to cover tidal or estuarial defences and the results will be published in 1992.

Rising sea level

9.17 As a result of increasing concentrations of greenhouse gases resulting from emissions due to human activities, the earth's surface and lower atmosphere are expected to warm (see also Chapter 3 on the global atmosphere). This warming is likely to continue, even if emissions of greenhouse gases were to cease, because of past increases in greenhouse gases and lags in the climate system. One of the major consequences of this warming will be a rise in the mean sea level globally. In addition there may be an increase in the frequency and intensity of coastal storms.

9.18 Over the past century the mean sea level is estimated to have risen by around 10 to 15 cm and the rate of rise is thought likely to increase over the next hundred years or so. Estimates of the likely effects of global warming [3] suggest additional rises in the sea level of perhaps 20 cm by 2030, with a likely range of 10 to 30 cm. Rising sea levels could have adverse effects, particularly coastal flooding and erosion, unless action is taken. In certain parts of the country (notably the South East) the rise in the sea relative to the land may be greater than this owing to subsidence. Since the last ice age, the British Isles has been readjusting in the north and west following glacier load removal, which has resulted in a gradual uplift. However, gradual subsidence has occurred on the margins of the North Sea basin in the east and south east. Estimates vary but it may be of the order of 1 mm per year, and such subsidence has caused the loss of numerous villages from low-lying east coast areas since the compilation of the Domesday Book. In particular regions other geological processes may occur which need to be monitored in the long term.

9.19 Figure 9.2 shows the areas of GB where a sea level rise could have a significant effect (shaded areas indicate land which is less than 5 m above sea level). Many of these include major conurbations or high grade

Table 9.3	NRA fluvial[1], estuarial and sea defences, by region[2], 1989 - 90 England and Wales		
			km
NRA region	Non - tidal fluvial	Tidal/ estuarial	Sea defences
Northumbria	307	19	8
Yorkshire	2,645	111	17
North West	9,145	171	64
Welsh	270	100	109
Severn Trent	6,878	294	40
Anglian	11,624	1,072	324
Thames	9,975	193	0
Southern	191	139	142
Wessex	4,264	112	29
South West	2,195	512	7
Total	**47,494**	**2,724**	**739**

Notes:
1. Relating to rivers
2. Figures given are approximate only.

Source: NRA

agricultural land. Major road and rail links situated near the coast would also be at risk. Several power stations are also situated on low lying land. In Northern Ireland, such areas are confined to narrow coastal strips and no significant areas exist inland.

9.20 The effects of sea level rise may be exacerbated by possible increases in the incidence of storms and thus wave activity, the greatest impact being likely on the exposed western coasts facing the Atlantic. A recent report [4] has suggested that the north east Atlantic has become notably rougher over the last 25 years.

9.21 As well as direct effects such as coastal erosion or the flooding of coastal areas, higher mean sea levels could also have an impact on underground water resources. The zone of mixing of sea water with freshwater in rivers is dynamic and a rise in sea level can cause it to move upstream. A similar effect can occur between freshwater contained in rocks under the land and salt water in sea floor sediments, causing intrusion of salt water beneath the land. This would adversely affect some abstractions along the lower reaches of rivers for domestic and irrigation purposes. These abstraction points would have to be moved upstream or become intermittent to avoid abstracting saline water. See also Chapter 6 on inland water resources and abstraction.

9.22 Rising sea level could also affect coastal habitats, particularly coastal wetlands and saltmarshes. The extent to which ecosystems are likely to be affected would depend on the rate of the sea level rise and the ability of ecosystems to adjust; and the extent to which habitats are prevented from migrating inland by coastal defences.

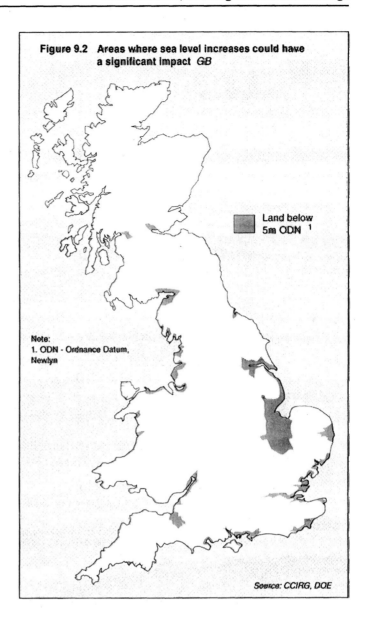

Figure 9.2 Areas where sea level increases could have a significant impact *GB*

Land below 5m ODN [1]

Note:
1. ODN - Ordnance Datum, Newlyn

Source: CCIRG, DOE

References and further reading

1. Jones, D.K.C. & Lee, E.M., (1992). Landsliding in Great Britain; a Review for the Department of the Environment. HMSO.

2. Department of Environment, (1991). Ground Movement in Ventnor, Isle of Wight. Geomorphological Services Ltd.

3. Department of the Environment, (1991). The Potential Effects of Climate Change in the United Kingdom. HMSO.

4. Carter & Draper, (1988). Nature 1988 vol 332, issue 6164 pp 494.

National Rivers Authority, (1991). NRA Facts 1990. NRA.

Boorman, L.A., (in press). The Environmental Consequences of Climate Change on British Salt Marsh Vegetation. Wetlands Ecology and Management vol.2.

Perry, A.H., (1977). Environmental Hazards in the British Isles. George Allen and Unwin.

10 Wildlife

- In the UK, there are an estimated 30,000 species of animals and 32,000 species of plants (10.1).

- The factors which affect the size of populations of species are generally complex and, in many cases, the reasons for changes are not fully understood. Changes may be due to weather, disease or habitat change, both natural and resulting from man's activities (10.3).

- The most important changes that have occurred to most species relate to change in the habitat (or habitats) which they depend upon. A number of habitats have diminished over time and some are now at risk. A major factor has been change in agricultural land use (10.3, 10.49).

- The area of moorland in GB has diminished by about 20 per cent since the 1940s. This has affected the population size of some birds of prey (10.6, 10.7).

- The area of lowland heath has diminished more markedly due to agricultural conversion, afforestation, urban or industrial development and neglect. Populations of some reptiles and invertebrates have been severely affected (10.9, 10.11).

- 40 per cent of the area of unintensified grassland in England and Wales has been lost since 1932; semi-natural grassland has been the most seriously affected (10.15). These losses have led to losses of orchid populations, of invertebrate populations, and to significant declines in some bird populations (10.16).

- Rivers and fens in lowland Britain are becoming increasingly affected by low water levels due to abstraction (10.27). The number of ponds is declining, but gravel pit habitat is increasing. Newts have been adversely affected by these changes, while populations of some water birds have increased (10.28).

- Afforestation has affected over one million hectares of upland GB in recent years. Populations of comparatively uncommon open-ground birds have declined, while some forest birds have increased. Afforestation often destroys existing semi-natural vegetation and invertebrate communities (10.55).

- Pesticides have had a serious impact on bird populations in the past 50 years, but following the banning of the most persistent and toxic of these, many bird populations have recovered (10.62). Otter populations may also have been affected, but now appear to be recovering (10.65).

- Air pollution, by both sulphur and nitrogen compounds, has affected communities of lower plants. Decreases in the level of sulphur in some areas in the 1980s have led to the recovery of some populations of lower plants (10.68). Acidification of streams has affected water birds and invertebrates (10.70).

- The two main bird monitoring schemes show that about 31 per cent of the 105 species covered regularly are increasing in population size, while 34 per cent are decreasing. The remainder show no consistent trend in numbers or are stable (10.77).

☐ **Of nineteen commoner species of butterfly monitored (mostly on protected areas), none decreased in abundance between 1976 and 1989, but three had increased. There were declines in the distribution of some rarer species (10.83).**

10.1 Wildlife, like air, water, land and soil, is an important natural resource. Changes in wildlife populations also act as indicators of other environmental changes. In the UK, there are an estimated 30,000 species of animals and 32,000 species of plants, including fungi and algae. The conservation of wildlife habitats is of primary importance in species protection. Many species are confined to single habitats, and it is therefore often easier to consider habitats rather than species when describing the wildlife resource. Not all species can be described in this chapter, so coverage is restricted to a selection of illustrative examples. Some species, particularly birds and mammals, require a range of habitats to fulfil their needs.

Table 10.1   Distribution of upland areas GB

Altitude (m)	Main land type	GB	sq km Scotland	England	Wales	% of GB
123-244	Marginal agricultural land	54,324	19,944	28,869	5,511	23.9
245-610	Hill pasture and moorland	47,315	27,030	12,363	7,922	20.8
611-914	Mountain range	5,263	4,645	394	224	2.3
> 915	High mountains	402	394	2	6	0.2

Source: Ratcliffe and Thompson (1988)

10.2 The first part of the chapter describes some of the habitats which have changed significantly or are at risk. These descriptions include some species which occur primarily in these habitats and which have been affected by habitat change. The first part concludes with descriptions of the status of some species which need a combination of several habitats, and cannot readily be ascribed to one. The second part of the chapter considers the effects on wildlife of some influences caused by man's activities which affect many habitats. The third part describes a few example species for which there are good time-series monitoring results.

Habitats and species

10.3 Most habitats and species in the UK have been affected by man's activities, with the result that most habitats are either man-made or semi-natural. Populations of species in these habitats are also much changed. Much of the change occurred in the distant past when the primeval woodlands were cleared to make way for agriculture. The clearances may have added initially to the range of habitats present and therefore increased biodiversity. However, in recent decades the pace of change has been increasing, and changes have generally been to simplify habitat diversity, for instance by removing hedgerows or creating monocultures. As a consequence, the extent of some habitat, such as deciduous woodland, wetlands and traditionally maintained farmland, has declined. There has been a corresponding decline in species that require these habitats. However, there are many influences, in addition to habitat change, which affect populations of species. These influences are generally complex and, in many cases, not fully understood.

10.4 Habitats are generally classified by type of vegetation, together with physical characteristics eg altitude, soil type or other characteristics such as scree, shingle, etc. At the broadest level, there are several types such as grasslands, woodlands and peatlands. These can be further subdivided: for example, grassland can be acidic, calcareous or neutral; woodland may be sub-divided into semi-natural and plantation woodland. There have been several full classifications of UK habitats, for example the Nature Conservancy Council habitat survey handbook. See references and further reading.

Moorland

10.5 Moorland is land with predominantly semi-natural upland vegetation consisting mainly of species characteristic of grassy

Uplands   *Box 10.1*

Hills, moors and mountains, collectively termed the uplands, represent the largest extent of natural and semi-natural wildlife remaining in Britain (see Table 10.1). The uplands are predominantly areas of low growing shrub heaths, grasslands and peatbogs. More than half, some 54,000 sq km - nearly 24 per cent of the GB land area, is largely marginal agricultural land. At altitudes above 245 metres there are a further 47,000 sq km (around 21 per cent) of mainly hill pasture and moorland, and 6,000 sq km (2.5 per cent) is classed as mountain range and high mountain.

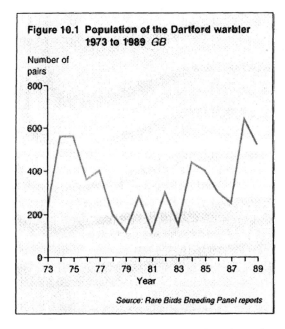

Figure 10.1 Population of the Dartford warbler 1973 to 1989 *GB*

Number of pairs

Source: Rare Birds Breeding Panel reports

and shrubby heaths, including heather. In 1985, 29 per cent of moorland in England and Wales was thought to be grazed at a level compatible with maintaining existing heather in good condition, compared with 70 per cent in 1975 [1]. This fall has been caused by overgrazing owing to rising sheep numbers and has implications for the upland wildlife. Overgrazing results in heather moorlands changing to grasslands, which form poorer habitats for wildlife species such as the merlin and hen-harrier [2]. The richest moorland for wildlife involves a mosaic of heather and grassland. Overall, upland heath (which is predominantly heather) in England and Wales fell from about 4,500 sq km in 1947 to 3,650 sq km in 1980.

10.6 Results from the National Countryside Monitoring Scheme for Scotland [3] indicate that the area of heather moorland cover declined from around 15,400 sq km in the 1940s to just over 12,600 sq km in the 1970s, a net loss of almost 20 per cent. Almost two thirds of this overall loss is accounted for by afforestation, and the remainder by extensive sheep and deer grazing combined with poor burning practices.

10.7 Moorlands, and uplands in general (see also Box 10.1 on Uplands), are home to some birds of prey which have generally shown an upward population trend in recent years [4,5] following earlier severe declines. These recoveries have been in part due to a reduction in illegal persecution as well as decreased use of organochlorine pesticides. An exception is the merlin, a small bird of prey requiring a particular mosaic of habitat types over extensive areas of open ground, usually heather moorland, which has shown a marked decline since the 1950s in GB [6].

Lowland heath

10.8 Heaths are areas of fairly short, open vegetation, typically covered in heather and gorse, although in some areas grasses and lichens predominate. Such areas develop on nutrient-poor soils where grazing or burning prevents the encroachment of trees.

10.9 British lowland heaths cover about 60,000 hectares. However, their total area has decreased markedly over the last 200 years, and this loss has accelerated during the present century. Traditional grazing and burning practices have been discontinued and heathland has been subject to agricultural conversion, afforestation, urban or industrial development and neglect [7].

10.10 The remaining British lowland heath occurs mainly in clusters, of which the most extensive are in southern Hampshire (including the New Forest), south east Dorset, the Brecklands around Thetford, eastern Suffolk (the Sandlings), western Surrey and the Lizard area in Cornwall. Similar, but smaller, heaths are scattered through the lowlands of southern and eastern Scotland [7].

Table 10.2	Sand lizard: number of colonies, 1970 to 1980	
	Number	
Area	**1970**	**1980**
Merseyside	7	2
Wealdon Heaths (SW Surrey)	7	1
S E Dorset (East of Wareham)	81	15
Purbeck and Dorset (West of Wareham)	71	26
Total	**166**	**44**

Source: Ginn (1983)(9)

10.11 Conservation of lowland heath is important for some of Britain's reptiles and

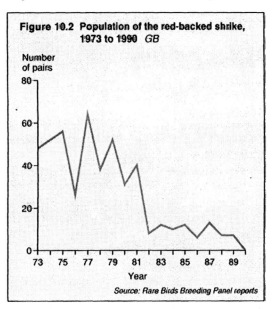

Figure 10.2 Population of the red-backed shrike, 1973 to 1990 *GB*

Number of pairs

Source: Rare Birds Breeding Panel reports

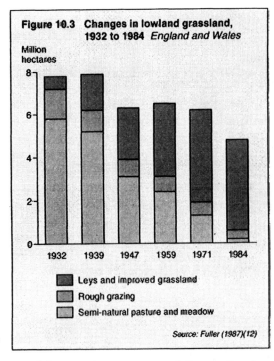

Figure 10.3 Changes in lowland grassland, 1932 to 1984 *England and Wales*

Million hectares

- Leys and improved grassland
- Rough grazing
- Semi-natural pasture and meadow

Source: Fuller (1987)(12)

amphibians. Accurate population estimates are difficult to obtain, but it is clear that loss and fragmentation of heath has led to declines in populations of some species. There are now estimated to be only around 2,000 adult smooth snakes, Britain's rarest reptile. This species prefers dry heathland habitat

Figure 10.4 Distribution of the green winged orchid; pre and post 1970 *UK*

- ○ Before 1970 (490)
- ● 1970 onwards (200)

Source: ITE Biological Records Centre

but climate as well as habitat is thought to be an important factor in its survival[8]. Sand lizard colonies are estimated[9] to have declined by over 70 per cent in four heathland areas in England between 1970 and 1980 (see Table 10.2).

10.12 Heathland habitats are also important to the survival of some invertebrates. Many invertebrates are, however, difficult to census accurately. For example, only two colonies of the heath-dwelling ant *(Formica pratensis)* have been found in recent decades, both in the area of Wareham, Dorset. One colony, on Morden Bog National Nature Reserve, is thought to have died out around 1980, and the other was last observed around 1988. The species may now be extinct in Britain[10].

10.13 Populations of some heathland birds, such as the Dartford warbler, show no clear trend (see Figure 10.1). The Dartford warbler, unlike most insect-eating birds, does not migrate to warmer countries for the winter and so is very susceptible to cold weather. Fluctuations in population are linked to heavy mortality during hard winters. A clearer trend can be seen in respect of the red-backed shrike, shown in Figure 10.2, which was a relatively widespread bird earlier this century occupying various habitats. As its population began to dwindle it became restricted to heathland, and its long term decline was marked with a significant fall in the early 1980s. The causes of the decline of red-backed shrike are not understood[11] and it has become rarer in much of north western Europe. This species has now ceased to breed regularly in the UK.

Lowland grassland

10.14 Grassland requires regular mowing or grazing to prevent the development of scrub and its progression to woodland. In most of lowland Britain this function is carried out mainly by domesticated livestock but wild mammals (eg rabbits, deer) also graze. Lowland grassland can be categorised into three main types depending on the degree of change brought about by agriculture:

- unintensified (agriculturally unimproved) pasture, including coastal machair (sandy, grassy land just above the high-water mark at a sandy shore), unintensified commons, rough grazing and meadows;

- semi-intensified grassland; and

- leys (land sown to grass for a year or more, grown in rotation with annually cultivated crops) and intensified grassland.

10.15 In England and Wales, the amount of lowland grassland not intensified for agriculture, has decreased by almost 40 per cent since the first comprehensive survey in 1932 (see Figure 10.3). The area of semi-natural grassland has declined to around 0.2 million hectares, compared with 5.8 million in 1932. Some has been converted into agriculturally improved grassland by application of herbicides and fertilisers and re-seeding, but most of the area has been ploughed and replaced by arable crops and grass leys. The area of rough grazings on commons and upland fringes has also reduced significantly, by 70 per cent over the same period [12]. Scotland has a smaller proportion of low lying land and a wetter climate in comparison to England. Much of its remaining agriculturally unimproved lowland grassland is poorly drained with characteristic wetland species.

10.16 Unintensified grassland is a scarce and effectively irreplaceable resource. Meadows and pastures include some of the most diverse habitats in western Europe, providing a habitat for over 400 species of flowering plant, as well as butterflies, grasshoppers, crickets and ground-nesting birds. Some species of plants are confined to unintensified semi-natural grassland of great age, owing to their poor powers of dispersal or to their being unable to compete with the vigorous grass strains used in re-seeded fertilised pasture. One such "ancient grassland indicator" is the green-winged orchid *(Orchis morio)*. Although still found throughout its former range, a large proportion of its colonies has been lost since the 1940s as suitable grassland has been reduced (see Figure 10.4).

10.17 Changes in vegetation structure can also affect wildlife. The wart-biter bush cricket *(Decticus verrucivorus)*, for example, one of Britain's rarest Orthoptera (grasshoppers, crickets etc) requires a mosaic of bare ground for egg laying [13], short flowery turf for feeding and tussocks to provide cover from predators,

Table 10.3	Number of calling corncrakes, 1978 to 1991 *Northern Ireland*	
	Co. Fermanagh	Rest of Northern Ireland
1978 [1]	26	143
1988	65 to 73	57 to 61
1990	35	10
1991	19	2

Note: 1. Underestimate due to incomplete coverage.

Source: Hamill and McSherry, (1990)(20)

with an additional requirement for warmth since it is vulnerable in cold weather [14]. With such exacting requirements, its populations are liable to fluctuate widely and, on small sites, are prone to recurrent local extinction [15].

10.18 The water level of wet grasslands is perhaps the most important factor determining the quality of the habitat for wading birds. During the breeding season, a high water level (ie the water table within 20 cm of the surface), and shallow surface pools are essential. These conditions provide ideal feeding conditions for redshank, which forage in the surface pools, and snipe and curlew which require moist soil for probing [16,17].

10.19 The population of breeding waders of wet grasslands declined in the 1920s and 1930s, and there have been further reductions in recent years, as a result of land drainage for agriculture. Lapwing, redshank and snipe have been particularly affected. The population of snipe, for example, has shown a marked downward trend on lowland farmland (see Figure 10.5). Lapwings declined significantly by 37 per cent on lowland grassland sites between 1982 and 1989 [18].

10.20 Corncrakes are also dependent on low-intensity farming, being particularly associated with crofting or similar systems. They now nest in the UK only in traditionally managed meadows, and the UK population (still nationwide early in the century) is now confined to west and north west Scotland and Northern Ireland. In Northern Ireland a recent trend away from traditional mowing for hay to an earlier silage cut is thought to have caused the failure of most nesting pairs, resulting in a rapid decrease in population (see Table 10.3). If the trend were to continue, the species would be extinct in Northern Ireland by the mid 1990s [19,20,21].

Ancient woodland

10.21 Although most kinds of woodland are relatively rich wildlife habitats, ancient woodlands are the richest. Today, ancient

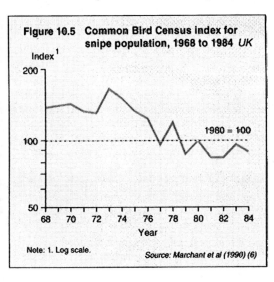

Figure 10.5 Common Bird Census index for snipe population, 1968 to 1984 *UK*

Note: 1. Log scale.

Source: Marchant et al (1990) (6)

Types of peatland   *Box 10.2*

The vegetation and underlying peat of peatlands show great variation reflecting differences in origins, topography, climate, base status and hydrology, especially in the relative influence of groundwater and rainfall. All peatlands are termed mires, with a primary distinction between fens, which are fed by groundwater as well as rainwater, and bogs, which are fed only by rainwater. Three main types of bog can be recognised:

- Raised bogs form as isolated units of domed peat within an otherwise non-peat landscape, usually as a result of the filling of lakes and pools with peat, following the development of stagnant conditions. All raised bogs were fens at earlier stages of development, but as the peat accumulates, the groundwater ceases to supply nutrients and the vegetation is fed only through rainfall. Rain-fed vegetation and peat is relatively acid and deficient in nutrients, and species such as Sphagnum mosses and those typical of wet acid moorland dominate.

- Blanket bogs generally form on flat or sloping ground with high rainfall and cool climatic conditions. They are very variable and some have also developed from fens. In the north of Scotland, domed lenses of peat which seem to have similar origins to raised bogs occur within blanket bog landscapes.

- Intermediate bogs are isolated units of peat which have some characteristics of raised bogs, but more generally resemble blanket bogs.

---

woodlands cover 2.6 per cent of England, 2.7 per cent of Wales and 1.9 per cent of Scotland, with most in the south east of England, the southern Welsh borders and the central Scottish Highlands. Although a clear distinction is difficult to make, it is estimated that around 40 per cent of ancient woods are plantations and 60 per cent are semi-natural woods [22]. See also the section on ancient and semi-natural woodlands in Chapter 5.

10.22 Simply preventing the felling of ancient woodlands is not sufficient to protect the full range of their wildlife. In lowland Britain, many woods were managed by coppicing, which involved the periodic cutting of the scrub layer, typically hazel, to promote a regular supply of small diameter logs and branches. This produced open, sunny glades sheltered from wind, and supported a rich butterfly fauna. It may be appropriate for a proportion of a coppiced wood to be allowed to regenerate to mature forest, but not for most as this would lead to the loss of these open areas and to the diminution of a number of species, including butterflies. Nightingales too, have declined since coppicing ceased to be a widespread method of woodland management [23,6].

10.23 Many invertebrates associated with wood decay ('saproxylic' species) are now rare because they require ancient woods in which there has been a continuous succession of ancient trees. Five British forests and deer parks have been identified as of potential international importance for their saproxylic invertebrate faunas: Abernethy Forest, Inverness; Epping Forest, Essex; Moccas Park, Hereford and Worcester; the New Forest,

Hampshire; and, Windsor Forest and Great Park, Berkshire [24].

Peatland

10.24 The term peatland covers a number of recognisably different habitats, each supporting different assemblages of wildlife. Peat is partially decomposed plant material that builds up where accumulation exceeds the rate of decay, usually because of a combination of waterlogging and low temperatures. Peatlands are often referred to as "mires" or "bogs" and have been classified according to their physical characteristics, their current vegetation, or by using other criteria such as chemistry. Box 10.2 describes the various types of peatland.

10.25 The total area of peat in the UK is about 1.5 million hectares, of which over 90 per cent is blanket bog. The largest areas are in the uplands and in the north of Scotland and Northern Ireland. The largest single complex is in Sutherland (0.4 million hectares). In Northern Ireland there are about 170,000 hectares of peatland [26]. The total area of lowland peat in Britain is about 65,000 hectares, of which 34,500 hectares is in England, 2,500 hectares in Wales and 28,000 hectares in Scotland [25]. An inventory of peatland in Northern Ireland was published in 1990 [26]. An equivalent for Britain is presently at a draft stage.

10.26 Because of the conditions found on the different types of peat, and in particular the acidity and low nutrient status, they are of high value as wildlife habitats with a number of specialised species which are not found elsewhere. Large areas have, however, been damaged or lost over the last few hundred years as a result of draining and conversion to agriculture and forestry, and on a smaller but still significant scale by peat cutting for fuel, the horticultural and landscaping industries as well as for domestic use in gardens.

Lowland freshwater

10.27 Low rainfall, compounded by increased groundwater abstraction, has affected many English wetland sites in recent years, particularly in the south and east. The

National Rivers Authority (NRA) has identified twenty sites as priorities for action on over-abstraction and one of these, the Redgrave and Lopham Fens Site of Special Scientific Interest (SSSI), is of major nature conservation interest. A further 24 fens in East Anglia which are of nature conservation importance are vulnerable to groundwater abstraction [27]. In the Redgrave and Lopham Fens SSSI, water abstraction is due to cease by 1994 and in the meantime water is being artificially fed to parts of the fen in an effort to maintain the aquatic wildlife, in particular the fen raft spider *(Dolomedes plantarius)*.

10.28 Field ponds were once a vital part of most farms, but the decline and intensification of farming has made ponds largely redundant and many have been filled in. An example of declining field ponds in Cambridgeshire is given in Figure 10.6. A survey carried out in 1988 compared modern Ordnance Survey maps with the previous editions of the same map. Current pond numbers were estimated at 291,000, but the survey suggested that Britain has lost about 182,000 ponds since 1945 [28].

10.29 Great crested newt *(Triturus cristatus)* are found at about 18,000 sites in GB, mostly in England and Wales, but these sites are decreasing at a rate of about 2 per cent every 5 years, mostly through pond

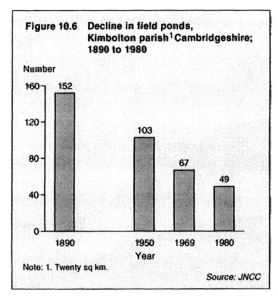

Figure 10.6   Decline in field ponds, Kimbolton parish[1] Cambridgeshire; 1890 to 1980

Note: 1. Twenty sq km.

Source: JNCC

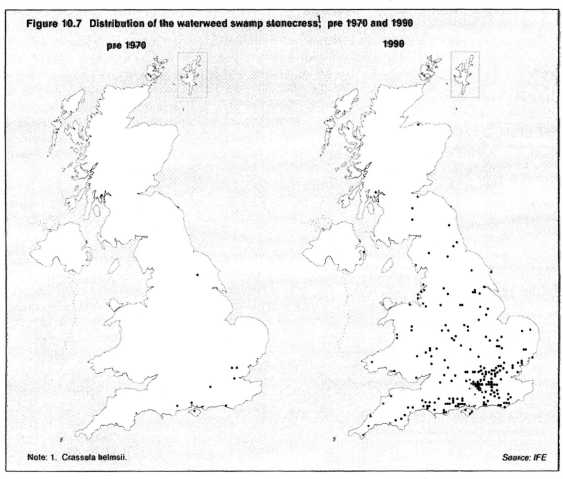

Figure 10.7   Distribution of the waterweed swamp stonecress[1] pre 1970 and 1990

pre 1970

1990

Note: 1. Crassula helmsii.

Source: IFE

destruction [29]. A current additional threat to pond-dwelling wildlife is an invasive waterweed, swamp stonecress *(Crassula helmsii)*, thought to have been introduced to this country from Australia by the aquarium trade. It is tolerant of a wide range of conditions in ponds, lakes and ditches, and is more vigorous than most native waterweeds. It is spreading rapidly in Britain (see Figure 10.7).

10.30 New aquatic environments, such as flooded gravel pits, have helped some species such as the great crested grebe. The species almost became extinct in the latter part of the last century, a trend partly attributed to human persecution, but recovered after a series of bird protection Acts were introduced. Populations have been increasing and recent increases were greatest in those counties with the largest areas of newly created wetlands such as gravel pits and reservoirs. The most recent national survey carried out in the mid 1970s, estimated the UK population at around 7,000 adult birds [30,6].

Estuaries

10.31 There are 163 estuaries in the UK. Estuaries provide a range of habitats. Inter-tidal expanses of mud or sand can grade landwards, in the absence of interference from man, into such habitats as saltmarsh, reedbed, grassland or even woodland. Sand dunes and shingle structures are frequently associated. These habitats support many communities of plants and animals including large numbers of wintering wildfowl and waders [31].

10.32 Land claim in estuaries has implications for these wildlife habitats. For example, encroachment of cord grass *(Spartina anglica)*, which was formerly planted extensively in England and Wales to stabilise tidal mud flats, has an adverse effect on wildlife by reducing estuarine habitat and by lowering the species diversity of some saltmarshes. These effects seem to have been particularly important in the case of the dunlin, where cord grass encroachment is thought to have caused a reduction in numbers over-wintering in Britain [32].

10.33 A recent survey of estuaries in GB identified 123 cases of land claim at 45 GB estuaries which were either in progress now or had planning consent. At 32 estuaries the land claims affected inter-tidal and sub-tidal areas, and at the remaining 13 the claims were in respect of surrounding land. The after-uses of the land claimed include rubbish tipping and spoil disposal (almost half of the current cases), transport schemes, marinas and water-based recreation, housing, sea defences, and agriculture. The size of claims range from small (less than 5 hectares) to very large (over 50 hectares). The total area under current claims in intertidal areas is estimated to be around 0.5 per cent of the resource.

Sand dunes

10.34 Around 7 per cent of the UK coastline is bordered by dunes, covering an area of around 56,000 hectares. These are widely distributed and most are relatively small. There are however a few large sites such as Culbin Sands in Scotland, covering approximately 4,000 hectares. There are around 120 sand dune sites in GB designated as SSSIs (SSSIs are covered in more detail in Chapter 5). Northern Ireland has two sand dune sites designated as National Nature Reserves (NNRs) at Murlough (282 hectares) and Magilligan Point (174 hectares).

10.35 Sand dunes have been seen as a threat to man, for instance through encroachment of sand onto agriculturally productive land, or through dune erosion leading to coastal inundation. Considerable effort has been spent on stabilisation of sand dunes through planting, resulting in the loss of natural dune habitats and dune flora and fauna. As new forests grow, or the dunes develop into coarse grassland and scrub, many of the species which are specially adapted for survival in unstable environments die as their natural habitat is lost.

10.36 Dune "slacks" (winter-flooded hollows between dunes) are particularly rich in flora and fauna. Development, and also the recent years of summer drought, have contributed to a reduction in flooding and areas of

Table 10.4	Sites of the highest national significance as sand dune habitat, larger than 1,000 ha [1] *GB*	
Site	**Location**	**Interest**
Braunton Burrows	Devon	Self contained, largely intact system. Species rich calcareous vegetation (NNR,MOD).
Morrich More	Highland Region	Self contained, largely intact prograding system. Northern dune heath vegetation (MOD).
Penhale Dunes	North Cornwall	Calcareous dune grassland and slacks (part MOD)
Sefton Coast	Merseyside	Rich calcareous dune remnant of a much larger area. Major recreational site. (part NNR, part MOD).
South Uist Machair	Western Islands	Almost continuous cultivated machair and dune, on the western shore, South Uist.
Torrs Warren	Dumfries and Galloway	Largely intact acid dune heath system (MOD)

Note: 1. Sites which are wholly or partly recognised as National Nature Reserves are shown (NNR). Sites used by the Ministry of Defence are shown (MOD).

Source: Doody (1989)(33)

standing water, which may result in the depletion or loss of rare plants and animals. One species especially affected by the drying out of such pools (where they would normally stay flooded in spring and early summer), is the natterjack toad. The most important of the 40 recently extant native populations of natterjacks are found on dunes, and dune "slacks" are vital for their survival [33].

10.37 There is no information about the overall loss in natural dune habitat in the UK, although there are examples of developments affecting some of the more important dune systems. These include afforestation, golf courses, military installations, housing and static caravans, industry, recreation, agriculture and sand extraction. Afforestation has affected an estimated 14 per cent, around 8,000 hectares, of the total dune area [34]. An example of major afforestation is at Culbin Sands where, over a period of about 140 years, the site was changed from one of the largest open dunes systems in the UK to an area of closed pine plantation. Some uses however, for instance golf courses and Ministry of Defence training areas, have effectively prevented more damaging developments which cover or radically alter the dune surface. Nevertheless, in some areas continued intensive recreational use has destroyed the vegetation covering the dunes and exposed the sand to erosion by water and wind, sometimes causing serious destabilisation.

10.38 Other dune sites are largely intact and there are several locations where these extend to over 1,000 hectares, such as Braunton Burrows in north Devon (see Table 10.4). Almost all of these are wholly or partly recognised as NNRs or are sites used by the Ministry of Defence.

Marine

10.39 Marine ecosystems around the UK coast are the most diverse of any European state bordering the Atlantic. They support in excess of 850 algal, 6,210 invertebrate (excluding insects), 310 fish, 190 bird and 30 mammal species [35]. Marine habitats in British territorial seas extend over an area of seabed equivalent to 70 per cent of the land surface of GB. An even larger area of sea falls within the UK's continental shelf and wider fishery limits.

10.40 There are a number of surveys of the marine environment and the populations of organisms supported by it (eg references [36, 37]).

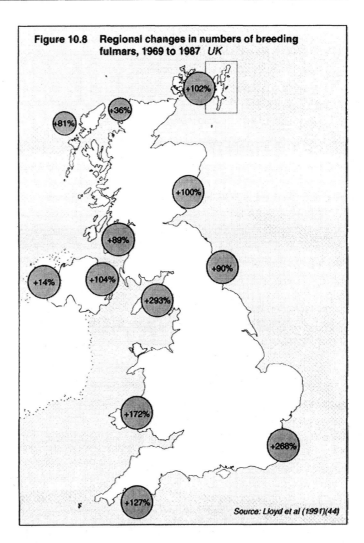

Figure 10.8 Regional changes in numbers of breeding fulmars, 1969 to 1987 *UK*

Source: Lloyd et al (1991)(44)

10.41 Grey seals and common seals are surveyed regularly in GB. The largest grey seal colonies are in the Outer Hebrides and Orkney. The total British population was estimated at 85,000 animals in 1990 and more than 90 per cent of these are associated with colonies in Scottish waters [38]. Common (harbour) seals are mainly found around the Scottish coast and in the Wash. More than 3,000 seals were found dead around the UK in 1988; most of these were common seals which had died as a result of infection with a distemper virus. The worst affected area was the Wash, where counts of common seals fell from 3,900 in 1988 to 2,000 in 1989 [39]. The estimated 1990 total of common seals in Britain is 25,000 [38].

10.42 The distribution of seabirds at sea in British and European waters has been mapped since 1979 [40,41]. This work has enabled the identification of areas of particular importance and aided in providing informed advice to both government and the offshore oil industry. For example, activities that are likely to increase the risk of oil pollution can now be carried out in areas and at times when fewest birds are present[42,43].

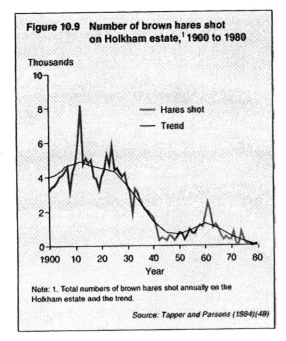

Figure 10.9 Number of brown hares shot on Holkham estate,[1] 1900 to 1980

Thousands

Hares shot
Trend

Year

Note: 1. Total numbers of brown hares shot annually on the Holkham estate and the trend.

Source: Tapper and Parsons (1984)(48)

10.43 In addition, numbers of breeding seabirds on the coasts of GB and Ireland were fully censused in 1969-70 and again in 1986-87 [44]. These surveys confirmed the outstanding importance of these coasts for the seabirds of Europe. Numbers for eighteen species were counted reliably in both surveys. Of these, fourteen species had increased overall and two had decreased. These trends however were not uniform on all British and Irish coasts. For example, fulmars increased everywhere, but in particular in south east England and on Irish Sea coasts (see Figure 10.8). The reasons behind these increases are more difficult to determine, but undoubtedly the continued provision of large quantities of offal and undersized fish from fishing fleets has supported the expansion of species that scavenge on this waste.

10.44 In 1986, fourteen Marine Consultation Areas (MCAs) in Scotland were identified as areas of particular nature conservation importance [45]. The areas have no statutory locus but are known to bodies with which Scottish Natural Heritage have consultations on marine conservation issues. There are currently twenty nine MCAs in Scotland. In 1992, the Department of the Environment (DOE) circulated a draft list of MCAs for England and Wales for comment.

Widespread species

10.45 Some species require a range of habitats to fulfil their needs. Evaluation of their status requires specific surveys of species, or species groups. Many such surveys have been undertaken, particularly on birds and mammals. The following section covers a few examples of widespread species.

10.46 Table 10.5 shows estimated populations of 228 species of British birds grouped by numbers of breeding pairs (or the number of territories or breeding females where this is more appropriate to a species' breeding habits). About a tenth of the species are very common (over 500,000 pairs); almost a half are relatively scarce (less than 5,000 pairs). The most abundant species of British breeding birds, with over three million breeding pairs, are the chaffinch, blackbird, house sparrow, starling, robin, blue tit, wren and pheasant.

Table 10.5 Abundance of British breeding birds

	Number of species
Number of pairs	British breeding birds
500,000 to 5 million	27
50,000 to 500,000	49
5,000 to 50,000	45
500 to 5,000	35
50 to 500	32
0 to 50	40
Total	**228**

Source: Batten et al (1990)(2)

10.47 The badger population in GB has been estimated in a recent study [46] at approximately 250,000 adults, with an annual production of 105,000 cubs. Hole blocking was found to be widespread and affected 16 per cent of badger setts. Further similar surveys will be carried out to examine the brown hare and bat distributions.

10.48 Reptiles and amphibians are widely dispersed across GB and records of both groups have been collected. Total population estimates for these groups are difficult to derive however, due to the secretive habit of most species. About 150 sites are known to hold four or more amphibian species [47].

Figure 10.10 Reduction in numbers of hares with increasing field size

4 fields with separate crops would produce 63 hares

Doubling the size of one field reduces the number of hares to 26

Reducing to 2 crops yields only 12 hares

Only two hares will live where fields are this size

Note: This schematic diagram shows a 1 km square on a typical farm and how the number of hares is reduced with the simplification of the farming pattern.

Source: Barnes and Tapper (1983)

spread influences on life

cultural change

Changes in agricultural land use have significant changes in Britain's rural ape (see also Chapter 5 on land use nd cover). The extent of many habitats, rticular wetlands and lowland heath, en reduced to provide more productive ltural land. Hedgerows and woodlands been removed or have become less rtive of wildlife through changes in gement. Many of these changes pose s of varying severity to a number of s.

An example of the effect of increased ize is the decrease in the population of own hare since the turn of the century, ling to records of numbers shot (see 10.9). Intensification of agriculture, in ular the increase in field size, is thought a major cause [48]. Although the average er of hares shot increases with the rtion of arable farmland in an area, do best in the old "patchwork quilt" ng of small fields and diverse crops (see 10.10).

The grey partridge has suffered from ffects of agricultural intensification. has been an 80 per cent decline of dges on farmland since the early s [6]. Important factors in this decline e decrease in the amount and quality ting habitats (hedgerows), making the s more susceptible to predators; and a tion in the main invertebrate food e for chicks, resulting from increased ide use. More chicks survive if the use rochemicals is restricted on perimeter around arable fields [49].

The Common Bird Census (CBC) was commissioned to examine changes in ers of common farmland birds. ular concerns when the CBC started the threats posed by new herbicides pesticides, and agricultural sification. Birds were seen as a nient way of monitoring the general t on wildlife. The decline of the corn g over the last two decades (see Figure) is thought to be mainly attributed to duction in traditional rotations of cereals vestock, and earlier harvesting under rn farm regimes [6]. The corn bunting ommon in the last century but has had success since reaching a high point in arly 1970s. Since then the population eclined and is now one third of the size in the early 1970s. The distribution of

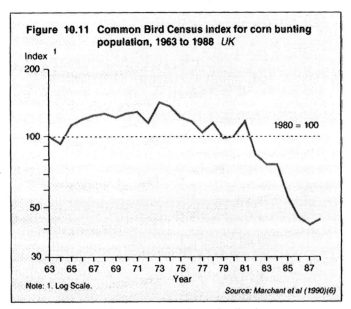

Figure 10.11 Common Bird Census index for corn bunting population, 1963 to 1988 *UK*

Index [1]

1980 = 100

Note: 1. Log Scale.

Source: Marchant et al (1990)(6)

the species has also contracted markedly, by some 40 per cent, over the last twenty years (see Figure 10.12). The lack of song posts and invertebrate food for chicks, provided by

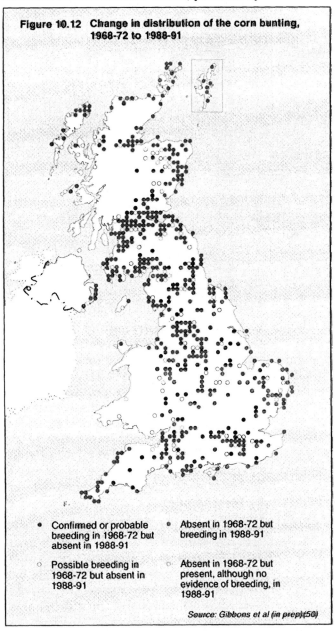

Figure 10.12 Change in distribution of the corn bunting, 1968-72 to 1988-91

- Confirmed or probable breeding in 1968-72 but absent in 1988-91
- Absent in 1968-72 but breeding in 1988-91
- Possible breeding in 1968-72 but absent in 1988-91
- Absent in 1968-72 but present, although no evidence of breeding, in 1988-91

Source: Gibbons et al (in prep)(50)

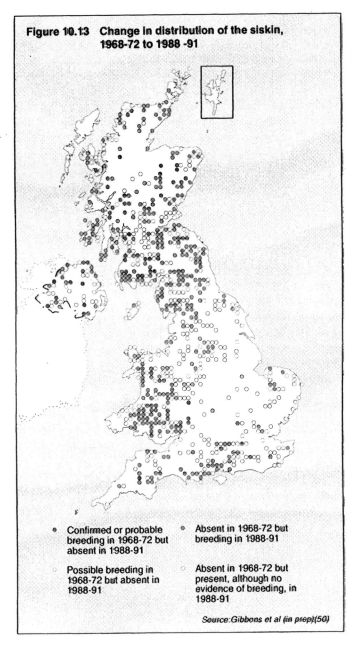

Figure 10.13 Change in distribution of the siskin, 1968-72 to 1988-91

- Confirmed or probable breeding in 1968-72 but absent in 1988-91
- Possible breeding in 1968-72 but absent in 1988-91
- Absent in 1968-72 but breeding in 1988-91
- Absent in 1968-72 but present, although no evidence of breeding, in 1988-91

Source:Gibbons et al (in prep)(50)

hedges and other boundaries, have also been suggested as contributing to the decline [50].

10.53 Linear features such as hedgerows, roadside verges and streamsides occupy less than 5 per cent of the land surface, but are of major ecological significance as reservoirs for species which are no longer able to survive in open fields. Hence in a field survey of one km squares, carried out in 1978, linear features contained a high proportion of all plant species recorded. In the lowlands, in particular, up to 50 per cent of all plant species in each square were found only in the linear features [51].

10.54 Headlands, hedgerows and banks form an exploitable habitat for common mammals such as rabbits, field mice and bank voles and they are also important for vascular plants and invertebrates. Although these may be considered pest species at times, they are of considerable importance in the diet of predators, both mammals and birds [52].

Afforestation

10.55 The planting of over one million hectares of new conifer forest in upland GB is thought to have reduced the size of a number of bird populations, though little information is available on the precise effects. However, a survey in 1987 of the effects of afforestation on wildlife in the peatlands of Sutherland and Caithness, estimated losses of about 900 pairs of golden plover (19 per cent), about 800 pairs of dunlin (17 per cent) and 130 pairs of greenshank (17 per cent) [53]. The survey also indicated that one of the few mainland nesting places of the red-necked phalarope has been degraded and deserted by the bird as a result of afforestation. All of these bird species are comparatively rare in Europe.

10.56 The maturing of conifer forest has nevertheless resulted in increases in populations of relatively common birds such as the siskin and the crossbill. Figure 10.13 shows the change in distribution of the siskin between the periods 1968-1972 and 1988-1991. It is estimated that at present there are around 440,000 siskin pairs compared with about 40,000 twenty years ago [6]. Forestry has caused high cyclical abundance in the vole population. These are a useful food source for some rare raptors (birds of prey) and the populations of goshawk and merlin have increased considerably. Also, the range of sparrowhawks, tawny owls and buzzards has expanded significantly. In addition to providing an increased food supply for these birds of prey, afforestation has increased their habitat and provided sanctuary from persecution [54].

10.57 Afforestation not only affects bird life, but also vegetation and animals other than birds. A study of the types of vegetation remaining within upland conifer plantation in north east England showed that 17 per cent of conifer stands had no further vegetation, while a further 40 per cent had extremely low species diversity and ground cover. Plant communities associated with the more extreme environmental conditions such as high altitudes and deep peats, were most sensitive to afforestation. Widespread communities composed mainly of cosmopolitan species were relatively resilient to the effects of plantations [55].

10.58 The effects of forestry can extend beyond land immediately planted. Mires adjacent to plantations have been examined in west Northumberland. An NNR in this area became completely surrounded by

stry and the wet peatland vegetation in
reserve changed to a heather/grass moor,
n some plant species becoming locally
nct. Causative factors in these changes
oably include changes in the hydrological
me of the mire due to water uptake into
plantation; changes in management of
ounding land (previously the mire and its
ounds was subject to slight grazing and
ing); and changes in the microclimate
to tree growth [56]. Conversion of moorland
commercial forestry is likely to have a
ked effect on invertebrate communities [57].

rsecution

59 The persecution of predatory birds,
of mammals, has occurred for many
turies and still continues either directly
ndirectly. A recent study [58] recorded a
l of 679 persecution incidents involving
killing of over 800 birds of prey and
s between 1979 and 1989. Shooting
trapping caused the deaths of 463 birds,
in another 145 cases, eggs and nestlings
e destroyed. Over 350 birds were illegally
soned using pesticides such as alpha
oralose, mevinphos and strychnine. These
res relate only to confirmed incidents;
al numbers of birds of prey and owls
d by persecution are thought to be higher.

60 The populations of several species
irds of prey have been reduced in the
t by persecution, in some cases leading
cal extinction. For one of these, the red
(a species vulnerable in world terms),
Joint Nature Conservation Committee
CC) and the Royal Society for the
tection of Birds (RSPB) are running a
gramme to re-establish it throughout
ain. The first successful breeding for over
years in England and Scotland occurred
1992. However, even some of the re-
oduced kites have been killed by illegal
oning.

61 Persecution continues of some species
are not predators. For example, even
ugh the digging of badger setts has been
al in GB since 1973, a survey [46] of
ger setts over the period 1985 to 1988
wed that 10.5 per cent of main breeding
s, and 9,000 setts in total, showed signs
igging in the previous 12 months. Most
this digging is for the illegal sport of
ger baiting. Further protective legislation
introduced in 1991. In Northern Ireland,
ger setts have been fully protected since
5.

sticides

62 The development and use of artificial
ticides for agricultural and other purposes,

based on compounds including chlorine (the
"organochlorines") increased rapidly after the
Second World War. Concern over the impact
of these chemicals on non-target wildlife
was highlighted when numbers of birds of
prey started to decline steeply [58]. These
persistent compounds tend to accumulate
through the food chain, occurring in greatest
concentrations in top predators. Effects were
observed through thinning of eggshells,
leading to eggs breaking in the nest, and
alterations in breeding behaviour. One of the
key chemicals was DDT, the derivatives of
which caused significant eggshell thinning in
many species of birds.

10.63 The populations of several species of
predators, notably the peregrine falcon and
sparrowhawk, declined also through increased
adult mortality following the use of the highly
toxic dieldrin and aldrin in cereal seed
dressings and sheep dip from 1956. Analyses
showed widespread contamination of wildlife
and the populations of these species fell
substantially in a few years; the peregrine
falcon declined from around 850 pairs in the
1930s to 360 pairs by 1962. Persecution
during the Second World War had caused a
severe drop in peregrine numbers prior to
dieldrin use, but there was a partial recovery
followed by further losses in the late 1950s
and early 1960s [59].

10.64 The phasing out of organochlorine
pesticides has resulted in increasing numbers
of birds of prey. The peregrine falcon has
now recovered to historically high levels,
with a minimum number of 1050 occupied
territories recorded in 1991. The latest
evidence suggests that the population has
risen to above the 1930 levels in some
areas [60], and the sparrowhawk has also
recovered from significant population
decline [61].

10.65 The otter is a protected species which
has shown a marked population decline since
the late 1950s. The decline may be due to a
number of factors including loss of habitat,
but there is some laboratory evidence that
organochlorine pesticides and PCBs can
cause reduced reproductive success in related
species. Two surveys, in 1977-79 [62] and
1984-86 [63], have shown that the otter too
has shown some recovery, helped by legal
protection (see Figure 10.14). Recovery and
movement of this species eastwards has been
slow due to the greatly reduced population
in the Midlands and the otters' lower mobility
compared with that of birds.

10.66 The use of chemicals in farming can
also have significant effects on other species.
For example, surveys suggest that the use of

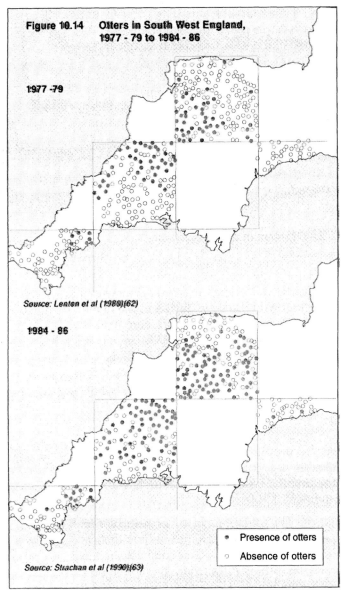

Figure 10.14   Otters in South West England, 1977 - 79 to 1984 - 86

1977 -79

Source: Lenton et al (1980)(62)

1984 - 86

• Presence of otters
○ Absence of otters

Source: Strachan et al (1990)(63)

the case of wildlife, cats and dogs was the deliberate illegal abuse of pesticides (chiefly strychnine, mevinphos and alphachloralose) in attempts to kill predators of game or livestock. The approved use of pesticides or careless misuse were the causes in most cases involving livestock or honeybees.

Pollution

10.68 Some lichens and bryophytes (mosses and liverworts) are sensitive indicators of air pollution and their distributions have been strongly influenced by sulphur dioxide (SO_2) and other chemicals causing acid rain. There is evidence that, as the level of pollutants becomes more diffuse because of the Clean Air Act and tall stacks spreading SO_2 further afield, the lichen flora of cities is gradually improving, while it has been suggested that pollution-sensitive species are being lost from hitherto little polluted country areas [66]. However, the decline of lichens in country

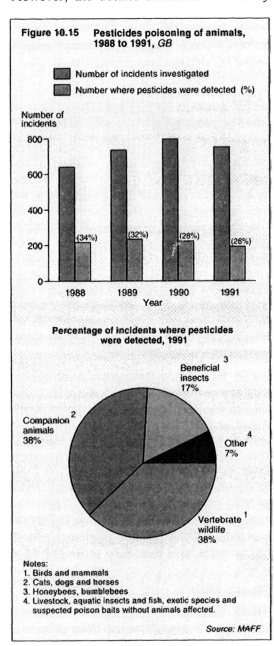

Figure 10.15   Pesticides poisoning of animals, 1988 to 1991, GB

■ Number of incidents investigated
■ Number where pesticides were detected (%)

Percentage of incidents where pesticides were detected, 1991

Beneficial insects 17% [3]

Companion animals 38% [2]

Other 7% [4]

Vertebrate wildlife 38% [1]

Notes:
1. Birds and mammals
2. Cats, dogs and horses
3. Honeybees, bumblebees
4. Livestock, aquatic insects and fish, exotic species and suspected poison baits without animals affected.

Source: MAFF

pesticides on cereals has reduced butterfly numbers on arable farmland. A 1984 survey of the number of species of butterfly on sprayed and unsprayed cereal field headlands on a Hampshire farm, found that for 13 out of the 17 species recorded, the numbers on unsprayed headlands were appreciably greater [64,65].

10.67 If an animal is reported dead or ill and pesticide poisoning is suspected, analyses are carried out to identify the presence of pesticide residues. The results of such investigations are reported to the Environmental Panel of the Advisory Committee on Pesticides and used in the evaluation of pesticide approvals. Figure 10.15 gives information about incidents investigated since 1988. In 1991, pesticides were found to be involved in 193 out of 752 cases investigated. Many of the remainder were attributable to disease, starvation or physical trauma. Misuse or abuse of pesticides was the cause of death in most confirmed cases. The most frequent cause of death in areas is probably attributable to a more complex mixture of factors. Research has shown that some pollution-sensitive species of bryophytes have made significant recoveries throughout southern England during the 1980s [66].

10.69 Deposition of nitrogen compounds may now be causing more problems for lichens than SO_2. Levels of deposition of nitrogen have increased significantly in recent years, resulting in the eutrophication of many lichen habitats and possibly changes in species mixture [67]. Lichens also have the ability to concentrate radionuclides and heavy metals, both of which become toxic at high levels [68] though several species are known to be particularly tolerant of high heavy metal concentrations, and are characteristic of rocks rich in heavy metals.

10.70 Pollution may also affect the breeding success of insects and birds. Dippers are birds which are dependant on invertebrates found in streams and rivers in upland areas. Studies have been carried out on the effects on dippers of water acidification, in Wales [69] and in Scotland [70]. Historical information suggests that the number of breeding dippers on the River Irfon in central Wales decreased between the 1950s and 1982, and that this was linked to increasing acidity of the river. The River Eddw, a "control river", showed no marked change in either pH or the density of breeding dippers. Further studies on the distribution of dippers in relation to stream chemistry showed fewer dippers in more acidic streams [71]. Dipper diet includes insects which are very sensitive to acidification [70,72], eg mayfly nymphs and certain caddisfly species which are scarce in acidic streams.

10.71 As well as changing the quantity of food available, it appears that acidity also changes the quality and that this affects the physiology and breeding performance of adult birds. Effects on dippers include delayed egg-laying, smaller clutches and thinner eggshells, lowered chick growth rate and reduced chick survival. As stream calcium concentrations decrease due to increased acidity so do invertebrates having an exoskeleton whose external body parts are mainly composed of calcium. Calcium availability to adult birds is reduced as their diet switches to these invertebrates not so dependant on calcium. This probably accounts for observed reductions in eggshell thickness and clutch size in acid streams [73]. In some cases however, the low calcium intake by female birds is seemingly compensated for by mobilising calcium from their skeleton [74].

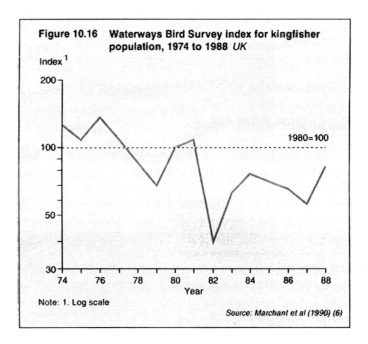

Figure 10.16   Waterways Bird Survey index for kingfisher population, 1974 to 1988 *UK*

Index [1]

Note: 1. Log scale

Source: Marchant et al (1990) (6)

10.72 Pollution may also poison birds directly by the accumulation of heavy metals and pesticides from their prey, or indirectly by reducing the food supply. The decline in the populations of kingfishers since the mid 1970s is attributed to several causes of which pollution is one. Other relevant factors are disturbance and loss of suitable breeding and feeding sites because of leisure activities, and water regulation through flood prevention schemes and water abstractions. Kingfishers are also especially vulnerable to severe winter weather [6]. Figure 10.16 shows the trend in the kingfisher population from 1974 to 1988.

10.73 The impact of oil pollution on seabirds may be significant, particularly near seabird colonies and at certain times of the year in other areas [75]. Summary details of incidents involving 50 or more birds in 1988-89 are given in Table 10.6.

10.74 Wildlife may be used to assess levels of pollution retrospectively by means of environmental specimen banks. These store specimens, under appropriate conditions, of plants, animals, birds and fish which can be examined for contaminants such as pesticide residues, heavy metals etc. There are several specimen banks in the UK: at Rothamsted, where specimens of soil and crops grown in the soil have been taken in experiments which have continued for more than 100 years; Burnham-on-Crouch (shellfish and fish) and the Institute of Terrestrial Ecology (animals and birds) [76].

143

Table 10.6 Oiling incidents involving more than 50 birds, 1988-89

Date	Location	Oil type	Minimum number of birds affected [1]	Species
August 88	Clyde approaches	Medium/heavy fuel oil, little weathering, and fresh heavy fuel oil	125	Guillemot: live but heavily oiled
December 88-January 89	South coast of England	Fresh heavy fuel oil	770	Mainly guillemots: also razorbills, eider, mallard, mute swans, great crested grebe, great northern diver and red-throated diver
May-June 89	Sutherland	Probably weathered crude	216	Mainly guillemots: also razorbills, puffin, fulmar, gannet, shag and Arctic skua
August 89	Mersey estuary	Fresh Venezuelan crude (pipeline fracture)	350	Cormorant, great crested grebe, little grebe, shag, heron, shelduck, mallard, guillemot, gulls spp.
September 89	Northumberland	Heavy fuel oil with some weathering	150	Almost all guillemot, but also puffin, razorbill, gannet
September 89	Whitby, Yorkshire	Heavy fuel oil with some weathering	70	Almost all guillemot, also puffin, fulmar, gannet
October 89	Solent, Hampshire	Syrian crude	400	Teal, mallard, shoveler, mute swan, and some waders
			2081	

Note: 1. Excludes live oiled waders, gulls and most wildfowl: includes all dead oiled birds and those live oiled species whose chances of recovery are very low (eg divers, auks, seaduck).

Source: RSPB

Time-series monitoring

10.75 Time-series monitoring of wildlife is important for assessing the success of conservation measures, but also has a wider role as an early warning of environmental problems (eg effects of acidification on dippers, discussed earlier). There are national monitoring schemes addressing several aspects. These have been established by several organisations usually co-ordinating the work of specialists, both amateur and professional. Many thousands participate in the larger schemes. In 1991, a survey was commissioned to investigate and report on data collected and held by public and voluntary organisations. The project looked at seven groups of species and has produced separate reports which examine data on rare breeding birds in GB [77]; wintering populations of wildfowl and waders in GB [78]; the Butterfly Monitoring Scheme [79]; the Rothamsted Insect Survey on moths in GB [80]; rare plants in GB [81]; squirrels in Crown Forests [82]; and national otter surveys [83]. Information from these and other studies is included in the following paragraphs on monitoring schemes and in Figure 10.17. An annual series of reviews of bird monitoring schemes is being published [84].

Birds

10.76 There are several schemes for monitoring breeding birds. Some of these schemes are well established: the Common Bird Census (CBC) began in the early 1960s and the Waterways Birds Survey in 1974. Other monitoring schemes have been in existence since the 1920s. Several other schemes monitor aspects of breeding performance and survival. Results of some of these schemes have already been integrated and published [6,84]. Numbers of some species of seabirds have been counted regularly since the late 1800s.

10.77 The CBC measures changes in breeding bird populations in the UK between years by means of annual breeding season censuses of target species holding territory on particular plots of land, usually woodland or farmland, but also a number of other habitats. Between 200 and 300 plots are

Figure 10.17 Monitoring of selected birds and insects

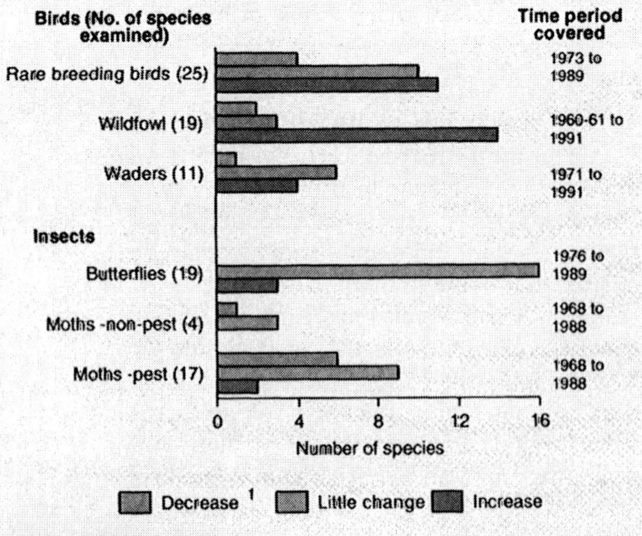

Note: 1. Species showing decrease:
rare breeding birds - red-backed shrike, marsh warbler, bittern, black-tailed godwit;
wildfowl - scaup, European white fronted goose;
waders - dunlin;
moths (non-pest) - cinnabar;
moths (pest) - heart and dart, turnip moth, magpie, rosy rustic, large yellow underwing, common rustic.

Source: Banwell and Crawford (1991)(77), (1992) (79,80) Kirby et al (1991) (31)

covered each year and more than 40,000 territories are mapped. The Waterways Bird Survey covers rivers, streams and canals, and surveys populations of breeding waterside birds using similar methods. These two schemes monitor birds with between a thousand and ten million breeding pairs in Britain. Of 105 species covered regularly by the two schemes, about 31 per cent are increasing in population size, while 34 per cent are decreasing. The remainder show no consistent trend in numbers or are stable [6].

10.78 The Rare Breeding Birds Panel collects information on breeding numbers and success of birds that nest in low numbers (less than 300 pairs) in the UK. Its results thus complement those of the CBC and the Waterways Bird Survey for land birds. The information comes mainly from voluntary county bird recorders, but specialist individuals or groups also contribute. Many of these rare birds are at the edge of their breeding ranges and are particularly sensitive to climatic or ecological threats. Between its inception in 1973 and 1989, the Panel has reported on 96 species which were either breeding in small numbers or were not breeding, but had shown the potential for doing so. Results for 25 of these species have been examined to identify trends over time [77]. There was evidence of significant decreases in breeding numbers of four of the species examined (red-backed shrike, marsh warbler, bittern and black-tailed godwit), but also of significant increases in 11 of the species (see Figure 10.17).

10.79 For all breeding birds, habitat loss, disturbance, persecution, nest-robbing and climatic influences are factors which may have contributed to decreases in numbers. Climatic change is also amongst the factors which may have increased breeding numbers, as well as protection and specific management of sites by conservation bodies, changes in land management by farmers, improved legislation and changes in normal behaviour by certain species [6].

10.80 Breeding seabirds are also monitored systematically (see later section on marine monitoring).

10.81 Wintering wildfowl have been counted each year since 1947 at about 1,400 wetland sites under the National Wildfowl Count. There have also been annual counts of wildfowl and waders at 112 estuarine sites in the UK since 1971 by the Birds of Estuaries Enquiry. Wintering population indices for 19 species of wildfowl and 11 species of wader are published annually. Changes have not been uniform in the period 1971-1991, but overall the wintering numbers of 3 of these species had fallen by more than 20 per cent (European white-fronted goose, scaup and dunlin), while numbers of 18 species had increased by more than 20 per cent [31] (see Figure 10.17). Reasons for increases in numbers include successful breeding seasons and improved protection in the UK and other countries. Likely reasons for the decline in numbers of dunlin include habitat loss (see earlier section on estuaries).

10.82 Cold weather is often the cause of fluctuations in the numbers of wildfowl and waders wintering in the UK [85,86]. Cold winters can cause many species to come further south and west than usual to avoid harsh conditions on the continent, leading to increased numbers in the UK. On the other hand in cold winters, species such as the pochard and teal leave the UK for southern Europe while others like the redshank and the shoveler are likely to die. UK wintering populations of birds that breed in or near the Arctic may fall as a result of adverse weather conditions during the breeding season. Such weather effects tend to cause year-to-year fluctuations, rather than longer term trends, which depend generally on the other factors mentioned above.

Invertebrates

10.83 The Butterfly Monitoring Scheme (BMS) has been operated since 1976. The BMS monitors about 100 sites, but with a concentration at present of monitoring sites on nature reserves and other protected or specifically managed sites; many monitored sites are in the south east of England. The scheme publishes annual indices for 29 species of butterfly. A study of the data for the main annual generation (some species have two generations per year) of 19 of the commoner species found that none had decreased appreciably over the period 1976 to 1989 [79]. Three species (the comma, the ringlet and the large skipper) had increased over the period (see Figure 10.17). The BMS, however, only records at a few selected sites, and cannot detect contractions in the overall distribution of butterflies. Evidence for the changes in distribution can be derived from the national mapping schemes run by the Biological Records Centre. Six of the eight species of fritillary butterfly in Britain used to be widespread, but have suffered major declines in distribution in the present century. The remaining two species have remained restricted in their distributions. Estimates of the decline in high brown *(Argynnis adippe)*, small pearl-bordered *(Boleria selene)* and pearl-bordered *(Boleria euphrosyne)* fritillaries show losses of breeding sites per decade running at 97 per cent, 41 per cent and 38 per cent respectively [87].

10.84 The Rothamsted Insect Survey (RIS) has run, with a few breaks, since 1933. RIS monitors numbers of moths attracted to light traps in GB. These traps operate virtually throughout the year at around 80 sites. The majority of sites are in England. Data on 21 species has been studied; 17 of the 31 species classed as "pests" are featured in the RIS Annual Summaries, plus unpublished data on four "non-pest" species over the period 1968 to 1988 [80]. The study's findings showed that of the 21 species examined, 2 have increased in abundance, 12 have remained stable and 7 show a significant decrease over the last two decades (see Figure 10.17). The 2 species which have increased (the March moth and the angle shades), and 6 of the 7 showing decreases (the heart and dart, the turnip moth, the magpie, the rosy rustic, the large yellow underwing and the common rustic) are "pest" species. One "non-pest" (the cinnabar) has decreased. The large yellow underwing and the angle shades are partially migrant, so numbers could be influenced heavily by events overseas.

Rare plants

10.85 The Biological Records Centre at the Institute of Terrestrial Ecology (ITE) holds distributional records of rare plants dating back to 1930. These records, and additional field surveys, provide the basis for two editions of the British Red Data Book for vascular plants, covering all information up to 1976 and to 1980 respectively [88,89]. About 320 species are included, and are those which occur in 15 or fewer 10 km squares in GB.

10.86 Data from the two Red Data Books were compared to illustrate possible changes in population distribution and the status of each species [81]. The distribution of around 60 per cent of rare species had not changed significantly. Twenty per cent of species had contracted in range, while the ranges of 16 per cent of species had spread - see Table 10.7 which also shows the status of species based on the IUCN classification and on "threat number". Many of the changes shown however, may have been due to changes in subjective judgement or to improvements in survey methods.

Mammals

10.87 The monitoring of mammals has a shorter history and is less intensive than for birds. Best data are available for squirrels, red deer, otters and seals. Information on squirrels, seals and otters is given below.

10.88 An annual survey of red and grey squirrels on Forestry Commission (FC) land has produced distribution maps of reported sightings in 10 km squares for both species annually since 1973 [90]. This information was used as the basis for a study [82] of changes in the distribution of red and grey squirrels (see Figure 10.18). The findings are restricted to FC land resulting in an uneven coverage over GB as a whole. In Scotland, the distribution of red squirrels has remained widespread, but it has declined rapidly in Wales and to a lesser extent in England. Grey squirrels spread in Scotland over the same period but remained fairly stable in England and Wales.

10.89 A major reduction in numbers of otters in GB in the 1950s was thought to be due to the introduction of organochlorine insecticide use in the countryside, although the otter had also long been hunted for sport and to protect fishing interests [91]. However, legal protection for the otter was introduced in England and Wales in 1978 and in Scotland in 1982, and (as mentioned earlier in this chapter) there are signs of a recent recovery due to a variety of factors. The National Otter Surveys were carried out in 1977-79 and 1984-86, and recorded presence or absence of otters on 600m stretches of rivers (see for example, Figure 10.14). Positive counts in both surveys showed that otters are most numerous in Scotland (in the later survey nearly two thirds of the Scottish sites had positive evidence of otters) and least numerous in England (one tenth of sites had positive evidence). Comparison of the two surveys suggests an overall increase in evidence for otter presence in GB, but with some regional variation [83]. Twenty five regions of GB were

Table 10.7 Indicators of changes in populations of species of rare plants, 1976 to 1980 *GB*

Method of examination	Total number of species covered	Percentage of species whose populations have;		
		decreased/ worsened	not changed	Increased/ improved
Number of 10 km squares in which species present [1]	318	20	64	16
Based on IUCN [2] classification of extinct, endangered vulnerable or rare	318	6	90	4
Based on "threat number" [3]	298	10	70	20

Notes:
1. Method involved comparing between years the number of 10km squares in which a species was present.
2. International Union for the Conservation of Nature and Natural Resources.
3. Number based on points system for six factors including population decline, attractiveness of plant, distribution, location and accessibility.

Source: Banwell & Crawford (1992)(81); Perring & Farrel (1977)(88), (1983)(89)

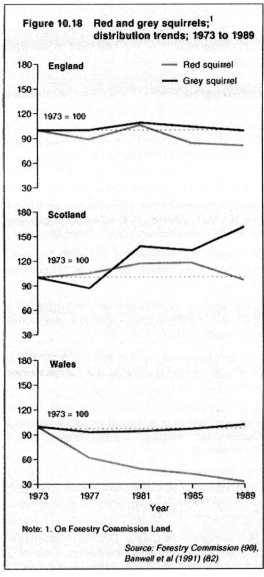

Figure 10.18   Red and grey squirrels;[1] distribution trends; 1973 to 1989

England
1973 = 100

Scotland
1973 = 100

Wales
1973 = 100

Red squirrel
Grey squirrel

Year
1973   1977   1981   1985   1989

Note: 1. On Forestry Commission Land.

Source: Forestry Commission (90),
Banwell et al (1991) (82)

examined for change. One region (Anglian Water region) showed a significant decline, 11 regions showed no significant change and 13 showed significant increases. The results also suggested that the recent increases in England and Wales were relatively larger than those in Scotland. An otter survey carried out in 1981 found positive evidence of otters in over 90 per cent of sites studied in Northern Ireland [92].

Marine

10.90 Monitoring of marine wildlife is undertaken for a wide range of purposes including evaluating effects of industrial activity (eg both onshore and offshore activities of the oil industry), management of Marine Nature Reserves (at Lundy and Skomer and five other possible Marine Nature Reserves - see Chapter 5 on land) and, for fisheries, fish landings and stock sizes of commercial species in order to recommend quotas under the EC's Common Fisheries Policy. One of the longest running marine monitoring projects is that based on the continuous plankton recorder, towed behind merchant ships around the British Isles and out into the Atlantic. This project monitors the variation in plankton communities in time and space and relate this to oceanographic and climatic changes.

10.91 The changing state of the offshore marine environment for birds is monitored under the Seabird Monitoring Programme. This programme assesses the numbers and breeding performance of seabirds at colonies throughout Britain and Ireland [93]. Virtually all species of seabird are monitored, but information for fulmars, shags, guillemots, kittiwakes and terns is the most comprehensive. Substantial information is available, for example, on the breeding performance of kittiwakes in recent years. Productivity of the kittiwake in the southern part of the North Sea was uniformly high during the period 1986 -1990, while in the Irish Sea and south west Britain productivity was much more variable. Breeding output in Shetland declined to a very low level in 1990, recovering in 1991 [93]. The decline in breeding performance was caused by the lack of suitably sized sand eels [94] and the reasons for this are the subject of a major research project now in progress.

10.92 MAFF maintain regular assessments of commercially exploitable stocks of fish and shellfish. They also participate in the regular monitoring of seal and cetacean stocks, and assist with some seabird surveys. MAFF also run some local programmes to check on the spread of pests, or environmental recovery following pollution control measures such as those imposed on Tributyltin (TBT) use. See Chapter 8 on the marine environment.

References and further reading

1. Felton, M., & Marsden, J., (1990). Heather Regeneration in England and Wales: Executive Summary. A Feasibility study for the Department of the Environment by Nature Conservancy Council, Peterborough.

2. Batten, L.A., Bibby, C.J., Clement, P., Elliott, G.D., & Porter, R.F., (1990). Red Data Birds in Britain. T. & A.D. Poyser, London.

3. Tudor, G. (in prep). Summary of Phase 1 of the National Countryside Monitoring Scheme for Scotland. Scottish Natural Heritage, Edinburgh.

4. Ratcliffe, D.A., (1990). Upland Birds and their Conservation. British Wildlife, 2: 1-12.

5. Ratcliffe, D.A., (1990). Bird Life of Mountain and Upland. Cambridge University Press, Cambridge.

6. Marchant, J.H., Hudson, R., Carter, S.P., & Whittington, P., (1990). Population Trends in British Breeding Birds. British Trust for Ornithology & Nature Conservancy Council, Tring.

7. Webb, N.R., (1986). Heathlands. Collins, London (New Naturalist Series).

8. Braithwaite, A.C., Buckley, J., Corbett, K.F., Edgar, P.W., Haslewood, E.S., Haslewood, G.A.D., Langton, T.E.S., & Whitaker, W.J., (1989). The Distribution in England of the Smooth Snake (Coronella austriaca Laurenti). Herpetological Journal, 1: 370-376.

9. Ginn, H., (1983). The Ecology and Conservation of Amphibian and Reptile Species Endangered in Britain. Nature Conservancy Council, London.

10. Falk, S., (1991). A Review of the Scarce and threatened Bees, Wasps and Ants of Great Britain. Research and Survey in Nature Conservation No. 35. Nature Conservancy Council, Peterborough.

11. Bibby, C.J., (1973). The Red-backed Shrike: a Vanishing Species. Bird Study, 20: 103-110.

12. Fuller, R.M., (1987). The Changing Extent and Conservation Interest of Lowland Grasslands in England and Wales: a Review of Grassland Surveys 1930-1984. Biological Conservation, 40: 281-300.

13. Cherrill, A.J., Shaughnessy, J., & Brown, V.K., (1991). Oviposition behaviour of the Bush-cricket Decticus Verrucivorus (L.) (Orthoptera: Tettigoniidae). Entomologist, 110: 37-42.

14. Cherrill, A.J., & Brown, V.K., (1990). The Habitat Requirements of the Wart-biter Decticus Verrucivorus (L) (Orthoptera: Tettigoniidae) in Southern England. Biological Conservation, 53: 145-157.

15. Cherrill, A.J., & Brown, V.K., (1990). The Life Cycle and Distribution of the Wart-biter Decticus Verrucivorus (L.) (Orthoptera: Tettigoniidae) in a Chalk Grassland in Southern England. Biological Conservation, 53: 125-143.

16. Green, R.E., (1991). Breeding Waders of Lowland Grasslands in England and Wales. In: Birds and Pastoral Agriculture in Europe, (eds. Curtis, D.J., Bignal, E.M., & Curtis, M.A.). Joint Nature Conservation Committee & Scottish Chough Study Group, Peterborough.

17. O'Brian, M., & Buckingham, D., (1989). A Survey of Breeding Waders on Grassland within the Broads Environmentally Sensitive Area. Royal Society for the Protection of Birds, Sandy.

18. O'Brien, M., (1991). Breeding Waders on Wet Lowland Grassland. pp87-89 in Stroud, D.A., & Glue, D., Britain's Birds in 1989/90: the Conservation and Monitoring Review. British Trust for Ornithology & Nature Conservancy Council, Thetford.

19. Stowe, T.J., & Hudson, A.V., (1988). Corncrake Studies on the Western Isles. RSPB Conservation Review, 2: 38-42.

20. Hamill, R., & McSherry, B., (1990). Report to RSPB and DOE(NI) on Corncrake Fieldwork in County Fermanagh in 1990. RSPB, Belfast.

21. Williams, G., Stowe, T.J., Newton, A.V., (1991). Action for Corncrakes. RSPB Conservation Review, 5: 47-53.

22. Peterken, G.F., & Allison, H., (1989). Woods, Trees and Hedges: a Review of Changes in the British Countryside. Nature Conservancy Council, Peterborough.

23. Warren, M.S., (1991). The Successful Conservation of an Endangered Species, the Heath Fritillary Butterfly Mellicta Athalia, in Britain. Biological Conservation, 55: 37-56.

24. Speight, M.C.D., (1989). Saproxylic Invertebrates and their Conservation. Council of Europe, Strasbourg. Nature and Environment series, No 42.

25. Burton, R.G.O., & Hodgson, J.M. (eds.), (1987). Lowland Peat in England and Wales. Soil Survey of England and Wales, Harpenden. Special Survey No 15.

26. Cruickshank, M.M., & Tomlinson, R.W., (1990). Peatland in Northern Ireland: Inventory and Prospect. Irish Geography, 23: 17-23.

27. Unpublished study by Birmingham University on behalf of Nature Conservancy Council and National Rivers Authority.

28. Swan, M.J.S., & Oldham, R.S., (1989). Amphibian Communities. Nature Conservancy Council Commissioned Research Report No 1020.

29. Whitten, A.J., (1990). Crested Newt pp 292-295 in Recovery: a Proposed Programme for Britain's Protected Species. Nature Conservancy Council, Peterborough.

30. Furphy, J.S., (1977). Census of Great Crested Grebes in Northern Ireland, Summer 1975. Irish Birds, 1: 56-58.

31. Kirby, J.S., Ferns, J.R. Waters, R.J. & Prys-Jones, R.P., (1991). Wildfowl and Wader Counts 1990-91. Wildfowl and Wetlands Trust, Slimbridge.

32. Goss-Custard, J.D., & Moser, M.E., (1988). Rates of Change in the Numbers of Dunlin, Calidris Alpina, Wintering in British Estuaries in relation to the spread of Spartina Anglica. Journal of Applied Ecology, 25: 95-109.

33. Denton, J.S., & Beebee, T.J.C., (1992). Pilot Investigation of Potential Sites for the Reintroduction of the Natterjack Toad Bufo Calamita. English Nature, Peterborough.

34. Doody, J.P., (1989). Management for Nature Conservation. Proceedings of the Royal Society of Edinburgh, 96B: 247-265.

35. Howson, C.M. (ed.), (1987). Directory of British Marine Fauna and Flora. Marine Conservation Society, Ross-on-Wye.

36. Bennett, T.L., Mitchell, R., & Copley, V.H., (in prep.). Coastwatch: a Survey of Coastal Habitats and Human Activities around the Coast of Great Britain. Joint Nature Conservation Committee, Peterborough.

37. Downie, A.J., & Davies, L.M., (1991). Synopsis of Survey Data held by the Marine Nature Conservation Review: a Contribution to the Great Britain Nature Conservation Resource Survey. JNCC Report, No 13.

38. Hiby, L., Duck, C., & Thompson, D., (1992). Seal Stocks in Great Britain: Surveys Conducted in 1990 and 1991. NERC News (January 1992): 30-31.

39. Hall, A.J., Pomeroy, P.P., & Harwood, J., (1992). The Descriptive Epizootiology of Phocine Distemper in the UK during 1988-89. Science of the Total Environment, 115.

40. Tasker, M.L., Webb, A., Hall, A.J., Pienkowski, M.W., & Langslow, D.R., (1987). Seabirds in the North Sea. Nature Conservancy Council, Peterborough.

41. Webb, A., Harrison, N.M., Leaper, G.M., Steele, R.D., Tasker, M.L., & Pienkowski, M.W., (1990). Seabird Distribution West of Britain. Nature Conservancy Council, Peterborough.

42. Tasker, M.L., & Pienkowski, M.W., (1987). Vulnerable Concentrations of Birds in the North Sea. Nature Conservancy Council, Peterborough.

43. Tasker, M.L., Webb, A., Harrison, N.M., & Pienkowski, M.W., (1990). Vulnerable Concentrations of Birds West of Britain. Nature Conservancy Council, Peterborough.

44. Lloyd, C., Tasker, M.L., & Partridge, K., (1991). The Status of Seabirds in Britain and Ireland. T. & A.D. Poyser, London.

45. Nature Conservancy Council, (1986). Initial Marine Consultation Areas: Scotland. Nature Conservancy Council Report, Edinburgh.

46. Cresswell, P., Harris, S., & Jefferies, D.J., (1990). The History, Distribution, Status and Habitat Requirements of the Badger in Britain. Nature Conservancy Council, Peterborough.

47. Swan M.J.S., & Oldham, R.S., (1991). Herpetile Sites: Interim Report No 2. Report by Leicester University to Nature Conservancy Council, Peterborough.

48. Tapper, S., & Parsons, N., (1984). The Changing Status of the Brown Hare (Lepus Capensis L.) in Britain. Mammal Review, 14: 57-70.

49. O'Connor, R.J., & Shrubb, M., (1986). Farming and Birds. Cambridge University Press, Cambridge.

50. Gibbons, D.W., Reid, J.B., Chapman, R.A., (in prep). The New Atlas of Breeding Birds in Britain and Ireland: 1988-1991. T. & A.D. Poyser, London.

51. Barr, C., Benefield, C., Bunce, B., Ridsdale, H., & Whittaker, M., (1986). Landscape Changes in Britain. Institute of Terrestrial Ecology, Abbots Ripton.

52. Baldock, D., (1990). Agriculture and Habitat Loss in Europe. Institute for European Environmental Policy, London.

53. Stroud, D.A., Reed, T.M., Pienkowski, M.W., Lindsay, R.A., (1987). Birds, Bogs and Forestry: the Peatlands of Caithness and Sutherland. Nature Conservancy Council, Peterborough.

54. Avery, M., & Leslie, R., (1990). Birds and Forestry. T & AD Poyser, London.

55. Wallace, H.L., Good, J.E.G., & Williams, T.G., (1992). The Effects of Afforestation on Upland Plant Communities: an Application of the British National Vegetation Classification. Journal of Applied Ecology, 29:180-194.

56. Lindsay, R.A., Charman, D.J., Everingham, F. O'Reilly, R.M., Palmer, M.A., Rowell, T.A., and Stroud, D.A., (1988). The Flow Country: the Peatlands of Caithness and Sutherland. Nature Conservancy Council, Peterborough.

57. Coulson, J.C., (1988). The Structure and Importance of Invertebrate Communities on Peatlands and Moorlands, and Effects of Environmental and Management Changes. pp 365-380 in Ecological Change in the Uplands. British Ecological Society, London (Special publication No 7 of the British Ecological Society).

58. Cadbury, J., (1991). Death by Design. The Persecution of Birds of Prey and Owls in the UK 1979-1989. Royal Society for the Protection of Birds & Nature Conservancy Council, Sandy.

59. Ratcliffe, D.A., (1972). The Peregrine Population in 1971. Bird Study, 19: 117-156.

60. Mead, C.J., (1992). The 1991 Peregrine Survey - Preliminary Results. BTO News, 180: 7.

61. Newton, I., (1986). The Sparrowhawk. Poyser, Calton.

62. Lenton, E.J., Chanin, P.R.F. & Jefferies, D.J., (1980). Otter Survey of England 1977-79. Nature Conservancy Council, London.

63. Strachan, R., Birks, J.D.S., Chanin, P.R.F. & Jefferies, D.J., (1990). Otter Survey of England 1984-86. Nature Conservancy Council, Peterborough.

64. Rands, M.R.W., & Sotherton, N.W., (1986). Pesticide Use on Cereal Crops and Changes in the Abundance of Butterflies on Arable Farmland in England. Biological Conservation, 36: 71-82.

65. Sotherton, N.W., Boatman, N.D., & Rands, M.R.W., (1989). The Conservation Headland

Experiment in Cereal Ecosystems. Entomologist, 108: 135-143.

66. Adams, K.J., & Preston, C.D., (1992). Evidence for the Effects of Atmospheric Pollution on Bryophytes from Local and National Recording. Pp 31-43 in: Biological Recording of Changes in British Wildlife (ed. Harding P.T.). HMSO, London.

67. Farmer, A.M., Bates, J.W., & Bell, J.N.B., (1992). Ecophysiological Effects of Acid Rain on Bryophytes and Lichens. Pp 284-313 in Bryophytes and Lichens in a Changing Environment (eds. Bates, J.W., & Farmer, A.M.). Clarendon Press, Oxford.

68. James, P.W., (1973). The Effect of Air Pollutants other than Hydrogen Fluoride and Sulphur Dioxide on Lichens. Pp 143-175 in: Air Pollution and Lichens, (ed. Ferry, B.W., Baddeley, M.S., & Hawkworth, D.L.). Athlone Press, London.

69. Ormerod, S.J., & Tyler, S.J., (1987). Dippers (Cinclus cinclus) and Grey Wagtails (Motacilla cinerea) as Indicators of Stream Acidity in Upland Wales. In Diamond, A.W., & Fillion, F.W., (eds.) The Value of Birds. ICBP Technical Publication, 6: 191-208. International Council for Bird Preservation, Cambridge.

70. Vickery, J., (1991). Breeding Density of Dippers (Cinclus cinclus), Grey Wagtails (Motacilla cinerea) and Common Sandpipers (Actitis hypoleucos) in relation to the Acidity of Streams in South West Scotland. Ibis, 133: 178-185.

71. Ormerod, S.J., Tyler, S.J., & Lewis, M.S., (1985). Is the Breeding Distribution of Dippers Influenced by Stream Acidity? Bird Study, 32: 33-39.

72. Ormerod, S.J., (1985). The Diet of Breeding Dippers (Cinclus cinclus) and their Nestlings in the Catchment of the River Wye, Mid Wales, a Preliminary Study by Faecal Analysis. Ibis, 127: 316-331.

73. Ormerod, S.J., Bull, K.R., Tyler, S.J., & Vickery, J.A., (1988). Egg Mass and Shell Thickness in Dippers (Cinclus cinclus) in relation to Stream Acidity in Wales and Scotland. Environmental Pollution, 55: 107-121.

74. Ormerod, S.J., (1991). Acidification and Freshwater Biotas. In Woodin, S.J., & Farmer, A.M., (eds.). The Effects of Acid Deposition on Nature Conservation of Great Britain. Focus on Nature Conservation, 26. Nature Conservancy Council, Peterborough.

75. Stowe, T.J., (1982). Beached Birds Surveys and Surveillance of Cliff Breeding Seabirds. Nature Conservancy Council Chief Scientist Directorate Report No 366. Royal Society for the Protection of Birds.

76. King, N., (1984). Environmental Specimen Banking in the UK; Do We Need To Go Any Further? Environmental Specimen Banking and Monitoring as Related to Banking. Proceedings of the International Workshop, Saarbruecken, Federal Republic of Germany, 10-15 May, 1982. Martinus Nijhoff.

77. Banwell, J.L., & Crawford, T.J., (1991). Key Indicators for British Wildlife: Rare Breeding Birds in Great Britain. Report to Department of the Environment by University of York.

78. Banwell, J.L., & Crawford, T.J., (1992). Key Indicators for British Wildlife: Wintering Populations of Wildfowl and Waders in Great Britain. Report to Department of the Environment by University of York.

79. Banwell, J.L., & Crawford, T.J., (1992). Key Indicators for British Wildlife: Butterfly Monitoring Scheme. Report to Department of the Environment by University of York.

80. Banwell, J.L., & Crawford, T.J., (1992). Key Indicators for British Wildlife: Rothamsted Insect Survey Moth Data. Report to Department of the Environment by University of York.

81. Banwell, J.L., & Crawford, T.J., (1992). Key Indicators for British Wildlife: Rare Plants in Great Britain. Report to Department of the Environment by University of York.

82. Banwell, J.L., Crawford, T.J., & Usher, M.B., (1991). Key Indicators for British Wildlife: the Red Squirrel (Sciurus vulgaris) and Grey Squirrel (Sciurus carolinensis) in Crown Forests. Report to Department of the Environment by University of York.

83. Banwell, J.L., & Crawford, T.J., (1992). Key Indicators for British Wildlife: National Otter Surveys. Report to Department of the Environment by University of York.

84. Stroud, D.A., & Glue, D., (1991). Britain's Birds in 1989-90: the Conservation and Monitoring Review. British Trust for Ornithology/Nature Conservancy Council, Thetford.

85. Prater, A.J., (1981). Estuary Birds in Britain and Ireland. T. & A.D. Poyser, Calton.

86. Owen, M., Atkinson-Willes, G.L., & Salmon, D.G., (1986). Wildfowl in Great Britain, Cambridge University Press, Cambridge.

87. Warren, M., (1992). Britain's Vanishing Fritillaries. British Wildlife, 3: 282-296.

88. Perring, F.H., & Farrell, L., (1977). British Red Data Books: 1 Vascular Plants 1st edition. Society for the Promotion of Nature Conservation, Lincoln.

89. Perring, F.H. & Farrell, L., (1983). British Red Data Books: 1. Vascular plants. 2nd edition. Royal Society for Nature Conservation, Lincoln.

90. Annual Reports on Squirrels from Forestry Commission, eg Pepper H.W., (1989). Squirrel Questionnaire 1988 Report. Forestry Commission, Farnham.

91. Jefferies, D.J., (1989). The Changing Otter Population of Britain 1700-1989. Biological Journal of the Linnean Society, 38: 61-69.

92. Chapman, P.J., & Chapman, L.L., (1982). Otter Survey of Ireland 1980-1981. Vincent Wildlife Trust, London.

93. Walsh, P.M., Sim, I., & Heubeck, M., (1992). Seabird Numbers and Breeding Success in Britain and Ireland, 1991. Joint Nature Conservation Committee, Peterborough (UK Nature Conservation No 6).

94. Wright, P.J. & Bailey, M.C., (1991). Biology of Sandeels in the Vicinity of Seabird Colonies at

Shetland. First Annual Report -April 1991. The Scottish Office Agriculture and Fisheries Department. Fisheries Research Services Report No 13/91.

Davidson, N.C., Laffoley, D.A., Doddy, J.P., Way, L.S., Gordon, J., Key, R., Drake, C.M., Pienkowski, M.W., Mitchell, R., & Duff, K.L., (1991). Nature Conservation and Estuaries in Great Britain. Nature Conservancy Council, Peterborough.

Nature Conservancy Council, (1990). Handbook for Phase I Habitat Survey - a Technique for Environmental Audit. Nature Conservancy Council, Peterborough..

Wyatt, G., (1991). A Review of Phase I Habitat Survey in England. Nature Conservancy Council, Peterborough.

Ratcliffe, D.A., & Thompson, D.B.A., (1988). The British Uplands: their Ecological Character and International Significance. In: Ecological Change in the Uplands, (eds Usher, M.B., & Thompson, D.B.A.). British Ecological Society Special Publication No 7. Blackwell, Oxford.

Rare Birds Breeding Panel Reports 1973 -1989. These are usually published two years after the year to which they refer in British Birds (eg Spencer, R. & the Rare Birds Breeding Panel, (1991). Rare Breeding Birds in the United Kingdom in 1989. British Birds, 84:349-392).

Barnes, R., & Tapper, S., (1983). Why We Have Fewer Hares. Game Conservancy Annual Review for 1982, 14: 51-61.

Fletcher, M.R., Hunter, K., Quick, M.P., Grave, R.C., (1992). Pesticide Poisoning of Animals 1991: Investigations of Suspected Incidents in Great Britain. Environmental Panel Report. MAFF, London.

11 Waste and recycling

☐ **Total waste arisings in the UK are estimated to be around 400 million tonnes of which mining and quarrying waste accounts for 27 per cent (Figure 11.1). About a third of total waste arisings is "controlled waste" (11.4).**

☐ **Special wastes in the UK (controlled wastes which are "dangerous to life") have increased in recent years, totalling 2.5 million tonnes in 1990-91 (Table 11.1).**

☐ **Landfill is the main method of disposal for controlled waste, including special waste (accounting for 85 per cent and 70 per cent respectively of controlled and special waste disposals) (Figures 11.2 and 11.3).**

☐ **About half of sewage sludge is used as a soil conditioner and nutrient on farm land and horticultural areas (Figure 11.4).**

☐ **Around 44,000 tonnes of hazardous wastes were imported into England and Wales in 1990-91 (Table 11.3). Over half (58 per cent) is disposed of by physical or chemical treatment and 29 per cent by incineration (Figure 11.7). Imports to Scotland in 1990-91 were less than one thousand tonnes (11.34).**

☐ **Over 50 per cent of household waste is estimated to be in principle recyclable (if composting of organic material is included) but only about 5 per cent is recovered (11.38).**

☐ **A greater weight of ferrous metal is recycled in the UK than all other materials combined. Around half of total consumption is recycled scrap (11.40).**

☐ **Recycling of paper and board has increased steadily in recent years. Over 30 per cent of total consumption is recycled scrap (Figure 11.11).**

☐ **The use of recycled glass has more than doubled since the early 1980s. 21 per cent of total consumption is recycled scrap (Figure 11.12). The number of participating councils in the bottle bank scheme increased from 158 authorities in 1981 to 448 in 1991 (11.44).**

☐ **About 10 per cent of the annual use of aggregates is derived from secondary materials (11.45).**

11.1 Around 400 million tonnes of waste per year is produced in the UK. This chapter covers the generation and disposal of most categories of waste, the recycling of waste and litter.

11.2 Reliable estimates of amounts of UK waste arisings (ie the amount of waste produced), are difficult to obtain. Very little waste is weighed, there are no standard conversion factors for converting volumetric measures to tonnes, and there is no widely accepted classification of wastes. Data are available in respect of wastes disposed of at sea but in general, annually updated information on arisings and disposal of waste is available only for Scotland. Steps are being taken to improve information by research and through new data collection initiatives. The estimates of total UK arisings given in this chapter are based on estimates derived from a variety of information sources and give a reasonable guide to arisings in any recent twelve month period from households and from commercial and industrial sectors. Estimated annual arisings of mining wastes, demolition and construction wastes, blast furnace and steel slag and power station ash in 1990 are taken from a recent study by

Arup Economics and Planning for the Department of the Environment (DOE)[1]. Estimates for quarrying waste have been provided by the Quarries Inspectorate, and the Ministry of Agriculture, Fisheries and Food (MAFF) have provided a revised estimate of waste from agricultural premises.

11.3 Some categories of waste are more appropriately described elsewhere in the report and are only covered briefly in this chapter. Waste disposal at sea (eg dredged material, some sewage sludge, some industrial waste, and wastes discharged into tidal waters) is covered in Chapter 8 on the marine environment. Liquid wastes which are discharged to inland water courses (eg industrial effluent and sewage) are covered in Chapter 7 on inland water quality. Information on radioactive waste, a very small percentage of all UK waste, is given in Chapter 13 on radioactivity.

11.4 The treatment and disposal of "controlled waste", ie industrial, household and commercial waste, is regulated in GB by the Control of Pollution Act 1974 (CoPA). The waste provisions of CoPA are being extended by the progressive implementation of the Environmental Protection Act 1990

(EPA). Around 140 million tonnes of total annual estimated waste arisings, roughly one third, is termed controlled waste as defined by CoPA and the EPA (see Box 11.1).

11.5 In England and Wales, District Councils act as Waste Collection Authorities for household waste, and some commercial wastes. Waste Disposal Authorities (WDAs), generally County Councils in England, District Councils in Wales and the English Metropolitan areas (there are some single purpose disposal authorities in the latter areas), are responsible for arranging the treatment and disposal of the controlled waste collected by the collection authorities. WDAs also regulate waste management by licensing firms undertaking this work and monitoring their activities. In Scotland and Northern Ireland both these functions are carried out by District or Islands Councils. Under the EPA, the disposal and regulatory functions of WDAs are being separated. In England and Wales, Waste Regulation Authorities will be responsible for the licensing and monitoring of waste facilities, the registration of waste carriers, control of special waste and other statutory regulatory functions. However, the Government has proposed that

---

Controlled waste: how it is regulated Box 11.1

Control of Pollution Act 1974 (CoPA)

CoPA regulates the disposal of industrial, household and commercial wastes, termed controlled waste in GB. Controlled waste specifically excludes mine and quarry waste, wastes from premises used for agriculture, some sewage and radioactive waste. (Similar legislative provisions exist in Northern Ireland based on the same definitions and control requirements.)

Control of Pollution (Special Waste) Regulations 1980

These provide a "cradle to grave" consignment note system for special waste, ie any controlled waste which consists of or contains certain substances which makes it "dangerous to life", together with prescriptive medicines. The regulations define "dangerous to life" and list substances. (The Pollution Control (Special Waste) Regulations (NI) 1981 apply in Northern Ireland.)

Licensing arrangements for controlled waste

Anyone wishing to operate disposal, storage and treatment facilities has to have a licence from Waste Disposal Authorities (WDAs), unless the activity qualifies for exemption. WDAs may impose conditions in licences and have powers to require that the operations are carried out satisfactorily.

Environmental Protection Act 1990 (EPA)

The Act is being implemented progressively. Part II, which extends and largely supersedes the waste provisions of CoPA, will strengthen the licensing system. Holders of licences will no longer be able to cancel them unilaterally; their surrender must be accepted by the Waste Regulation Authority (WRA), who will not do so until they are satisfied no further pollution is likely. WRAs will be able to check whether proposed licensees are fit and proper persons to hold a licence and will make charges related to the cost of regulation. Part II also introduced a Duty of Care from 1 April 1992 on all producers and holders of waste, to take reasonable steps to prevent illegal disposal of their waste.

Planning and Compensation Act 1991

Under this Act English counties are required to prepare waste local plans to inform their development control decisions on waste treatment, recycling and disposal facilities.

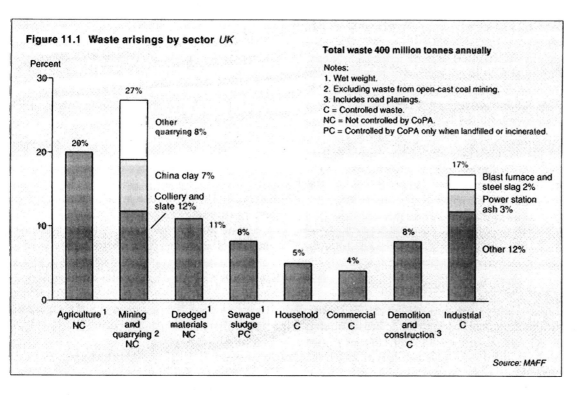

Figure 11.1 Waste arisings by sector *UK*

Total waste 400 million tonnes annually

Notes:
1. Wet weight.
2. Excluding waste from open-cast coal mining.
3. Includes road planings.
C = Controlled waste.
NC = Not controlled by CoPA.
PC = Controlled by CoPA only when landfilled or incinerated.

Source: MAFF

waste regulation should become the responsibility of a national Environment Agency when it is set up. In Scotland, District and Island Councils will remain the collection, disposal and regulation authorities, although again the division of functions would be affected by the establishment of a Scottish Environment Protection Agency.

Waste arisings

11.6 Figure 11.1 shows the estimated annual arisings of waste in the UK by main source. As explained earlier, the figure of around 400 million tonnes for total waste arisings is the best estimate available in the absence, except in Scotland, of regular data for most sources. In England and Wales, yearly estimates are available only for sewage sludge and dredged materials. Most estimates given in this chapter for individual sources do not refer to a specific year but may be regarded as valid for any 12 month period in recent years; they have been rounded to the nearest million tonnes, except where figures exceed 50 million tonnes, when they have been rounded to the nearest 10 million tonnes. The absence generally of annually updated data means that few trends can be presented in respect of arisings.

Controlled waste

11.7 Controlled waste, regulated by CoPA, includes household and commercial waste, some sewage sludge, demolition and construction waste, and industrial waste, but specifically excludes wastes from mining and quarrying, wastes from agricultural premises, and radioactive waste.

Household and commercial waste

11.8 Household waste is estimated at 20 million tonnes annually (5 per cent of total arisings), and commercial waste at 15 million tonnes (4 per cent). In England and Wales, information about these categories has, in the past, been provided mainly by the Chartered Institute of Public Finance and Accountancy (CIPFA) from surveys of local authorities. However, this series is no longer considered reliable owing to a falling off in responses by authorities in recent years. Currently DOE and CIPFA are working to improve the flow of information from authorities and in future surveys there will be greater emphasis on obtaining data relevant to environmental protection and recycling. A separate survey is also being mounted by Warren Spring Laboratory for DOE, to provide information on the composition of household waste, and results are expected during 1993. In Scotland, the Scottish Office Environment Department collects statistics annually from Scottish local authorities which reported 2.2 million tonnes of household waste and 2 million tonnes of commercial waste in 1989.

Sewage sludge

11.9 Sewage sludge is the by-product of the treatment of raw sewage. Sludge disposed of to landfill sites or incinerated is regulated by CoPA or Part I of the EPA; sewage sludge disposed of at sea or used for agricultural purposes is controlled by other means. Sewage sludge arisings (wet weight) are estimated to be between 30 and 40 million tonnes annually in the UK (roughly 8 per cent of total estimated arisings). The dry weight equivalent is around 1 million tonnes. The residual liquid effluent from sewage treatment plants is discharged into inland rivers and tidal waters after being treated to a standard specified in the consent held by the discharger. Such discharges are controlled by the National Rivers Authority in England and Wales, the River Purification Authorities in Scotland and the Department of the Environment for Northern Ireland (DOE(NI)). (See also later sections of this chapter and the references to sewage sludge in Chapters 7 and 8 on inland water quality and pollution and the marine environment).

Demolition and construction wastes

11.10 Demolition and construction wastes, including asphalt road planings, are estimated at 32 million tonnes annually (about 8 per cent of total arisings). Materials such as concrete, masonry, steel, timber, and cement resulting from demolition, and similar materials and damaged items from construction work, are estimated at 24 million tonnes in the Arup report[1], although other recent surveys suggest a higher estimate. Asphalt road planings removed from road surfaces prior to resurfacing or reconstruction amount to between 7 and 8 million tonnes annually.

Industrial waste

11.11 Industrial waste arisings form a major proportion of wastes regulated by CoPA and account for 17 per cent of waste arisings. Industrial waste includes power station wastes and blast furnace and steel slag which together amounts to 19 million tonnes annually (5 per cent of total arisings). A further 50 million tonnes a year (12 per cent) is generated by a wide variety of industrial activities and includes such substances as residues from food manufacture, horticulture, containers and packaging.

Wastes not subject to waste disposal licensing: regulation and guidance
Box 11.2

Wastes from agricultural premises

Ministry of Agriculture, Fisheries and Food (MAFF)/Welsh Office Agricultural Department (WOAD) and Scottish Office Agriculture and Fisheries Department (SOAFD) Codes of Good Agricultural Practice: Codes of Good Agricultural practice are being produced for England and Wales by MAFF/WOAD for the Protection of Water, Air and Soil. Each of the Codes will give practical guidance to farmers on, amongst other things, the storage and disposal of farm wastes to prevent pollution in the respective media. The Water Code was published in July 1991, the Air Code in July 1992 and the Soil Code is in preparation. SOAFD published in March 1992, a Code of Good Practice for the Prevention of Environmental Pollution from Agricultural Activity and similar guidance is issued in Northern Ireland.

Control of Pollution (Silage, Slurry and Agricultural Fuel Oil) Regulations 1991: these regulations, which apply in England and Wales, were made under the Water Act 1989 and set minimum construction standards for new or substantially altered farm waste storage installations. Similar regulations have been made in Scotland.

Sewage sludge (and other wastes)

The Food and Environment Protection Act 1985: the disposal of sewage sludge (and other wastes) by dumping at sea is controlled by means of licences issued under the Food and Environment Protection Act 1985 (Part II) by MAFF in England and Wales, SOAFD in Scotland and DOE(NI) in Northern Ireland. The UK will cease dumping sewage sludge into the sea by the end of 1998 - see also Chapter 8 on the marine environment. (Sewage sludge disposed of to landfill sites or incinerated is dealt with as controlled waste, regulated by CoPA.)

Mining and quarrying wastes

Conditions attached to Planning Permissions: the landfilling of mine and quarry waste (overburden, waste rock etc), whether generated at the site or brought in from elsewhere, is controlled by planning conditions with the mineral planning authority as the competent authority. In cases where the wastes are likely to result in discharges coming under the terms of the EC directive on the protection of groundwater (80/68/EEC) the National Rivers Authority is a statutory consultee.

Regulation of certain operations

Environmental Protection Act 1990: Part I of the EPA controls a variety of operations involving wastes, including most incineration, some recovery processes, the burning of refuse derived fuels (RDF), many waste combustion processes and glass processes using waste glass (cullet).

Special waste

11.12 Many everyday commodities are the result of complex industrial processes producing wastes which can contain substances which are dangerous to life. These wastes are classed as special wastes in the UK. They are controlled under CoPA but are subject to additional regulations designed to track their movement from production to disposal (see Box 11.1). Annual figures for arisings are obtained from the consignment notes which must be used by producers, carriers and disposers to notify Waste Regulation Authorities (WRAs) in England and Wales, of the movement and disposal of special waste. The figures are collated by DOE from returns made by WRAs. In Scotland and Northern Ireland, the Scottish Office Environment Department (SOEnD) and DOE(NI), respectively, collect special waste statistics.

Table 11.1 Special waste arisings: 1986-87 to 1990-91 UK				
				Thousand tonnes
Year	England	Wales	Scotland[1]	N.Ireland
1986-87	1,500	94	30	5
1987-88	2,070	..	59	12
1988-89	1,762	60	66	17
1989-90	2,146	80	71	18
1990-91	2,352	76	94[2]	20[2]

Notes:
1. Figures on calendar year basis.
2. Provisional figure

Source: DOE, SOEnD, DOE (NI)

11.13 Estimated arisings in the UK in 1990-91 were 2.5 million tonnes. Table 11.1 gives estimates of special waste arisings by UK country for the last five years. The figures should however be treated with caution in view of incomplete returns by local authorities in England and Wales in some years. Special wastes dealt with in-house do not have to be reported to WRAs and are not included. Special wastes arisings have increased in recent years but there have been large year to year fluctuations, particularly in England. These may be partly due to contaminated soil arisings and one-off disposal of dangerous materials.

Waste not subject to waste disposal licensing under CoPA

11.14 The following paragraphs cover wastes not regulated by CoPA, amounting to around 260 million tonnes annually, about two thirds of total estimated arisings. Box 11.2 gives information about legislation and guidance for the main types of wastes.

Waste from agricultural premises

11.15 Previous published figures have estimated waste from agricultural premises at around 250 million tonnes annually, based mainly on estimates of excreta from all livestock (ie both housed and grazing animals). It is now recognised that a more realistic assessment includes only housed animal excreta which requires handling. This category of agricultural waste is estimated at 80 million tonnes annually (20 per cent of total arisings). In addition there is a relatively small amount of other agricultural waste such as straw and plastics but estimates are not generally available for these categories. The potentially most significant is straw, but most is incorporated into the soil or used as animal bedding.

11.16 Some agricultural waste can, if properly applied, be a beneficial residue and recycled back to the land as a fertiliser or soil conditioner. Some wastes, however, are potentially harmful; animal slurries are up to one hundred times and silage effluent up to two hundred times more polluting than raw sewage. Improper handling of these wastes may result in water pollution. Most such incidents reported arise from problems of storage rather than from use on the land. The National Rivers Authority reports on water pollution incidents annually (see also Chapter 7 on inland water quality).

11.17 In 1989 the Farm and Conservation Grant Scheme was introduced. This provides grants for providing, replacing or improving facilities for the storage, treatment and disposal of agricultural wastes and silage effluent.

Mining and quarrying waste

11.18 Mining and quarrying waste is estimated to be around 107 million tonnes annually (27 per cent of estimated total waste). Mining wastes are the by-products of the extraction process and may include soil, rock and dirt some of which may be hazardous owing to contamination and require special treatment and disposal. The major mining wastes in GB include colliery spoil and china clay and slate waste from quarrying.

Dredged materials

11.19 Dredged materials were estimated at 43 million tonnes (wet weight) in 1991 (11 per cent of waste arisings). The equivalent dry weight is 23 million tonnes. Dredged materials are disposed of at sea and are discussed in Chapter 8 on the marine environment.

Sewage sludge

11.20 Another major category of waste not controlled by CoPA is the proportion of sewage sludge which is disposed of by methods other than landfill or incineration. See also other sections on sewage sludge in this chapter.

Waste disposal

11.21 The UK is almost entirely self sufficient in the disposal of waste from all sources but there is some international movement of metallic containing residues and other industrial by-products. Controlled waste may only be treated or disposed of under a licence (unless the treatment or disposal is exempt) from the WDA. From April 1993, licensing will be under the EPA and will encompass any storage, treatment or disposal of controlled waste (unless exempt). The number of waste disposal licensed facilities by type is shown in Table 11.2.

Landfill

11.22 About 85 per cent of controlled waste currently goes to landfill sites (see Figure 11.2). Regulatory authorities impose appropriate licence conditions so that landfill sites have caused little serious pollution. However, leachates arising from waste in these sites may contaminate surface or ground waters (see also Chapter 7 on inland water quality), and the breakdown of biodegradable wastes may result in the production of methane, a greenhouse gas. About a quarter of total UK emissions of methane is estimated to come from this source (see also Chapter 3 on the global atmosphere). Methane is also of concern because of the potential dangers of explosion if accumulations build up in confined spaces. Around 70 per cent of special waste goes to landfill sites. Figure 11.3 gives a rough breakdown of UK disposal routes for special waste which is probably

Table 11.2 Number of waste disposal facilities, by type UK	
Type of disposal facility	Number of disposal licences
Landfill	4,196
Civic amenity	559
Transfer [1]	936
Storage [2]	274
Treatment [3]	122
Incineration	212
Other	366

Notes:
1. Facilities licensed for receipt, sorting, consolidation and onward movement of waste.
2. Facilities licensed principally for storing waste remotely from final disposal.
3. Including physical, chemical and biological treatment and solidification.

Source: DOE, SOEnD, DOE (NI)

typical for any 12 month period in recent years. More accurate figures are not available.

11.23 Landfilling costs are expected to increase following the introduction of further controls under the EPA. The terms of a waste management licence will require control of emissions of liquids and gases from landfill sites, both during the filling process, and for a number of years following closure.

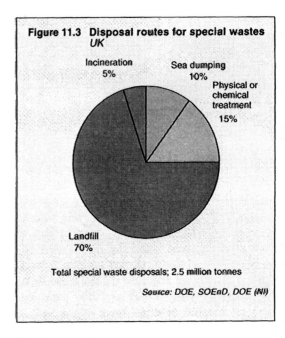

Figure 11.3 Disposal routes for special wastes UK

Incineration 5%
Sea dumping 10%
Physical or chemical treatment 15%
Landfill 70%

Total special waste disposals; 2.5 million tonnes

Source: DOE, SOEnD, DOE (NI)

Incineration

11.24 About 4 per cent a year of controlled waste is incinerated, including almost all clinical waste and 3 million tonnes a year of municipal waste. Incineration sterilises and reduces the volume of material for final disposal by up to 90 per cent, and its weight by about two thirds. There are also opportunities to recover energy in the form of heat or electricity by the incorporation of suitable boiler plant. Incineration however is a capital intensive disposal option and in most cases more costly than landfilling.

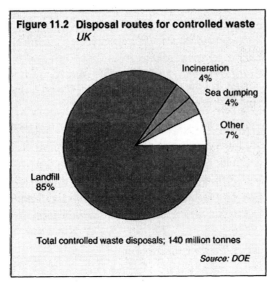

Figure 11.2 Disposal routes for controlled waste UK

Incineration 4%
Sea dumping 4%
Other 7%
Landfill 85%

Total controlled waste disposals; 140 million tonnes

Source: DOE

11.25 New plant standards under Part I of the EPA include controls over smoke, acid gases, heavy metals and residual trace organic species (see Box 11.2). The majority of existing plant is old and will require modification to improve combustion conditions and post combustion cleaning of waste gases to satisfy future requirements. This applies to all 34 current mass burn municipal waste incinerators and to most of the current 900 or so hospital waste incinerators. A significant rationalisation of incinerator stock is anticipated by the mid 1990s.

11.26 The rationalisation is expected to lead to the development of a smaller number of larger facilities to deal with wastes for which incineration is the preferred or only option. For other wastes, the development of replacement incineration capacity will depend on the cost of alternative disposal routes, but the economics favouring incineration may be improved by incorporating energy recovery. The five largest municipal waste incinerators have energy recovery facilities.

Other disposal methods

11.27 Some liquid industrial waste and power station ash is disposed of at sea, but the UK and the other North Sea countries are in the process of phasing out most forms of sea disposal (see Chapter 8 on the marine environment). About 4 million tonnes of power station ash is used in construction, about 5 million tonnes of blast furnace slag is used in construction and road building and about 1 million tonnes of municipal waste is recovered for reuse or recycling (see the later section in this chapter on recycling).

Disposal of sewage sludge

11.28 Figure 11.4 shows the disposal routes for sewage sludge in 1990. Almost half was spread on farm land and horticultural areas as a soil conditioner, and about a quarter was disposed of at sea. Around a fifth went to landfill sites or was incinerated. Disposal methods need to take account of contamination with heavy metals, organic chemicals, oils, etc, depending on the source of the effluent and the effectiveness of treatment. Limits on the maximum permitted concentration of seven potentially toxic elements in soils to which sludge is applied

---

EC Directives concerning waste disposal
Box 11.3

Directive 91/156/EEC amending Framework Directive 75/442/EEC on the reduction, recycling and safe disposal of waste, has to be implemented by 1 April 1993. The Directive stipulates that:

- there must be competent authorities responsible for implementing the Directive;
- they must produce waste disposal plans;
- disposal and recovery facilities must have a permit (unless they fall within permitted exemptions);
- the polluter pays principle should apply; and
- member states are to encourage recycling.

The Toxic and Dangerous Wastes Directive (78/319/EEC) is to be replaced by a new Hazardous Wastes Directive (91/689/EEC) which has to be implemented by 12 December 1993.

The Transfrontier Shipment of Hazardous Wastes Directive (84/631/EEC), as amended, introduced a control system for movements of hazardous waste between member states and to and from third states. The Directive is to be replaced by a new EC Regulation implementing the international convention signed in Basel in 1989 which provides a framework for world wide control of transfrontier shipment of hazardous wastes.

Directives 89/369/EEC and 89/429/EEC introduced requirements for controls over municipal waste incinerators (implemented through Part I of the EPA). The EC has published a draft directive on controls over incinerators for hazardous waste including clinical waste, sewage sludge and a broad range of industrial and chemical wastes. It proposes minimum performance requirements for incinerators to be adopted throughout member states.

Directive 84/360/EEC on the combating of air pollution from industrial plants requires prior authorisation and the use of the Best Available Technique Not Entailing Excessive Cost (BATNEEC) for specified incinerators. A draft Integrated Pollution Prevention and Control Framework directive also applies to certain incinerators.

New directives on landfill and packaging are currently under discussion.

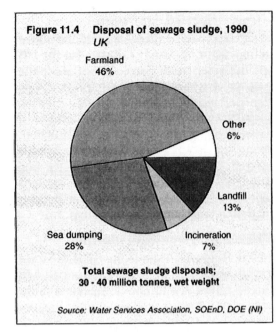

Figure 11.4   Disposal of sewage sludge, 1990 UK

Farmland 46%
Other 6%
Landfill 13%
Incineration 7%
Sea dumping 28%

Total sewage sludge disposals;
30 - 40 million tonnes, wet weight

Source: Water Services Association, SOEnD, DOE (NI)

are set out in an EC Directive. The UK will end sea disposal of sewage sludge by the end of 1998, after which land based methods will have to be used. Some form of disposal will be needed for an estimated additional 300,000 tonnes (dry weight) of sludge by 1998. (See also references to sewage sludge disposal in Chapter 8 on the marine environment).

Energy from waste

11.29 It has been estimated that the potential energy value in household, industrial and agricultural waste could, if realised, meet about ten per cent of the UK's electricity needs. Currently, only part of this potential is used, largely from energy recovery from municipal waste incineration and landfill gas recovery[2]. Electricity and heat provided annually by the combustion of refuse and refuse derived fuel is estimated to equal approximately the amount that would be generated by 150,000 tonnes of oil. Landfill gas contributes the equivalent of energy generated by around 100,000 tonnes of oil a year[3].

Problem materials

11.30 Some local authorities and industries organise collection of wastes which may pose serious environmental problems eg small batteries, old tyres, used oil from motor vehicles and chlorofluorocarbons (CFCs) in refrigerators. In some cases the materials collected may be recycled, but the costs of doing so may be too high to make this commercially viable.

European Community legislation on waste

11.31 The European Community (EC) is also involved in setting standards for the disposal and movement of waste (see Box 11.3).

Imported hazardous wastes

11.32 In addition to dealing with waste produced in this country, the UK also treats and disposes of hazardous wastes imported from other countries. The Transfrontier Shipment of Hazardous Waste Regulations 1988 implement an EC Directive (see Box 11.3). Under the regulations, shipments of hazardous waste must be notified in advance to the competent authorities, and must be accompanied by a consignment note giving details of the holder, producer, destination and nature of the waste. This applies to shipments to or from non EC countries as well as those between member countries. Full details of imports to the UK have been available since October 1988 from WRAs who are notified of individual consignments to the UK.

Table 11.3 Imports of hazardous waste, by type; 1990-91 England & Wales

Type of waste	Tonnes
Inorganic or organic acids	10
Alkalis	2,939
Toxic metal compounds	185
Inorganic compounds	758
Organic materials (exc PCBs)	8,745
PCBs/PCB contaminated waste	3,928
Polymeric materials and precursors	490
Fuels, oils and greases	885
Fine chemicals & biocides	1,361
Miscellaneous chemical waste	406
Filter materials, sludge & rubbish	6,027
Interceptor wastes	1,318
Miscellaneous wastes	16,861
Animal & food wastes	43
Total	**43,953**

Note: In 1990, 954 tonnes of hazardous waste were imported into Scotland.

Source: DOE, SOEnD

11.33 Total imports of hazardous wastes to England and Wales in 1989-90, the first complete year for which records are available, were just under 35,000 tonnes. Imports in 1990-91 were about 44,000 tonnes. Table 11.3 shows the weight of imports by type in 1990-91. Imports to Scotland were 954

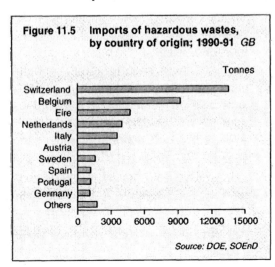

Figure 11.5   Imports of hazardous wastes, by country of origin; 1990-91 GB

Tonnes

Switzerland
Belgium
Eire
Netherlands
Italy
Austria
Sweden
Spain
Portugal
Germany
Others

0   3000   6000   9000   12000   15000

Source: DOE, SOEnD

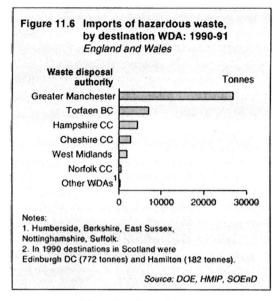

Figure 11.6 Imports of hazardous waste, by destination WDA: 1990-91
England and Wales

Notes:
1. Humberside, Berkshire, East Sussex, Nottinghamshire, Suffolk.
2. In 1990 destinations in Scotland were Edinburgh DC (772 tonnes) and Hamilton (182 tonnes).

Source: DOE, HMIP, SOEnD

within the UK. All other waste is disposed of in the UK.

Recycling

11.36 Re-using useful waste materials helps to conserve natural resources, and helps also to reduce production costs and the amount of energy needed for production and transport. Extraction of re-usable materials cuts down on the volume of waste which has to be transported to landfill sites or incinerated. Recycling (which unlike re-use, involves collection and separation of materials from waste, and subsequent processing to produce marketable products) also helps to conserve natural resources. As the costs of traditional disposal methods increase, recycling schemes will become increasingly attractive economically.

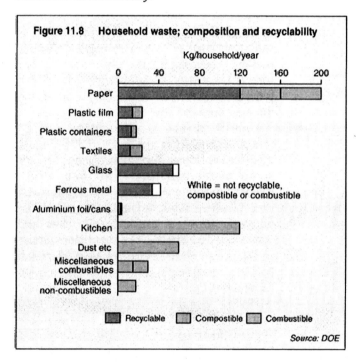

Figure 11.8 Household waste; composition and recyclability

White = not recyclable, compostible or combustible

Source: DOE

tonnes in 1990. Combined GB imports came from 28 countries, 10 of which each accounted for 1,000 tonnes or more, as shown in Figure 11.5.

11.34 Figures 11.6 and 11.7 show for 1990-91 imports of hazardous waste to England and Wales by destination WDA and the main disposal methods used. About 65 per cent of imported wastes were physically or chemically treated (including solidification) at specialist facilities before being disposed of. Almost 30 per cent was incinerated. 135 tonnes were sent direct to landfill (included in the other routes shown in Figure 11.7). Imports to Scotland in 1990 amounted to 954 tonnes of which 772 tonnes went to Edinburgh DC and 182 tonnes to Hamilton DC.

11.35 Hazardous waste amounting to 525 tonnes was exported from Northern Ireland to Finland for high temperature incineration in 1990-91. These arisings have since been reduced and re-directed to disposal facilities

11.37 Currently in the UK there are high recovery levels for some materials such as scrap metal, and recycling methods are becoming well established for others. Some sectors of industry recover significant amounts of scrap from in-house processes for their own use, or sell it to other industries as raw material or for processing before re-use. See later sections in this chapter.

11.38 As far as household waste is concerned, although at least 50 per cent, by weight, is estimated to be recyclable in principle (if composting of organic material is included), only about 5 per cent is recovered. The Government has set a target of recycling 50 per cent of the recyclable element of household waste by the year 2000, (usually taken to mean 25 per cent of all household waste). Figure 11.8 illustrates what typically UK household waste is composed of and

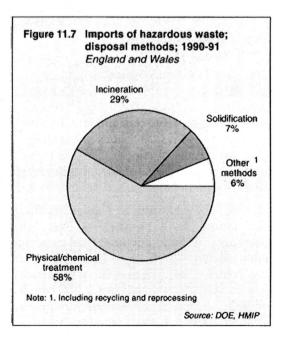

Figure 11.7 Imports of hazardous waste; disposal methods; 1990-91
England and Wales

Incineration 29%

Solidification 7%

Other [1] methods 6%

Physical/chemical treatment 58%

Note: 1. Including recycling and reprocessing

Source: DOE, HMIP

161

what proportions are recyclable or potentially degradable.

11.39 Arrangements by waste collection and waste disposal authorities to separate collection of valuable materials such as glass, waste paper and metals, from the mainstream household collection are likely to spread as a result of the implementation of recycling plans, which local authorities are required to prepare under the EPA. DOE has issued guidance to authorities on recycling[4], and it is also currently monitoring kerbside collection schemes in Sheffield, Adur, Leeds and Cardiff as part of a three year programme. A report on the first two years of the Sheffield experiment was published in October 1991[5]. The final report is due to be published in the autumn of 1992.

Ferrous metals

11.40 The reclamation of ferrous scrap is well established and a greater weight of ferrous metal is recycled in the UK than all other materials combined. Over 8 million tonnes of ferrous scrap is used in the UK representing 46 per cent of the total consumption of ferrous metal used in manufacture. Over 3 million tonnes was exported in 1990. Figure 11.9 shows home and export sales of ferrous scrap metal between 1980 and 1990. Included in the total amount of ferrous metal recovered are small amounts extracted from domestic waste at civic amenity sites and waste transfer stations, or other places where waste is handled before its final disposal. Much larger amounts arise from the scrapping of 2 million obsolete or crashed vehicles, and 6 million units of "white goods" (washing machines, cookers, freezers, refrigerators). Over 1,000 million steel cans are recycled each year and are collected mainly by

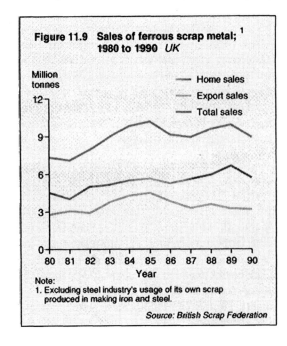

Figure 11.9 Sales of ferrous scrap metal; [1]
1980 to 1990 UK

Million tonnes

Home sales
Export sales
Total sales

Note:
1. Excluding steel industry's usage of its own scrap produced in making iron and steel.

Source: British Scrap Federation

magnetic extraction from household waste. In areas where magnetic extraction is not available consumer collection is important. The "Save-a-Can" scheme run by British Steel Tinplate operates over 400 can banks around the country, collecting both steel and aluminium cans.

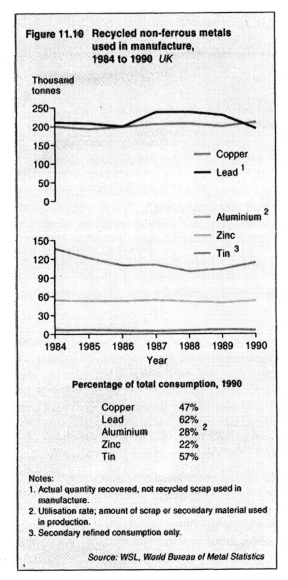

Figure 11.10 Recycled non-ferrous metals used in manufacture, 1984 to 1990 UK

Thousand tonnes

—— Copper
—— Lead [1]

—— Aluminium [2]
—— Zinc
—— Tin [3]

Percentage of total consumption, 1990

Copper	47%
Lead	62% [2]
Aluminium	28% [2]
Zinc	22%
Tin	57%

Notes:
1. Actual quantity recovered, not recycled scrap used in manufacture.
2. Utilisation rate; amount of scrap or secondary material used in production.
3. Secondary refined consumption only.

Source: WSL, World Bureau of Metal Statistics

Non ferrous metals

11.41 Figure 11.10 shows the trends in the recycling of aluminium, copper, lead, tin and zinc in the UK from 1984 to 1990, and the percentage the amount recycled represented in terms of total 1990 consumption. In 1990, approximately 95,000 tonnes of aluminium, 111,000 tonnes of copper, 20,000 tonnes of lead and 25,000 tonnes of zinc scrap were exported.

11.42 The Aluminium Can Recycling Association has a network of over 350 merchants willing to purchase used cans and there are over 1100 sites where the public may return cans. The weight of aluminium cans recycled has grown from 1,200 tonnes in 1989 to around 6,000 tonnes (336 million cans) in 1991. A new recycling plant at

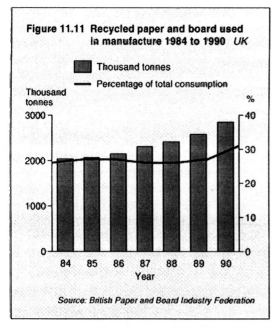

Figure 11.11 Recycled paper and board used in manufacture 1984 to 1990 *UK*

Source: British Paper and Board Industry Federation

Warrington with a capacity to deal with 50,000 tonnes of cans was commissioned in 1991.

Paper and board

11.43 Paper and board recovery is increasing and in 1990 2.8 million tonnes was re-used (31 per cent of total consumption in the UK), see Figure 11.11. The scope for recycling is determined by market conditions, the capacity of mills to deal with reclaimed stocks, and the level of cheap imports. The Government has agreed with publishers and newsprint producers a target for increasing the proportion of recycled paper used in newspapers and magazines from 27 per cent in 1990 to 40 per cent by the year 2000.

Glass

11.44 The weight of glass recycled has more than doubled since the early 1980s. In 1990

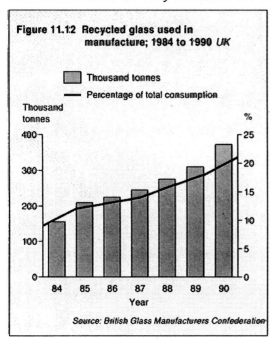

Figure 11.12 Recycled glass used in manufacture; 1984 to 1990 *UK*

Source: British Glass Manufacturers Confederation

372,000 tonnes was recycled representing 21 per cent of UK consumption, see Figure 11.12. The number of local authorities participating in the British Glass Manufacturers Confederation bottle bank scheme increased from 158 authorities in 1981 to 448 in 1991. Figure 11.13 shows the rate of growth in the number of local authority collection sites from 6 in 1977 to over 5,800 in 1990. 178,000 tonnes of glass was collected in 1990, see Table 11.4.

Table 11.4 Bottle bank sites and glass collected[1]; 1990 *UK*

| | Number of sites | | Tonnage | Kilograms |
	Public	Commercial	collected	/head
England [2]	5,003	1,285	154,250	3.2
Wales	196	11	1,080	0.4
Scotland	630	2,268	18,000	3.5
N. Ireland [3]	202	..	5,100	3.2
Total	**6,031**	**3,564**	**178,430**	**3.1**

Notes:
1. Figures for tonnage collected and estimates of amount collected per head based on reporting councils.
2. Excludes Isle of Man, Guernsey and Isles of Scilly.
3. Separate figures for public and commercial sites are not available.

Source: British Glass Manufacturers Confederation, DOE (NI)

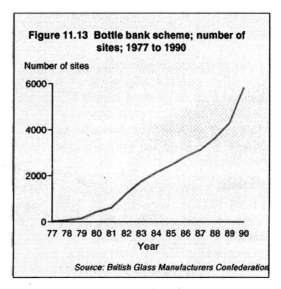

Figure 11.13 Bottle bank scheme; number of sites; 1977 to 1990

Source: British Glass Manufacturers Confederation

Mineral and construction

11.45 In the late 1980s approximately 10 per cent of total aggregates used annually were derived from secondary materials. Larger proportions of mineral and construction wastes are potentially usable as secondary aggregates in construction projects including roadworks, but the relative prices of primary and secondary aggregates, and the cost of transporting wastes to where they could be used, discourage increased use.

Mining and quarrying wastes

11.46 Waste resulting from china clay production in Devon and Cornwall is a significant local problem with existing large and often unsightly stockpiles. Around 27 million tonnes of waste is produced *every* year and around 1.5 million tonnes annually is used in local industries. Less than half of the annual china clay waste arisings has the potential to be used as significant secondary aggregates (ie china clay sand amounting to 12 million tonnes annually). Transportation costs to the areas of high demand for aggregates, such as the south east, limit further use of the waste. This is also true of slate waste of which around 6 million tonnes is produced annually, mainly in north Wales but also in Scotland, the Lake District and Cornwall[1].

Construction and demolition wastes

11.47 Demolition and construction wastes re-used by the industry largely comprise crushed concrete and masonry and amount to around 11 million tonnes annually (roughly a third of total arisings). Of this, approximately 1 million tonnes is recycled to produce graded aggregate. The main uses are for levelling sites and as fill, and also in road construction. Utilisation is often greatest at or close to the site of demolition in areas where the costs of haulage and landfill are high. Total arisings include asphalt road planings of between 7 and 8 million tonnes annually. A large proportion (perhaps 80 per cent) goes into some form of secondary use, such as footpaths, but virtually none is currently recycled for use on road surfaces. The new Department of Transport Specification for Highways will permit the use of up to 10 per cent of asphalt road planings in road construction[6].

Plastic

11.48 Plastic recycling is still being developed. Although recycling of production waste is well established, collection and other technical difficulties prevent greater recycling of plastic waste discarded by consumers. More than 70 firms are currently engaged in recycling in the UK, and 200,000 tonnes a year is recycled of which around 7,000 tonnes a year is post consumer scrap. A high proportion of the latter is polythene film; in 1991 more than 3,750 tonnes of polythene was collected from farms for recycling under a free collection scheme established jointly by the National Farmers Union and the plastics industry.

Litter

11.49 The EPA introduced measures to deal with the problem of litter. The Act places a duty on local authorities, schools and colleges, statutory transport undertakers, such as British Rail, and government departments, to keep their public land clear of litter and refuse.

The local authority can extend the duty to owners of other areas, such as shopping centres and car parks, by designating them as "litter control areas". Individual citizens can take any body under the duty to court if it fails to keep its land litter-free. The Act also allows local authorities to serve notices on certain types of shop owners who allow their frontages to become persistently littered.

11.50 DOE will be able to assess the effectiveness of the litter control measures. Regular surveys will examine the levels of street cleanliness throughout the country, based on the standards set in a Code of Practice. The survey completed in March 1992 reported an overall improvement of more than 11 per cent compared with one year earlier. The survey results will be published in autumn 1992.

11.51 The Tidy Britain Group (TBG) is a registered charity which carries out a range of anti-litter activities, focusing principally on education and campaigning. The Group also undertakes research on litter, provides advice to local authorities and others on the implementation of the EPA, and collects information on the implementation of the legislation.

References and further reading

1. Arup Economics and Planning, (1991). Occurrence and Utilisation of Mineral and Construction Wastes. HMSO.

2. Department of the Environment, (1992). Waste Management Paper No 1 (second edition) - A Review of Options. HMSO.

3. Department of Energy, (1992). UK Digest of Energy Statistics, Annex B: Renewable Sources of Energy. HMSO.

4. Department of the Environment, (1991). Waste Management Paper No 28 - Recycling. HMSO.

5. Civic Amenity Waste Disposal Project, (1991). Sheffield Recycling City - The First Two Years. Putteridge Bury, Hitchen Road, Luton, Bedford.

6. Department of Transport, (1992). Manual of Contract Documents for Highway Works, Vol I, Specification for highways. HMSO.

Department of the Environment, (1991). A Guide to the Environmental Protection Act 1990.

United Nations Environment Programme, (1991). Environmental Data Report, 3rd edition. Basil Blackwell, Oxford.

Department of the Environment, (1991). Waste Management - The Duty of Care, A Code of Practice. HMSO.

Her Majesty's Inspectorate of Pollution, (1991). Fourth Annual Report 1990-91. HMSO.

Warren Spring Laboratory, (1991). Market Barriers to Materials Reclamation and Recycling. WSL.

Department of the Environment. Environmental Protection Act 1990, Part I; Processes Prescribed for Air Pollution Control by Local Authorities; Secretary of State's Guidance series relating to the waste sector. HMSO.

Her Majesty's Inspectorate of Pollution, (1991). Guidance Note for Inspectors, Waste Disposal Industry Sector (IPR5). HMSO.

12 Noise

☐ **Noise is a pollutant which can irritate and annoy, and is one of the environmental pollutants identified as a health hazard by the World Health Organisation (12.1).**

☐ **In a study of noise outside dwellings in England and Wales, about half the dwellings in the sample were exposed to a daytime level exceeding that suggested by WHO as desirable to prevent significant community annoyance (12.9).**

☐ **Local authority Environmental Health Officers received almost three thousand complaints per million people in England and Wales in 1989-90, and over nine hundred per million in Scotland in 1990 (12.15). Both these rates have more than doubled since 1980, although this does not necessarily reflect a corresponding increase in the annoyance caused by noise (Figure 12.3).**

☐ **In 1989-90, over 60 per cent of complaints received by Environmental Health Officers in England and Wales related to domestic premises and around 25 per cent to industrial/commercial premises. In Northern Ireland, the pattern was similar (around 60 per cent and 15 per cent respectively) (12.17). In Scotland in 1990, around 50 per cent of complaints related to industrial/commercial premises and around 20 per cent to domestic premises (Figure 12.4).**

☐ **A survey of complaints about domestic noise in England and Wales between mid 1984 and mid 1986 showed that noise from amplified music systems and barking dogs accounted for two thirds of domestic noise complaints (Figure 12.8).**

☐ **Although aircraft movements have increased in recent years at Heathrow and Gatwick airports, both the number of people and the area affected by aircraft noise have declined, largely as a result of improvements in jet engine design and the banning of aircraft without a noise certificate (Figure 12.6).**

☐ **Since 1974, 58 Noise Abatement Zones have been designated in England and Wales (12.41).**

12.1 Noise is a pollutant which can irritate and annoy, and is one of the environmental pollutants identified as a health hazard by the World Health Organisation (WHO)[1].

12.2 The pollutant effects of noise are difficult to quantify because people's tolerance to noise levels and different types of noise varies considerably. Distinct variations in noise intensity and noise levels can occur from place to place (even within the same general area), and from one moment to the next. Similarly there may be large variations during each day, week, or year.

12.3 Annoyance is perhaps the most common adverse effect of noise on the general public and thousands of complaints are made every year about many different types of noise. There is evidence of a clear relationship between degrees of individual annoyance and noise levels; for example it has been demonstrated that less use is made of private gardens and public parks when there is too much noise[2].

12.4 Prolonged exposure to high noise levels may have an adverse effect on health. It may cause some permanent loss of hearing, particularly for people who have an

---

How noise levels are measured Box 12.1

Noise, often defined as unwanted sound, is caused by small pressure fluctuations in the air. The human ear is a very sensitive detector, and the range of sound pressures that can be heard is extremely large. To reduce the range of numbers involved, sound levels are measured in decibels (dB), a logarithmic scale. With this scale, 0 dB corresponds to the quietest audible sound (the threshold of hearing), whilst 140 dB represents a sound at which pain would probably occur. An increase in sound level of 3 dB anywhere on the scale corresponds to a doubling of the sound *intensity*, whereas an increase of 10 dB is perceived as an approximate doubling of the *loudness* of the sound.

The usual form of measurement is dB(A) where the 'A' refers to the frequency weighting used in sound meters enabling them to mimic human hearing. Normally, noise levels vary from around 30 dB(A) (quiet), to around 100 dB(A) (loud). It is widely accepted that continuous exposure to levels above 85 dB(A) can result in some permanent loss of hearing. The risk of disability increases as the noise level and/or the duration of exposure increases.

Because noise levels fluctuate, there are different measures of sound levels (L):

- L_{Aeq} measures the average energy of the noise level;
- L_{A90} measures the background noise persisting when intermittent sounds are not heard, that is, the level exceeded 90 per cent of the time in the period measured. This is often used in the assessment of industrial noise;
- L_{A10} measures the louder end of the noise range in a given period, that is, the level exceeded only 10 per cent of the time.

Depending on the circumstances either L_{Aeq} or L_{A10} can have the greater value, but both are always greater than L_{A90}, provided they relate to the same period of time.

For the purposes of noise insulation regulations for road traffic the measure $L_{A10,18h}$ is used, which corresponds to the arithmetic average of the 18 hourly values of L_{A10} determined over the period 6 am until midnight on a normal working day. See also Box 12.4.

Prior to 1990 the Noise and Number Index (NNI), was used for measuring daytime exposure to aircraft noise. In September 1990, a change of index to Leq (16 hour) dB(A) was announced. This index takes account of the duration of noise from each aircraft movement. The sixteen hour day used for the index is 0700-2300 BST. To enable a comparison to be made between the indices the Department of Transport has published both NNI and Leq contours for Heathrow, Gatwick and Stansted airports for 1988 and 1989.

---

occupational contact with noise, such as employees on factory production lines, workers involved in general construction, and those working on road repairs. In addition to noise induced hearing loss, exposure to high noise levels may result in ear damage which may be manifested by tinnitus (ringing in the ears). Noise may also contribute to or aggravate stress related health problems, including raised blood pressure, and to minor psychiatric disease in the community. Exposure to noise during sleep may alter normal sleep patterns. Conclusive evidence of any general adverse effects of noise on health is however limited to noise induced hearing loss and tinnitus[3].

Measuring noise

12.5 There are two main ways of assessing noise; measuring sound pressure levels and recording the discomfort or annoyance that noise causes.

Assessment of noise levels

12.6 The level of noise is measured in decibels (dB) and Figure 12.1 shows illustrative noise levels and their decibel equivalents. A continuous and reliable monitoring system for measuring noise levels in the UK does not exist. Any national system

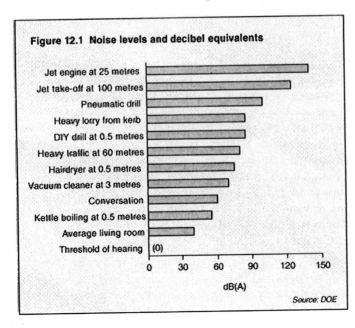

Figure 12.1 Noise levels and decibel equivalents

Jet engine at 25 metres
Jet take-off at 100 metres
Pneumatic drill
Heavy lorry from kerb
DIY drill at 0.5 metres
Heavy traffic at 60 metres
Hairdryer at 0.5 metres
Vacuum cleaner at 3 metres
Conversation
Kettle boiling at 0.5 metres
Average living room
Threshold of hearing (0)

0   30   60   90   120   150

dB(A)

Source: DOE

would be difficult to set up, and because of the nature of noise such a system would not give a representative profile of the range of levels and intensities of noise to which individuals are exposed.

12.7 Different measurements are used to assess noise from different sources. Consequently, most assessments of noise levels in the UK have been made in studies relating to a particular source or local area. In some circumstances, statutory obligations, for instance under the Control of Pollution Act 1974 (CoPA) and the Environmental Protection Act 1990 (EPA), require local authorities to measure the levels of different types of environmental noise. Information about the measurement of noise levels is given in Box 12.1, and the relevance of these measures to OECD and WHO international general guidelines on exposure to noise are given in Box 12.2.

Measured noise levels in the UK

12.8 A number of studies have been undertaken to measure noise levels from different sources and in different areas of the UK. The most up to date survey is the National Noise Incidence Study (NNIS) carried out by the Building Research Establishment (BRE) in 1990. Sound level measurements were taken outside 1,000 dwellings in England and Wales over a 24 hour period. Some preliminary results from the study were available at the end of 1991, and further results will be published by BRE in 1992.

12.9 Initial findings of the survey are that 7 per cent of dwellings in the sample were exposed to noise levels above 68 dB $L_{A10,18h}$. This is the level of noise at which dwellings would qualify for provision of sound insulation against noise from new roads. About half the dwellings in the sample were found to be

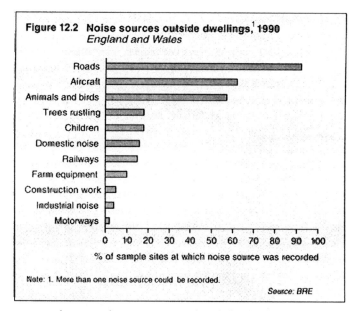

Figure 12.2 **Noise sources outside dwellings,**[1] **1990**
England and Wales

% of sample sites at which noise source was recorded

Note: 1. More than one noise source could be recorded.

Source: BRE

exposed to a daytime noise level ($L_{Aeq,16h}$) exceeding 55 dB(A). WHO suggest that general daytime outdoor noise levels of below 55 dB L_{Aeq} (equivalent to 57 Leq (16 hour) dB(A) for aircraft noise) are desirable to prevent any significant community annoyance. See Box 12.1 for an explanation of noise levels.

12.10 Figure 12.2 shows the percentage of measurement sites where the different sources of noise were recorded during the survey. Road traffic noise was noticeable outside over 90 per cent of the dwellings, although only about 5 per cent of the sample faced main ('A') roads. Aircraft noise was noticeable outside over 60 per cent of dwellings.

12.11 These findings are broadly supported by a survey which has been funded each year since 1986 by the Department of the Environment (DOE), and carried out by students at the Open University. The size of the sample used is small and therefore may not be representative of the UK as a whole,

Guidelines on exposure to noise   Box 12.2

A number of studies undertaken in OECD countries on the effects of noise and the behaviour that it induces have shown that, to comply with desirable limits for indoor comfort, the outdoor level of noise should not exceed 65 dB (day time L_{Aeq}). In the case of new residential areas, the outdoor levels should not exceed 55 dB (day time L_{Aeq}). These levels often serve as the reference for defining noise black spots where the levels of exposure to noise exceed 65 or 70 dB (daytime L_{Aeq}) and grey areas where the noise level is between 55 and 65 dB(A)[2].

The World Health Organisation suggests that general daytime outdoor noise levels of below 55 dB L_{Aeq} (equivalent to 57 Leq (16 hour) dB(A) for aircraft noise) are desirable to prevent any significant community annoyance[1].

The 1973 Planning and Noise circular[5] refers to noise from roads, aircraft and industry, and gives guidance on how to take exposure to noise into account when making planning decisions. The draft Planning Policy Guidance on Planning and Noise, describes new mechanisms and guidelines for local planning authorities to adopt when determining planning applications. When it is published in its final form it will replace the 1973 Circular.

but the results provide some illustration of the levels and types of noise to which the UK population is exposed. The main sources of noise identified were very consistent over the period, and similar to the NNIS results illustrated in Figure 12.2. Road traffic was mentioned as the source of some noise at over 90 per cent of sites, and it was found to be the main noise source at over 60 per cent of sites. Noise level measurements were also taken at each site, and a summary of the results has been published by DOE[4]. The results for the five years from 1986 to 1990 are very similar and show no significant change in measured levels of noise over the period.

Complaints about noise

12.12 An indication of annoyance from different types of noise can be given by looking at the number of complaints made about the various sources of noise. Such figures should however be treated with caution, and should not be regarded as a statistical measure of annoyance. People tend to be disturbed by, and react differently to, different types of noise. Some people may complain about an occasional very loud noise, whereas others may complain about a moderate but constant noise, and some people are more inclined to complain than others. Also, more than one complaint may stem from the same place or specific source of noise, and sometimes an organised campaign designed to draw attention to a particular source may result in a large number of complaints.

12.13 Although transport is a major source of noise, people are less inclined to complain about it.

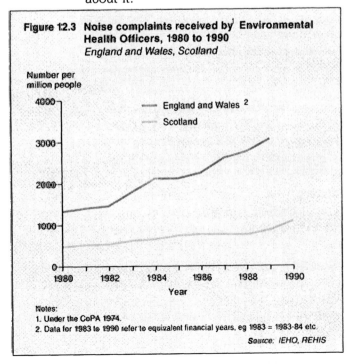

Figure 12.3 Noise complaints received by Environmental Health Officers, 1980 to 1990
England and Wales, Scotland

Number per million people

— England and Wales [2]
— Scotland

Notes:
1. Under the CoPA 1974.
2. Data for 1983 to 1990 refer to equivalent financial years, eg 1983 = 1983-84 etc.

Source: IEHO, REHIS

12.14 Complaints about noise are generally made to local authority Environmental Health Officers (EHOs), who have powers under EPA in England and Wales and CoPA in Scotland to limit or stop noise from premises, including land (see Box 12.3 for a description of the legislation). In Northern Ireland, the Pollution Control and Local Government (NI) Order 1978 applies. EHOs do not have powers over road, rail or aircraft noise and complaints about noise from these sources should be made to the relevant transport authority or airport manager. The number of complaints made to local authority EHOs about transport noise therefore understates the total number of complaints made about noise from transport sources.

12.15 Figure 12.3 shows the total number of complaints per million people received by EHOs between 1980 and 1990. In England and Wales in 1989-90 almost three thousand per million were received, and over nine hundred per million for 1990 in Scotland. Both these rates have more than doubled since 1980, although this does not necessarily reflect a corresponding increase in the annoyance caused by noise.

12.16 Complaints to EHOs per million people are much lower in Scotland than in England and Wales, perhaps because Scotland is generally much less densely populated than England and Wales and because there are different procedures for dealing with complaints in Scotland. The Civic Government (Scotland) Act 1982 enables a complainant to seek a court order if animals are causing noise and disturbance to local residents. Also, disturbance associated with the playing of musical instruments and sound amplification systems can be dealt with directly by the police under the 1982 Act, whereas in England and Wales only EHOs have such powers.

12.17 Figure 12.4 shows the proportion of the noise complaints received by EHOs about different sources of noise in England and Wales during 1989-90, and for Scotland in 1990. It includes complaints made to EHOs about transport and other noise, which should be made to the relevant authority. Figure 12.4 therefore shows only some of the total number of such complaints. There is a marked contrast in the respective pattern of complaints as a result of the factors described above. In Northern Ireland in 1989-90, the pattern of complaints received was similar to that in England and Wales, with about 60 per cent of complaints relating to domestic premises, and around 15 per cent to industrial and commercial premises.

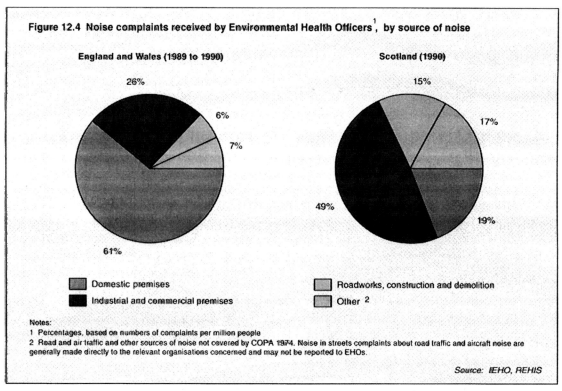

Figure 12.4 Noise complaints received by Environmental Health Officers[1], by source of noise

England and Wales (1989 to 1990)

26%
6%
7%
61%

■ Domestic premises
■ Industrial and commercial premises

Scotland (1990)

15%
17%
49%
19%

■ Roadworks, construction and demolition
□ Other [2]

Notes:
1 Percentages, based on numbers of complaints per million people
2 Road and air traffic and other sources of noise not covered by COPA 1974. Noise in streets complaints about road traffic and aircraft noise are generally made directly to the relevant organisations concerned and may not be reported to EHOs.

Source: IEHO, REHIS

12.18 Table 12.1 shows the action taken over noise complaints in England and Wales in 1989-90 and for Scotland in 1990. Taking these together, EHOs received a total of about 100,000 complaints about noise, relating to about 50,000 specific sources. During the same period EHOs confirmed over 25,000 cases as a nuisance (although these may not relate to noise complaints made in the same year). In most cases the nuisance is remedied informally, so formal action is necessary in only a small proportion of cases. EHOs issued almost 3,300 noise abatement notices in the period and made about 300 prosecutions for contravention of such notices.

Table 12.1	Noise complaints[1] and prosecutions, England and Wales, 1989-90 Scotland 1990	
	England & Wales	Scotland
Complaints received	97,798	4,629
Specific sources complained of	49,232	--
Nuisances confirmed	25,177	1,644
Nuisances remedied informally	24,998	2,134
Abatement notices issued	3,180	116
Prosecutions for contravention	284	3
Convictions	233	-

Notes: 1. Complaints received by Environmental Health Officers under CoPA 1974.

Source: IEHO, REHIS

Sources of noise

12.19 The following sections give more information about the sources of noise, covering transport, domestic, industrial/

commercial premises, and other sources for which data are available. Trend information based on numbers of complaints must be regarded with caution because of the difficulties with these data described in the previous section on noise complaints. The rapid growth in road and air traffic in recent years has further complicated the difficulty of looking at noise pollution data since information is neither comprehensive nor necessarily representative of the overall situation for individual countries in the UK.

Transport noise

12.20 Transport noise is perhaps the most familiar background noise and may result in complaints to transport authorities or EHOs. Some data about transport noise complaints submitted to local authority EHOs are available, although these are not representative of overall trends in transport complaints[4]. Sound level measurements of motor vehicles may also be taken by EHOs, but these measurements can only be used to help identify offending vehicles and support prosecutions.

Motor vehicle noise

12.21 Information is available about noise offences relating to motor vehicles. However, the introduction of the Vehicle Defect Rectification Scheme (VDRS) in 1986 (which requires motorists to remedy defects or submit their vehicle for scrap, rather than be reported for prosecution) has significantly affected trends in numbers of offences dealt with, which fell from 11,400 in 1986 to 7,900 in 1990. Between 1987 and 1990,

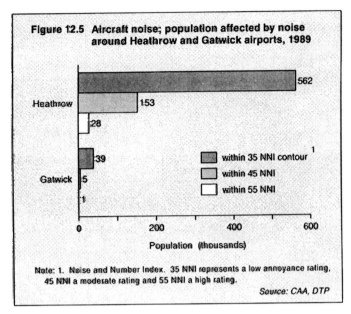

Figure 12.5 Aircraft noise; population affected by noise around Heathrow and Gatwick airports, 1989

Heathrow
562
153
28

Gatwick
39
5
1

within 35 NNI contour [1]
within 45 NNI
within 55 NNI

Population (thousands)
0   200   400   600

Note: 1. Noise and Number Index. 35 NNI represents a low annoyance rating, 45 NNI a moderate rating and 55 NNI a high rating.

Source: CAA, DTP

about a quarter of a million VDRS notices were issued each year in England and Wales. Notices specifically involving noise offences cannot be separately identified within this total.

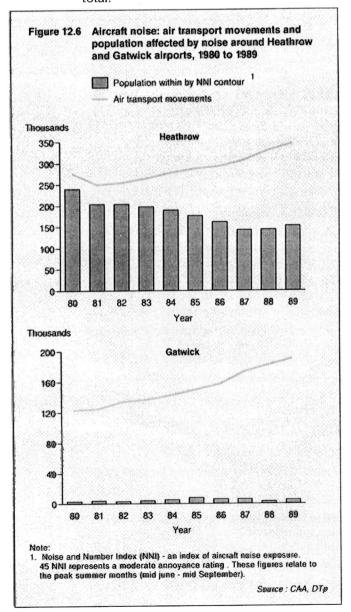

Figure 12.6 Aircraft noise: air transport movements and population affected by noise around Heathrow and Gatwick airports, 1980 to 1989

Population within by NNI contour [1]
Air transport movements

Thousands

Heathrow

350
300
250
200
150
100
50
0
80 81 82 83 84 85 86 87 88 89
Year

Thousands

Gatwick

200
160
120
80
40
0
80 81 82 83 84 85 86 87 88 89
Year

Note:
1. Noise and Number Index (NNI) - an index of aircraft noise exposure. 45 NNI represents a moderate annoyance rating . These figures relate to the peak summer months (mid june - mid September).

Source : CAA, DTp

12.22 National standards for noise limits for new vehicles are controlled by European Community (EC) legislation, described in Box 12.4.

Aircraft noise

12.23 Aircraft noise measurements are carried out annually around Heathrow, Gatwick and Stansted by the Civil Aviation Authority Department of Safety, Environment and Energy, and results are reflected in the aircraft noise exposure contours published by the Department of Transport.

12.24 Although a change in the aircraft noise index was announced in September 1990 (see Box 12.1), the data presented in this report are all in terms of the Noise and Number Index (NNI) for consistency. Using noise contours it is possible to estimate populations, and areas, exposed to differing levels of aircraft noise. Figure 12.5 shows populations within three NNI levels for the two major airports in the UK, Heathrow and Gatwick, in 1989, the latest year for which data are available. The levels shown are 35 NNI, representing a low annoyance rating; 45 NNI a moderate rating; and 55 NNI a high rating. These data indicate that around 600 thousand people were exposed to aircraft noise within the 35 NNI contour around these two airports in 1989, and that nearly 30 thousand people were exposed to levels of 55 NNI or more, mostly around Heathrow airport.

12.25 Even though aircraft movements have increased substantially in recent years, the area and number of people affected by aircraft noise at Heathrow and Gatwick have declined significantly. These reductions in noise exposure result largely from improvements in jet engine design. Since 1986 all subsonic jets on the UK register have required a noise certificate, and all foreign registered aircraft followed suit in 1988. Agreement has also now been reached to phase out the so-called "Chapter 2" types of subsonic airliners, the next noisiest class to those already banned. Other measures to minimise noise at UK airports are incorporated in air traffic control procedures and restrictions upon night movements (see Box 12.4).

12.26 Individual trends in air transport movements and population affected by 45 NNI or more for Heathrow and Gatwick airports are shown separately in Figure 12.6. At Heathrow, air transport movements increased by a quarter between 1980 and 1989. However, both the population and the area affected by aircraft noise were more than a quarter lower in 1989 than in 1980.

At Gatwick, air transport movements increased by over a half during the 1980s. The area and population affected by 45 NNI or more, was at its highest in 1985, but has fallen in the late 1980s despite a continuing increase in air transport movements.

Railway noise

12.27 Whereas road traffic usually generates a fairly uniform sound level, interspersed with frequent peaks generated by individual vehicles, railway noise is usually characterised by relatively short periods of noise followed by longer periods when sound exposure returns to the local ambient or background level. There are also considerable differences in the character of the noise generated by railways compared to that generated by road traffic. The differences affect the way in which noise needs to be measured and assessed, and the degree of disturbance caused. Noise from railways should be assessed using the equivalent continuous noise level scale L_{Aeq} (see Box 12.1).

12.28 The Department of Transport's publication "Railway Noise and the Insulation of Dwellings"[6] looks at evidence from studies that have considered the effects of road and railway noise on communities. It concludes that during the day railway noise is probably less annoying than road noise, and that while railway noise causes more disturbance to communication than road noise, it causes less sleep disturbance.

Noise from domestic, industrial/ commercial premises, and construction noise

12.29 Figure 12.7 shows trends over the last decade in complaints per million people received by EHOs on account of domestic, industrial/commercial and construction site noise. In England and Wales, most of the complaints were about domestic noise in every year during the decade; in Scotland, most complaints were about industrial/ commercial premises.

Domestic

12.30 The number of complaints about noise from domestic premises has doubled since 1980 in England and Wales, although it is not possible to say whether domestic noise levels themselves have actually increased. Table 12.2 shows the pattern of complaints reported to EHOs about domestic noise over the last decade, and the legal measures employed by EHOs to deal with such noise. In England and Wales, the number of complaints more than doubled, the number of specific sources complained of increased by slightly less - around 80 per cent, and the

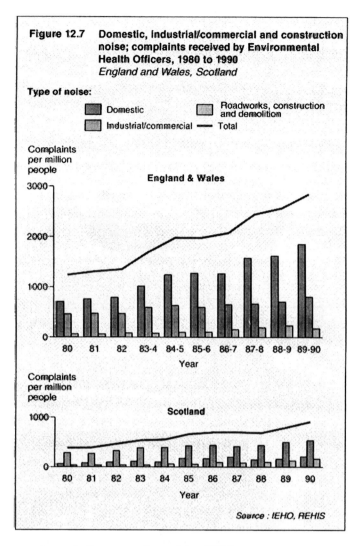

Figure 12.7 Domestic, industrial/commercial and construction noise; complaints received by Environmental Health Officers, 1980 to 1990
England and Wales, Scotland

number of these confirmed by EHOs as a nuisance increased by just over 50 per cent. Noise abatement notices were issued in

Table 12.2 Complaints about noise from domestic premises [1] and abatement notices issued, 1980 to 1990
England and Wales, Scotland

England & Wales	80	85-86	86-87	87-88	88-89	89-90
Complaints received	31,076	56,414	46,803	59,132	59,061	62,416
Number of sources complained of	18,047	35,636	29,223	33,516	35,748	33,106
Nuisances confirmed	10,011	17,780	13,626	14,485	16,788	15,354
Abatement notices issued	951	3,156	1,763	2,636	1,865	1,238
Percent of sources considered to be a nuisance	55.5	49.9	46.6	43.2	47.0	46.4
Percent of sources for which an abatement notice was issued	5.3	8.9	6.0	7.9	5.2	3.7
Scotland	**80**	**86**	**87**	**88**	**89**	**90**
Number of complaints	269	844	999	718	797	1,041
Number of sources complained of	..	..	..	..	..	..
Nuisances confirmed	..	245	180	208	164	251
Abatement notices issued	3	16	12	5	12	6

Notes: 1. Complaints received by Environmental Health Officers under COPA 1974

Source: IEHO, REHIS

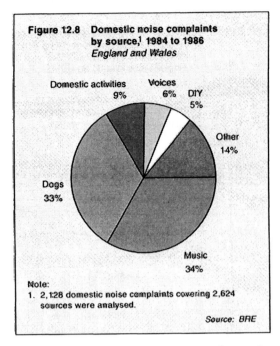

Figure 12.8 Domestic noise complaints by source,[1] 1984 to 1986 England and Wales

Domestic activities 9%
Voices 6%
DIY 5%
Other 14%
Dogs 33%
Music 34%

Note:
1. 2,128 domestic noise complaints covering 2,624 sources were analysed.

Source: BRE

12.32 BRE undertook a survey of complaints about domestic noise in England and Wales between mid 1984 and mid 1986[7]. Figure 12.8 shows the proportions of different sources of domestic noise complaints. Noise from amplified music systems and barking dogs accounted for two thirds of complaints.

Industrial/commercial premises

12.33 The trend in the number of complaints relating to industrial and commercial premises made to EHOs over the last decade is shown in Figure 12.7 for England and Wales and for Scotland. The number of complaints per million people has grown by 75 per cent in England and Wales and by 89 per cent in Scotland since 1980. In Scotland, noise from this source results in more complaints than any other source of noise. However, complaints from this source as a proportion of the total subject to CoPA (Section 58) have fallen in both England and Wales by one quarter, and in Scotland by one fifth, since 1980.

Roadworks, construction and demolition

12.34 Complaints per million for noise from roadworks, construction and demolition activities have increased almost threefold in England and Wales and by more than fourfold in Scotland. The proportion of total complaints made subject to CoPA (Section 58) has hardly changed in England and Wales

respect of just under 4 per cent of specific sources complained of in 1989-90; over the decade this proportion varied from under 4 per cent (in 1989-90) to almost 9 per cent (1985-86).

12.31 In Scotland, the number of complaints received about domestic noise nearly quadrupled from 269 in 1980 to a peak of over 1,000 during 1990, despite lower levels in 1988 and 1989 than in the middle of the decade.

Control of noise by local authorities: *Box 12.3*

The Control of Pollution Act 1974, the Environment Protection Act 1990, and the Town and Country Planning Act 1990

The Control of Pollution Act 1974 (CoPA) and the Environmental Protection Act 1990 (EPA) include most legislation for environmental noise control in GB. In Northern Ireland the Pollution Control and Local Government (NI) Order 1978 is almost identical to CoPA. These Acts give responsibility to local authorities to deal with complaints about noise from a variety of sources (mainly neighbourhood noise), and also give local authorities powers to protect people against the adverse effects of noise pollution. Aircraft noise is specifically excluded.

Under the 1974 and 1990 Acts, local authorities have powers to control noise nuisance, to establish noise abatement zones, and to limit construction site noise. EPA contains some tightening up of the noise control regime defined under CoPA. Penalties for breaches of abatement notices by commercial and non-commercial enterprises were increased under EPA, and the Act also clarifies the duties of local authorities in responding to noise complaints.

During 1990 the Noise Review Working Party, comprising representatives from local authorities, government departments and a number of other organisations with expertise on dealing with environmental noise, reviewed existing legislation and provided a report to the Secretary of State for the Environment (the Batho Report[9]). It considered aspects of noise control with particular reference to CoPA. The Working Party found that the increase in complaints about many types of noise during the 1980s suggested a need to revise sections of CoPA. One of the recommendations made by the Working Party was that the duty of local authorities to investigate complaints should be made clearer and this was subsequently done in EPA.

Powers under the Town and Country Planning Act 1990 allow planning authorities to control the location of potentially polluting development. This includes the effects of noise sources on new and existing noise sensitive premises.

since 1980, but in Scotland such complaints accounted for 18 per cent of the total made in this group in 1990, double the proportion in 1980.

Noise in streets and other noise

12.35 Complaints about noise in streets, and noise from sources other than those already covered, have also increased substantially in recent years (see Table 12.3). Other noise in

this case includes a wide variety of activities not controlled by CoPA, including that generated by roadside car maintenance, sporting activities and refuse collection.

Limiting exposure to noise

12.36 Noise is controlled by setting limits on the emission of noise at source (eg vehicle regulations, planning permission, noise abatement notices and entertainment

---

Other United Kingdom legislation   Box 12.4

The commonly used legislation covering environmental noise control in the UK is covered in Box 12.3. Outside these Acts the main UK legislation covering noise, particularly transport noise, is covered below.

Road traffic noise

Motor vehicle noise limits are regulated through the Motor Vehicle Type Approval, and Construction and Use Regulations. The current maximum limits are 84 dB(A) for the most powerful goods vehicles, 77dB(A) for cars, and 82 dB(A) for the largest motorcycles. The Road Vehicles Construction and Use Regulations also contain provisions to control excessive noise, applicable both to the condition of the vehicle silencer and to the driver of the vehicle.

There is at present no metered noise testing of road vehicles in the UK, either in the annual roadworthiness test ("MOT") or at the roadside, although the annual test does include a check on the condition of a vehicle's silencer.

The Noise Insulation Regulations of the Land Compensation Act 1973 impose a duty on the highway authority to offer a grant to provide noise insulation for householders if they experience increased noise levels from a new or substantially upgraded road, (eg from a single to a dual carriageway) and the resulting noise level reaches or exceeds 68 dB ($L_{A10,18h}$). The Act also provides for compensation where the value of properties fall after road opening as a result of increased noise from road traffic and other factors such as dust and fumes.

In Scotland, councils have the power under Section 56A of the Countryside (Scotland) Act 1967 to make by-laws to control certain vehicles and aircraft to prevent disturbances of quiet countryside areas by engine noise. However, there is no record of this power being used.

Aircraft noise

Section 78 of the Civil Aviation Act 1982 enables the Secretary of State for Transport to specify requirements at designated aerodromes for limiting or mitigating the effect of noise or vibration of aircraft landing or taking off. Measures may include night restrictions and requirements to follow noise preferential departure routes. At other airports local management may adopt similar measures.

Reductions in noise nuisance from aircraft have largely come about as a result of improved jet engine design. Standards in the UK are implemented through the Air Navigation (Noise Certification) Order. Civil aircraft are only permitted to use UK airports if they have a noise certificate which is issued if noise levels on take-off and landing meet agreed international standards.

Railway noise

The Department of Transport is working on regulations that will set the levels for eligibility for noise insulation alongside new railway lines.

Entertainment noise

Responsibility for controlling places used for public music, dancing and similar entertainments is vested in local authorities under the Local Government (Miscellaneous Provisions) Act 1982 and the London Government Act 1963. Local authorities can also adopt powers under the Private Places of Entertainment (Licensing) Act 1967 to require the licensing of private events which involve music and are promoted for private gain.

Occupational noise

This is controlled in the UK through the Noise at Work Regulations 1990 under the Health and Safety at Work Act 1974.

Table 12.3	Noise in streets and other noise; complaints received by Environmental Health Officers, 1980 to 1990 *England and Wales, Scotland*					
					Number per million people	
England & Wales	**1980**	**85-86**	**86-87**	**87-88**	**88-89**	**89-90**
Noise in streets	31	45	55	52	69	60
Other noise	17	27	54	48	35	59
Scotland	**1980**	**1986**	**1987**	**1988**	**1989**	**1990**
Noise in streets	25	32	27	23	24	25
Other noise	6	6	4	6	7	8.7

Source: IEHO, REHIS

licences), and by keeping noise and people apart (eg noise insulation and planning). Some data on noise abatement notices were explained in the section on complaints about noise, and shown in Table 12.1. Many of the powers for controlling noise are vested in local authorities. The main legislation used to control noise in the UK is given in Box 12.3.

12.37 Where new development is involved, planning can ensure that noise-sensitive development is located at an adequate distance from a source of noise (see Box 12.3). Site lay-out and building design can also help minimise exposure to noise for noise-sensitive activities. Where appropriate, planning conditions can also be imposed. These may, for example, limit the hours of a noise-making activity; require noise barriers; or specify an acceptable noise limit.

12.38 The main regulations used to limit noise emissions from vehicles are briefly explained in Box 12.4. Noise from transport sources can be reduced by various measures. For example, earth mounds and noise barriers are often incorporated into the design of a new road. On railways, jointed track has been progressively replaced over the last two decades by continuously welded track, eliminating the familiar wheel-on-joint noise of the traditional railway.

12.39 CoPA and EPA do not cover many types of noise, including aircraft noise and other transport noise. Box 12.4 explains other UK legislation relevant to the control of noise, and Box 12.5 some of the EC legislation which is applicable.

12.40 The limitation of noise from public and private entertainments is carried out by local authorities using their licensing powers (see Box 12.4).

Noise abatement zones

12.41 The Control of Pollution Act 1974 enables a local authority to designate noise abatement zones (NAZs) within its area, and to specify the classes of premises to which an NAZ applies. The local authority is then obliged to measure the noise from any classified premises within a zone and to maintain a public register of the permitted noise levels. Since 1974, 58 NAZs have been designated in 37 authorities in England and Wales.

European Community legislation
Box 12.5

The European Community (EC) has produced a number of Directives aimed at reducing noise in the environment, which the UK must implement. EC Directives set standards for noise from vehicles, aircraft, lawnmowers, construction plant and for noise in the work place. They have a dual purpose; firstly to ensure that noise limits imposed by individual member states do not create barriers to trade, and secondly to progressively reduce noise levels for environmental reasons.

Examples are:

- Directive 84/424/EEC, sets limits on noise from cars, buses and lorries.

- Directive 87/56/EEC, sets limits on noise from motorcycles.

- Aircraft noise is covered by four Directives. 80/51/EEC and 83/206/EEC require member states to ensure that the relevant categories of civil aircraft registered in their territories are not used unless certificated in accordance with Annex 16 of the Convention on International Civil Aviation. Directive 89/629/EEC bans the addition of further Chapter 2 subsonic jet airliners to the registers of EC states to cap the overall number of such aircraft in Europe. Directive 92/14/EEC provides for their withdrawal from operating during the period 1995 to 2002.

- Directive 87/252/EEC sets noise limits for lawn mowers according to the cutting width of the mower.

- Directive 86/188/EEC aims to protect workers from the risks related to noise exposure at work. Member states had to implement this Directive by January 1990.

Insulation against noise

12.42 Some local authorities in England and Wales undertake sound insulation against road traffic noise. The number of local authorities undertaking traffic noise insulation work rose from 24 in 1987-88 to 45 in 1989-90, although over the period only around 1,000 dwellings were insulated each year[8].

12.43 For Heathrow and Gatwick airports noise insulation grant schemes have been established by the Department of Transport and funded by the airports. At all other airports it is the responsibility of the local airport management to adopt whatever noise amelioration measures may be appropriate.

12.44 In the past there have been no noise insulation regulations applying to new railways. However, with proposals emerging for new railway projects, including the rail link for the Channel Tunnel, the Government appointed a Committee to recommend national noise insulation standards for new railway lines. This Committee published its report in February 1991[6], and the Department of Transport is now working on the regulations that will set the relevant eligibility levels for noise insulation.

References and further reading

1. World Health Organisation, (1980). Noise; Environmental Health Criteria 12. WHO.

2. Organisation for Economic Co-operation and Development, (1991). The State of the Environment. OECD, Paris.

3. Taylor, S.M. & Wilkins, P.A., (1987). Health Effects from Transportation Noise Reference Book. Butterworths, London.

4. Department of the Environment, (1991). Digest of Environmental Protection and Water Statistics No 14. HMSO.

5. Department of the Environment/Welsh Office, (1973). Joint Circular on Planning and Noise. (DOE 10/73, WO 16/73). HMSO

6. Department of Transport, (1991). Railway Noise and the Insulation of Dwellings (The Mitchell Report). HMSO.

7. Utley, W.A. & Keighley, E.C., (1988). Community Response to Neighbourhood Noise. Clean Air Vol 18, No 3, p121-128. National Society for Clean Air.

8. The Institution of Environmental Health Officers, (1991). Environmental Health Report 1987-1990. IEHO.

9. Department of the Environment, (1990). Report of the Noise Review Working Party 1990 (The Batho Report). HMSO.

Mant, D.C. & Muir-Gray, J.A., (1986). Building Regulations and Health. BRE.

Civil Aviation Authority, (1985). United Kingdom Aircraft Noise Index Study. (Report DR 8402). CAA Cheltenham.

Civil Aviation Authority, (1990). The Use of Leq as an Aircraft Noise Index. (Report DR 9023). CAA Cheltenham.

Civil Aviation Authority, (annual). UK Airports: Annual Statement of Movements, Passengers and Cargo. CAA Cheltenham.

Department of Transport, (1991). Transport and the Environment. HMSO.

13 Radioactivity

☐ In the UK, about 87 per cent of the average amount of radiation to which the population is exposed each year is from natural sources. Artificial or man-made radiation contributes 13 per cent to the annual average radiation dose from all sources. Of this, most is accounted for by the use of radiation in medical practice (Figure 13.1).

☐ Radon, a radioactive gas which occurs naturally, accounts for a large proportion of natural radiation exposure and over half of the annual average dose received from all sources of radiation (Figure 13.1). It is estimated that about 100,000 houses in the UK may have average radon concentrations exceeding the action level of 200 Bq/m^3 (Figure 13.2).

☐ Together with fallout from nuclear weapons testing conducted in the past, the Chernobyl accident currently contributes 0.4 per cent to the UK annual average dose from all sources of radiation (13.25). Radioactive discharges from the nuclear power industry contribute less than 0.1 per cent to the annual average dose (13.27).

☐ In the UK, a total of around 260,000 people have jobs which bring them into contact with additional radiation, and about 60 per cent of these work with artificial or man-made radiation (13.30). The dose to all radiation workers currently averages less than 5 mSv annually; less than 3 per cent receive an annual dose above 5 mSv (Table 13.2). In UK law, the maximum annual dose limit set for workers is 50 mSv (13.36).

☐ Atmospheric radioactivity comes from a number of the major nuclear establishments in the UK, and includes the chemically inert gases argon-41 and krypton-85 (Figure 13.5). Between 1980 and 1990, the majority of discharges were from Sellafield, although these were well within the limits set under authorisation over the period (13.46).

☐ Sellafield has been the main source of radioactive liquid discharges between 1980 and 1990. There were substantial reductions in liquid discharges of both beta and alpha radiation over this period. Beta liquid discharges (excluding tritium) from Sellafield in 1990 were some 2 per cent of 1980 levels (Figure 13.6).

☐ Low level nuclear waste constitutes about 90 per cent by volume of all solid radioactive waste arising annually (13.55). The majority of radioactive waste stored is intermediate level waste held at Sellafield, although radioactive waste stocks are also held by other nuclear establishments (Table 13.4).

☐ In 1988, the Radioactive Incident Monitoring Network (RIMNET) was set up to enable detection of any abnormal increase in gamma radiation of the kind that might arise from an overseas nuclear incident, such as the Chernobyl reactor accident in 1986 (Figure 13.9).

13.1 We are all exposed to radiation. Radiation caused by human activities is generally what people think of when the subject is discussed, but most radiation has natural origins. Radiation from natural sources has always been present in the environment. Sources include cosmic rays from outer space, the earth's crust, the air, and food, which all contain minute quantities of radioactive material. About 87 per cent of

---

Radiation and radioactivity   Box 13.1

Throughout the text, all ionising radiation (the full scientific description) is referred to as simply "radiation".

Most species of atom can be characterised by the name of the element and the mass number: they are then called *nuclides*. Examples are lithium-7 and carbon-14. Some nuclides are stable, but many are not. The unstable nuclides are called *radionuclides* (radioactive nuclides), and *radioactivity* involves the spontaneous disintegration of the atomic nuclei with the emission of radiation.

This process of transformation or disintegration is known as *decay*, and results in the release of ionising radiation. The *activity* is the rate of decay, in terms of the number of nuclear disintegrations occurring per unit of time in a quantity of radioactive substance. The *decay chain* is a series of radionuclides, each of which disintegrates into the next radionuclide in the series until a stable nuclide is reached. The most common forms of radiation are alpha particles (the nucleus of a helium atom), beta particles (electrons) and gamma rays. All these forms of radioactivity are potentially dangerous to health, depending on the dose received (see Box 13.2 for an explanation of "dose"). *Irradiation* is the exposure of a body or substance to radiation.

The rate of decay of a radionuclide is characterised by its *half-life*, which is the time taken by a radionuclide to lose half its activity by decay, or the time it takes for half of the atoms of the radionuclide to disintegrate. Each individual radionuclide has a unique and unalterable half-life, but the half-life for different radionuclides can vary enormously from fractions of a second to millions of years. For example, argon-41 at 1.8 hours has a very short half-life, tritium at 12 years has a longer half-life, and uranium-238 at 4,500 million years has a very long half-life. As the amount of a radionuclide decreases the radiation emitted decreases proportionately.

---

the average amount of radiation to which the UK population is exposed each year occurs naturally. Artificial or man-made radiation has been introduced into the environment since the turn of the twentieth century, and currently makes up about 13 per cent of the annual average UK dose. Of this, the majority (12 per cent) is received from medical exposure, and the remaining 1 per cent is mostly received from fallout, occupational and miscellaneous exposure. Discharges from nuclear installations contribute less than 0.1 per cent to the

annual average dose. Figure 13.1 shows the contributions from the various man-made and natural sources to the annual average dose to an individual in the UK. A short explanation of the most common scientific terms used in discussing radiation is contained in Box 13.1.

13.2 This chapter draws together information about the natural and man-made sources of radiation; trends on discharges of man-made radioactive substances from sites undertaking practices giving rise to man-made radiation; what this means for the public; and measures to control exposure to radiation.

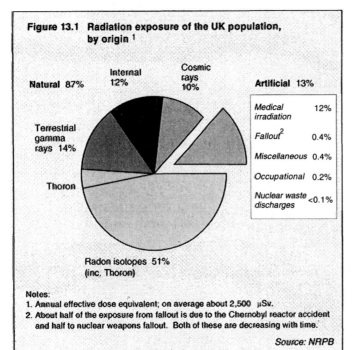

Figure 13.1 Radiation exposure of the UK population, by origin [1]

Natural 87%
Internal 12%
Cosmic rays 10%
Artificial 13%

Terrestrial gamma rays 14%

Thoron

Radon isotopes 51% (inc. Thoron)

Medical irradiation	12%
Fallout[2]	0.4%
Miscellaneous	0.4%
Occupational	0.2%
Nuclear waste discharges	<0.1%

Notes:
1. Annual effective dose equivalent; on average about 2,500 μSv.
2. About half of the exposure from fallout is due to the Chernobyl reactor accident and half to nuclear weapons fallout. Both of these are decreasing with time.

Source: NRPB

Table 13.1 Radiation exposure by source; average and collective dose *UK*

Source	Annual average dose (μSv)	Annual collective dose (manSv)
Natural		
Cosmic	250	14,000
Gamma Rays	350	19,600
Internal	300	16,800
Radon	1,200	67,200
Thoron	100	5,600
Artificial		
Medical	300	17,000
Miscellaneous	10	500
Fallout	10	550
Occupational	5	300
Discharges	< 1	30
Total (rounded)	**2,500**	**140,000**

Source: NRPB

Measurement of radioactivity and dose
Box 13.2

The International System of Units (SI) is used for measuring radioactivity and its effects. Common multiple and sub-units used include mega-, M (one million); kilo-, k (one thousand); milli-, m (one thousandth) and micro-, μ (one millionth).

The activity of an amount of a radionuclide, expressed by the rate at which spontaneous transformations occur in it, is measured in *becquerels* (Bq). One becquerel equals one disintegration per second. Measurements of radioactivity in air, water, rain, and food are made in becquerels, and concentrations of radioactivity (known as *activity concentrations*) in these media are expressed in becquerels per cubic metre (Bq/m^3), per litre (Bq/l), per kilogram (Bq/kg), etc, as appropriate. Assumptions are made about individual habits such as consumption of particular foods, in order to calculate how much radionuclide is taken into the body. This intake is then used to estimate the dose, measured in sieverts (see below).

Absorbed doses of radiation

The various sources of radiation to which the population is exposed give rise to the absorption of energy in the human body. The amount of radiation absorbed in body tissues is expressed in terms of the energy imparted by that radiation to a unit mass of matter. The unit of *absorbed dose* is the *gray* (Gy). Radiation is sometimes measured in terms of the *dose rate*, typically expressed in micrograys per hour (μGy/hr). However, equal amounts of radiation absorbed by the body do not necessarily have the same biological effects. Alpha radiation has more energy per unit and is more harmful than beta and gamma radiation.

The effective dose equivalent

In order to put all radiation on an equal basis with regard to the relative potential for causing damage to tissue and harm to health, the energy deposited in tissues is expressed as the *dose equivalent*, in *sieverts* (Sv). For gamma rays and beta particles, 1 Gy = 1 Sv; for alpha particles, which are much more damaging, 1 Gy = 20 Sv. The dose equivalent is the absorbed dose multiplied by a factor which takes account of the way a particular radiation distributes energy in living tissue, causing it harm.

The organs and tissues of the body differ in their susceptibility to harm from an equal dose equivalent of radiation. In order to assess the risk for the whole body, the dose equivalent for each of the major organs and tissues in the body is multiplied by a factor weighted in relation to the risk associated with that organ. The sum of these weighted dose equivalents is then called the *effective dose equivalent*. This allows a variety of distributions of dose equivalent in the body to be expressed as a single number, and is a broad indicator of the risk to health from any exposure to radiation.

Often it is useful also to have a measure of the total radiation dose to a group of people or a whole population. This is the mean average dose to an individual in the population multiplied by the number of people in that population, and is called the collective effective dose equivalent, or *collective dose*, expressed in man sieverts (manSv).

Summary

Absorbed dose	energy imparted by radiation to unit mass of tissue
Dose equivalent	absorbed dose weighted for harmfulness of different radiations
Effective dose equivalent	dose equivalent weighted for susceptibility of different tissues to harm
Collective dose equivalent	effective dose equivalent to a group from a source of radiation

---

13.3 Humans may be exposed to radiation by a number of different routes, called pathways. Pathways include inhalation, ingestion of radionuclides in food, and external radiation. The radiation dose to which different groups of people may be exposed is calculated from the measurement of radioactivity concentrations in environmental media such as air, freshwater, and seawater. Box 13.2 explains the measurement of radioactivity and dose. Table 13.1 shows the annual average individual dose and the annual collective dose for the UK population. The data in this table refers to the best information available in 1988[1].

13.4 Because of the potential danger to health of excessive exposure to radiation from whatever source (see Box 13.3), there are controls on human practices which release radioactivity and radiation into the environment.

Health effects of radiation

Box 13.3

Radiation gives rise to ionisation in the medium which it passes through. This can damage biological molecules within living tissue and a number of health effects can result.

Radionuclides that emit alpha particles are not hazardous unless they are taken into the body. The principal routes by which this may occur are by inhalation or ingestion. Radionuclides that emit beta particles are hazardous to superficial tissues, but not to internal organs unless they become incorporated in them. Gamma rays can pass through the body and so radionuclides that emit them may be hazardous whether the source of exposure is inside or outside the body.

Epidemiological studies[2] of populations exposed to moderate and high doses of radiation (such as the survivors of the atomic bombings of Hiroshima and Nagasaki) have shown that there is an increased risk of cancer for doses of 0.2 to 0.5 Sv upwards, (and a linear dose-response relationship is assumed, that is, the risk of cancer increases proportionally to increases in dose). Studies on animals also indicate that radiation increases the risk of hereditary disease, although this has not been demonstrated in studies of humans. The information on health effects has been reviewed by the United Nations Scientific Committee on the Effects of Atomic Radiation (UNSCEAR)[3]. A study published in October 1990 concluded that no observable effects of radioactivity on plants and animals in the wild can be demonstrated in the UK[4].

The main health concern about low doses of radiation is the increased likelihood of leukaemia and certain other cancers. There have been a number of studies into the incidence of leukaemia and other cancer rates in young people around nuclear installations in the UK. The Committee on Medical Aspects of Radiation in the Environment (COMARE) judged that these studies showed radioactive discharges from nuclear sites are most unlikely to explain the increased incidence of leukaemia and other cancers using current estimates of risk[5]. However, research continues in this and related areas as to the health effects of exposure to radiation.

In the UK, it is estimated that about 3 per cent of all cancer deaths may result from the effects of all types of radiation. Of a total of around 39,000 UK deaths from lung cancer each year (1990 data), current estimates suggest that 5 to 6 per cent may be partly attributable to radon, while over 80 per cent are caused by smoking[6]. See also Chapter 14 on environment and health.

13.5 In the UK, the National Radiological Protection Board (NRPB) advises the Government and others on all aspects of protection from radiation hazards, and undertakes research into the subject. The NRPB publication "Living with Radiation"[7] is a guide for the general public about the nature of radiation, its sources and effects, and the system of protection.

Sources of radiation

Annual average radiation dose of the UK population

13.6 The annual dose from radiation, averaged over the whole UK population, is about 2,500 microsieverts (μSv) in total, although individual doses can vary considerably depending on such factors as people's occupation or where they live.

Natural sources of radiation

13.7 Figure 13.1 shows that natural sources of radiation contribute 87 per cent of the total annual average radiation dose. There are a number of natural sources of radiation, which together give an annual average dose of 2,200 μSv, although there may be large variations about this average from one locality to another. The annual collective dose to the UK population from natural sources is about 120,000 manSv. Materials in the earth's crust are radioactive, and naturally radioactive materials are found in building materials and soils, and in the air, food and water. Cosmic radiation enters the atmosphere and reaches the earth from the sun and outer space.

Radon

13.8 Radon is a natural radioactive gas which accounts for over half of the annual average dose received in the UK from all sources of radiation. On average, the annual dose from radon decay products is estimated to be 1,300 μSv. Radon-222 and its isotope thoron (radon-220) form in rocks and soils by the radioactive decay of uranium and thorium. Radon from here on refers only to radon-222 which gives the larger radiation dose.

13.9 Radon forms in the ground and mixes with air, seeping into the atmosphere as radon gas. In the open air, it is quickly diluted to low levels but higher levels can collect in enclosed spaces such as buildings. The health hazard associated with radon results from the breakdown of atoms of radon gas to a series of minute radioactive particles called radon daughters or radon decay products.

When these are formed in the air, they may be inhaled and deposited in the lungs, where radiation from them can damage the lung tissues, and may increase the risk of lung cancer.

13.10 NRPB recommended in 1987 that exposure to radon daughters in the home should be limited. A programme of surveys, research and information provision[8] was started to discover dwellings in the UK with high radon concentrations, and to provide advice on design of new dwellings in affected areas of the UK.

13.11 A national survey[9] based on a sample of 2,300 homes (just over 1 in 10,000 from the UK stock of dwellings) determined the average concentration of radon gas to be 20 becquerels per cubic metre (Bq/m^3) in air per year, equivalent to an annual dose from exposure to radon daughters of about 1,200 µSv against the average dose from all sources of 2,500 µSv. There are wide variations from the average, and a few individual houses have received annual doses of 100,000 µSv (100 mSv) or more. The type of building construction and ventilation of the premises can affect indoor concentrations, and in some locations can cause considerable variations in radon levels.

13.12 An action level of 200 Bq/m^3 was recommended by the NRPB in January 1990[10]. This is about ten times the average for homes in the UK. About 100,000 houses in the UK may have average radon concentrations exceeding the action level. Figure 13.2 shows the estimated percentage of homes with radon concentrations above 200 Bq/m^3 in the UK, by county.

13.13 In 1990, NRPB also recommended the designation of affected areas, which were defined as areas where more than 1 in 100 homes were likely to have radon concentrations above the action level.

13.14 The counties of Cornwall and Devon were formally designated as an affected area in November 1990. About 12 per cent of homes in the area may have radon concentrations exceeding the 200 Bq/m^3 action level. This estimate is borne out by the results of many further measurements made in 1991 and 1992. Parts of Somerset, Northamptonshire and Derbyshire were similarly defined as affected areas in July 1992. About 1 to 2 per cent of houses in these counties may similarly exceed the action level. The designation of affected areas is based on a considerable volume of data. For other parts of the country, there are fewer data, but additional geographically based

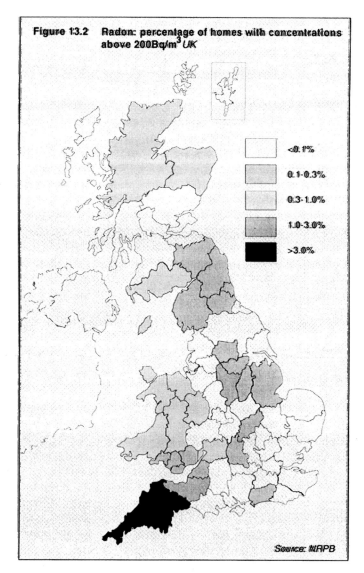

Figure 13.2 Radon: percentage of homes with concentrations above 200Bq/m^3 UK

<0.1%
0.1-0.3%
0.3-1.0%
1.0-3.0%
>3.0%

Source: NRPB

surveys have been commissioned to identify more precisely any additional areas where homes with radon concentrations above the action level are likely to be found.

13.15 The Institution of Environmental Health Officers (IEHO) has also undertaken radon screening measurements of houses for some local authorities. The IEHO surveys found moderately high levels of radon in certain limestone areas where the high permeability of this rock type enabled the rapid movement of radon to the surface[11].

Terrestrial gamma radiation

13.16 Terrestrial gamma radiation accounts for 14 per cent of the annual average dose received in the UK from all sources of radiation. The origins of this radiation are geological. The earth contains some very long lived radionuclides, in particular, uranium and thorium, together with their decay products. These can emit either alpha or beta particles, in some cases with gamma rays. The concentration of radionuclides in rocks in the UK varies with location, depending on the rock and soil type[12]. Building materials derived from radioactive rocks can emit radiation, and are predominantly the source of gamma rays indoors. However, such materials are often not representative of the underlying rock type at a particular location.

13.17 The exposure of the population to gamma radiation from rocks, soils and building materials in the UK has been assessed in two major surveys[9,12]. The annual average indoor dose was assessed as about 340 µSv per year, although the results of the surveys illustrate the variability of exposure in different areas of the UK. Annual average outdoor exposure adds only about 10 µSv to the annual average indoor dose, since on average people spend relatively little of their time outdoors. The overall average exposure from indoor and outdoor terrestrial gamma rays is therefore, about 350 µSv per year.

Internal radiation

13.18 Internal exposure to radiation accounts for some 12 per cent of the annual average radiation dose in the UK. About 40 per cent of internal exposure from ingested radionuclides is from sources of uranium and thorium radioactivity (in particular lead-210 and polonium-210). These kinds of natural radionuclides occur in soils and in the atmosphere, and are taken up by plants and so enter the food chain, passing into the human body by ingestion. Drinking water also contains traces of natural radionuclides. The concentration of radioactivity in foodstuffs depends mainly on the soils where they were produced, and is therefore not necessarily determined by the area in which a person lives. Surveys of the concentration of natural radionuclides in many different foods, combined with consumption patterns in the UK, suggest an annual average dose of 300 µSv[13].

Cosmic radiation

13.19 Cosmic radiation accounts for 10 per cent of the annual average radiation dose in the UK. Cosmic rays are formed when high-energy particles from outer space interact with the atoms in the earth's upper atmosphere and produce showers of secondary particles and gamma rays. The dose rate from cosmic rays depends on latitude and altitude, but since the UK is a low-lying area in middle latitudes, most of the population receives an annual average dose from cosmic sources of about 250 µSv. At flying altitudes the dose rate to cosmic radiation is about a hundred times greater than that at ground level, and so those who fly frequently receive a much higher dose.

For example, aircraft crew members receive about 2,000 µSv a year. A typical passenger receives an extra 4 µSv from cosmic rays for each hour of flying undertaken.

Man-made sources of radiation

13.20 Figure 13.1 shows that artificial or man-made radiation contributes 13 per cent, or about 325 µSv to the total annual average radiation dose of 2,500 µSv. Nearly all (12 per cent) of this amount is accounted for by the use of radiation in medical practice. The remaining 1 per cent comes from other sources of man-made or artificial radiation; fallout from nuclear weapons testing, discharges from nuclear power plants and nuclear reprocessing plants, occupational exposure and miscellaneous sources.

Medical

13.21 Radiation is used in medicine in two ways. Most commonly it is used as an aid to diagnosis in medical examination, or it can be used therapeutically to kill cancerous cells. Diagnostic X-ray procedures used in hospitals and in dental surgeries result in an annual average dose of around 300 µSv, over 90 per cent of the total from man-made sources. In 1988, the total annual collective dose from all diagnostic X-ray procedures was assessed as being about 17,000 manSv, whilst a more recent appraisal of medical exposures gave a value of about 20,000 manSv[14]. Some 36 million X-ray examinations take place annually in the UK. Each X-ray examination may give rise to exposures of only a few µSv.

13.22 When radiation is used for radiotherapy in the treatment of certain serious malignant illnesses, such as tumours, much higher doses are usually necessary. Personal exposure to radiation exposure from medical sources will therefore depend very much on the treatment given to each individual. The use of nuclear medicine has increased rapidly over the past two decades but is still much less common than the use of diagnostic X-rays. Nuclear medicine involves administering a medical drug, incorporating a gamma emitting radionuclide, to the patient. This technique enables a particular organ or tissue to be studied. About 1,000 manSv total annual collective dose is from nuclear medicine.

Fallout from nuclear weapons testing

13.23 Fallout from nuclear weapons tests accounts for about 0.2 per cent of the annual average dose in the UK from all sources of radiation. Testing of nuclear weapons in the atmosphere results in the injection of radioactivity into the stratosphere, which may subsequently return to earth by wet deposition or dry deposition. Collectively deposition, whether wet or dry, is called fallout.

13.24 Since the limited Nuclear Test Ban Treaty of 1963, the frequency of nuclear weapons testing in the atmosphere has decreased considerably. No atmospheric testing of nuclear weapons has been carried out since 1980, but a small amount of radioactive material continues to fall out. The three main radionuclides from this source which contribute to current doses, are carbon -14, accounting for half the total, strontium-90 and caesium-137. When fallout from the atmosphere is deposited, it can be taken up into the food chain, for example by cows grazing on contaminated herbage. Measurement of radioactivity concentrations in milk provide a useful indication of levels in total diet. Figure 13.3 shows the concentrations of strontium-90 and caesium-137 in milk in the UK since the 1960s. Concentrations of caesium-137 have fallen from an initial peak in 1964, with a second peak in 1986 following the Chernobyl reactor accident. The annual average dose in the UK from fallout declined from a peak of 140 µSv in 1963 to about 5 µSv at present, equivalent to a collective dose of 280 manSv.

Chernobyl

13.25 The Chernobyl nuclear reactor accident in April 1986 resulted in some environmental contamination from fallout in the UK, mainly in areas where heavy rainfall intercepted the radioactive cloud - south west Scotland, Cumbria, North Wales and Northern Ireland. In the UK, the most important contributor to overall dose from Chernobyl was via consumption of foodstuffs. As Figure 13.3 shows, the accident had most effect on average concentrations in Scotland and Northern Ireland. Fallout amounted to an annual average dose of 37 µSv in the first year after the accident, and doses are currently about 5 µSv on average. An average dose totalling 50 µSv is expected to be spread over the 50 year period following the accident. Together with nuclear weapons fallout, the Chernobyl accident contributed up to 10 µSv, which is 0.4 per cent of the total annual average dose in the UK, equivalent to a current collective dose of 550 manSv.

13.26 In 1987, following a UK review of the Chernobyl accident, the Government set up the Radioactive Incident Monitoring Network (RIMNET), under the National Response Plan for dealing with the consequences of overseas nuclear accidents.

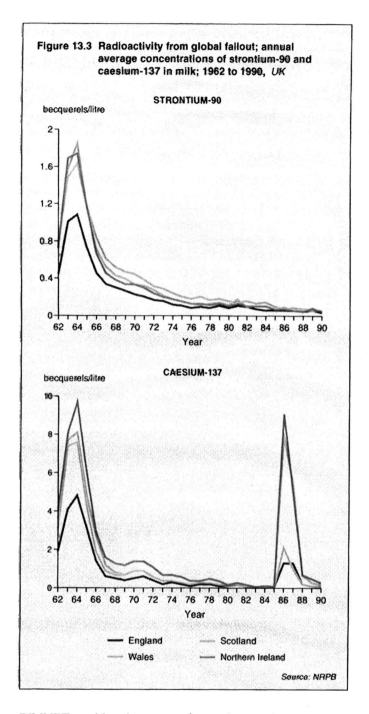

Figure 13.3 Radioactivity from global fallout; annual average concentrations of strontium-90 and caesium-137 in milk; 1962 to 1990, UK

Source: NRPB

RIMNET enables detection of any abnormal increase in gamma radioactivity of the kind that might arise from a nuclear accident. The network is covered briefly at the end of this chapter.

The nuclear power industry

13.27 Nuclear reactors have been used to produce electricity in the UK since the 1950s. Less than 0.1 per cent of the total annual average radiation dose to the public is due to discharges of radioactive substances by the nuclear power industry. The dose received by individual members of the public depends on a number of factors, such as the type of radionuclide released, location, and dietary and other habits.

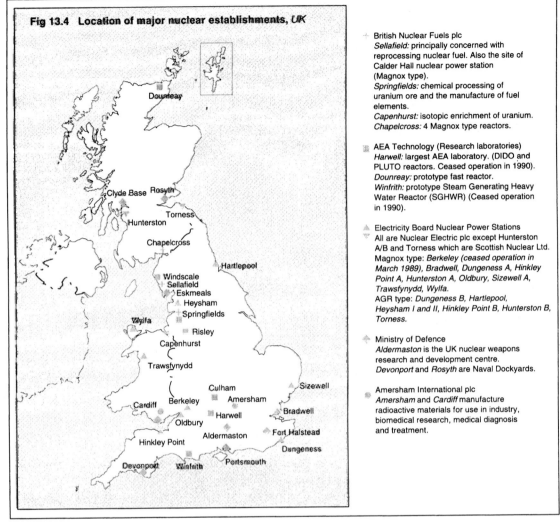

Fig 13.4 Location of major nuclear establishments, UK

British Nuclear Fuels plc
Sellafield: principally concerned with reprocessing nuclear fuel. Also the site of Calder Hall nuclear power station (Magnox type).
Springfields: chemical processing of uranium ore and the manufacture of fuel elements.
Capenhurst: isotopic enrichment of uranium.
Chapelcross: 4 Magnox type reactors.

AEA Technology (Research laboratories)
Harwell: largest AEA laboratory. (DIDO and PLUTO reactors. Ceased operation in 1990).
Dounreay: prototype fast reactor.
Winfrith: prototype Steam Generating Heavy Water Reactor (SGHWR) (Ceased operation in 1990).

Electricity Board Nuclear Power Stations
All are Nuclear Electric plc except Hunterston A/B and Torness which are Scottish Nuclear Ltd.
Magnox type: *Berkeley (ceased operation in March 1989), Bradwell, Dungeness A, Hinkley Point A, Hunterston A, Oldbury, Sizewell A, Trawsfynydd, Wylfa.*
AGR type: *Dungeness B, Hartlepool, Heysham I and II, Hinkley Point B, Hunterston B, Torness.*

Ministry of Defence
Aldermaston is the UK nuclear weapons research and development centre.
Devonport and *Rosyth* are Naval Dockyards.

Amersham International plc
Amersham and *Cardiff* manufacture radioactive materials for use in industry, biomedical research, medical diagnosis and treatment.

13.28 Figure 13.4 shows the location of the major current nuclear installations, including defence and research establishments. At some installations, such as older Magnox power stations, the most important exposure pathways to radiation are direct external irradiation (direct shine) from the plant itself, or external irradiation resulting from the release of gases such as argon-41. Consequently, the highest doses are received by people who live closest to the site. In other cases, different exposure pathways, such as external exposure to contaminated beach materials, or ingestion of radionuclides in foodstuffs could be more important, and may affect people further away from the site.

13.29 Included in the contribution to the annual average dose are small amounts of radioactive discharges to the atmosphere and surface water produced by some 1,000 hospitals, research and industrial establishments which use radioactive materials. Lists of premises currently authorised under the Radioactive Substances Act 1960 to discharge radioactive waste are available from Her Majesty's Inspectorate of Pollution (HMIP)[15] in England and Wales and Her Majesty's Industrial Pollution Inspectorate (HMIPI) in Scotland.

Occupational exposure

13.30 For the UK population as a whole, occupational exposure accounts for an annual average dose of about 5 µSv or 0.2 per cent of the annual average dose from all sources. However, those exposed to additional

Table 13.2 Occupational exposure to radiation, 1988 UK

Type of work	Number of workers	% of workers in dose range (mSv)			
		0-5	5-15	15-50	>50
Natural[1]					
aircraft crew	20,000	99.5	0.5	-	0
coal mines	81,500	99.4	0.5	-	0
non-coal mines	2,000	40	35	25	-
Total	**103,500**	**98.3**	**1.2**	**0.5**	**-**
Artificial					
Nuclear	47,669	90.4	8.2	1.4	0
Defence	11,824	95.9	3.9	0.3	0
Other industry	21,000	97.6	1.9	0.5	-
Education	13,000	100	0	0	0
Health	64,000	99.7	0.3	-	0
Total (rounded)	**157,500**	**96.3**	**3.2**	**0.5**	**-**
Total (rounded)	**261,000**	**97.3**	**2.4**	**0.5**	**-**

Note: 1. Excludes indoor employment in radon-prone areas

Source: NRPB

radiation at work may receive a dose above this average. A total of around 260,000 people have jobs which bring them into contact with additional radiation. Table 13.2 gives a summary of the distribution of doses from occupational exposure for 1988[16].

13.31 About 100,000 workers (40 per cent of the total exposed workforce) are known to be exposed to elevated levels of radiation from natural sources at work. Ninety eight per cent receive an annual average dose below 5,000 μSv. Aircraft crews and miners are the two occupational groups which commonly come into contact with natural radiation. The annual average dose to crew members of most aircraft is around 2,000 μSv. Crew members of Concorde, since it flies at greater altitude, experience higher annual doses of around 2,500 μSv. It is difficult to reduce exposure for this occupational group except by reducing the number of flying hours. Most coal miners are exposed to radiation from radon, an average of 1,200 μSv a year. The average dose from radon in non-coal mines (such as gypsum and tin mines) is generally higher; the most recent assessment, made in 1991, suggests an annual average dose from radon in non-coal mines of 4,500 μSv [17].

13.32 Almost 160,000 workers (60 per cent of the total exposed workforce) are known to be exposed to man-made radiation in the UK. These people are routinely monitored for occupational exposure to man-made radiation. On average, the annual dose is 900 μSv; 96 per cent of those working with artificial radiation receive an annual average dose below 5,000 μSv and less than 0.1 per cent receive an annual average dose above 15,000 μSv. Doses to those who work in education, medicine, dentistry and veterinary practice are generally low. For instance, the annual average dose to medical workers is about 200 μSv. There are some medical practices from which higher doses can be received, particularly the use of advanced medical treatment such as radiotherapy and nuclear medicine.

Miscellaneous sources

13.33 About 0.4 per cent of the annual average dose from all sources of radiation results from miscellaneous sources. This includes exposure to natural radionuclides released in emissions from industrial processes. The release of natural radionuclides into the atmosphere from coal burning is estimated to result in an annual average dose of 0.2 μSv.

13.34 Miscellaneous man-made sources include watches luminised with radioactive materials, television receivers, smoke detectors and gas mantles, amongst other items. Taken together artificial and natural radionuclides account for 3 per cent of the annual average dose from man-made sources, or 0.4 per cent of the annual average dose from all sources.

Protection from man-made radiation

13.35 The International Commission on Radiological Protection (ICRP), publishes international recommendations for protection against radiation. In the UK, NRPB advises the Government on the application of ICRP recommendations, which currently includes a maximum dose limit from man-made sources of 1 mSv per year for members of the public[18]. This excludes medical exposure, for which no recommendation is made. The requirements of radiological protection are based on three main elements, given in Box 13.4.

Occupational protection

13.36 Most employers who work with radiation in the UK are subject to control under the Ionising Radiations Regulations 1985 (IRR 85). These set legal maximum dose limits for all workers of 50 mSv in any one year, which has been the accepted standard since the 1960s, though the dose to radiation workers currently averages less than 5 mSv annually.

The main elements of radiological protection **Box 13.4**

- All practices giving rise to exposure to radiation must be justified; that is, the need for the practice must be established in terms of its overall net benefit.

- For any source, radiation exposure, or likelihood of exposure to individuals should be kept to levels which are as low as reasonably achievable (ALARA), economic and social factors being taken into account.

- The dose equivalent to all individuals should not exceed limits recommended for the appropriate circumstances by the International Commission on Radiological Protection (ICRP). ICRP dose limits do not apply to medical exposures.

13.37 NRPB has subsequently advised that exposure levels should not exceed an average effective dose of 15 mSv for an individual worker in any one year. Less than 1.5 per cent of the workforce in the nuclear industry currently receive doses exceeding 15 mSv per year (see Table 13.2).

13.38 Since 1976, NRPB has maintained a UK National Registry for Radiation Workers, which records how much radiation workers are exposed to in the course of their work, and of the eventual cause of their death. The first material published from the registry covered the histories of 95,000 individuals working for the main employers in the UK nuclear industry: British Nuclear Fuels plc (BNFL), Nuclear Electric, the UK Atomic Energy Authority, and the Ministry of Defence[19].

Protection of the environment and the UK population

13.39 On average, members of the public receive a dose of less than one μSv per year as a result of authorised discharges from nuclear facilities in the UK. Current recommendations from ICRP[20] and NRPB[21] are that the principal annual dose limit for members of the public should be one mSv. This limit refers to exposures from all controlled sources of artificial radionuclides

and radiation except those received for medical reasons. The NRPB recommended in 1987 that the dose received by an individual as a result of current discharges from a single site should not exceed 0.5 mSv.

13.40 It is not generally feasible to estimate doses to all population groups on the basis of measurements of activity concentrations made in various environmental media. A practical approach taken in the UK and other countries is to identify those groups of people who, because of where they live or their lifestyle, receive higher than average doses. These groups are the "critical groups". People living nearest to nuclear installations are often among those identified as a critical group. The principle of the approach is that if the critical groups are adequately protected from exposure to radiation and radioactivity, then the remainder of the population will also be adequately protected.

13.41 Doses resulting from normal operations can be controlled at source by imposing limits on the amounts of activity that may be discharged. In the UK, the powers to impose these controls are contained in the Radioactive Substances Act 1960. People who come into contact with radiation as a result of their work are not treated as part of any critical group, since they are covered by occupational exposure regulations.

Control of radioactive discharges

13.42 The disposal of radioactive waste in the UK is controlled by the Radioactive Substances Act 1960, as amended by Part V of the Environmental Protection Act 1990. Under the 1960 Act, all radioactive waste in whatever form, whether solid, liquid or gas, (except that excluded by order as being of no radiological significance) must be authorised before it can be disposed of. In England and Wales, HMIP is responsible for the registration of all premises using or keeping radioactive materials, and for authorising the accumulation and disposal of radioactive waste. In Scotland, HMIPI carries out these functions, and in Northern Ireland, the Alkali and Radiochemical Inspectorate is responsible.

13.43 The Secretary of State for Defence is responsible for the overall management of wastes arising at Crown premises occupied for military or defence purposes.

13.44 Applications for authorisations are considered against the objectives for radiological protection, which are given in Box 13.4. Authorisations usually limit the amount of radioactivity which may be

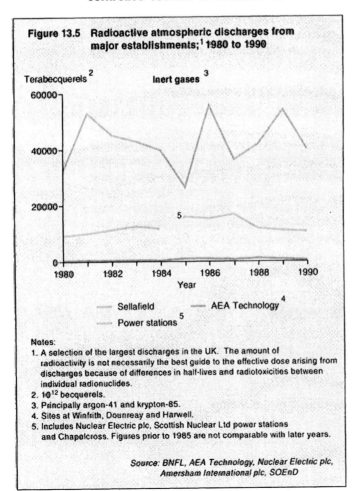

Figure 13.5 Radioactive atmospheric discharges from major establishments;[1] 1980 to 1990

Terabecquerels[2]   Inert gases[3]

Year

— Sellafield   — AEA Technology[4]
— Power stations[5]

Notes:
1. A selection of the largest discharges in the UK. The amount of radioactivity is not necessarily the best guide to the effective dose arising from discharges because of differences in half-lives and radiotoxicities between individual radionuclides.
2. 10^{12} becquerels.
3. Principally argon-41 and krypton-85.
4. Sites at Winfrith, Dounreay and Harwell.
5. Includes Nuclear Electric plc, Scottish Nuclear Ltd power stations and Chapelcross. Figures prior to 1985 are not comparable with later years.

Source: BNFL, AEA Technology, Nuclear Electric plc, Amersham International plc, SOEnD

discharged over specified periods of time, and require operators to use the best practicable means to limit the radioactivity of wastes discharged. Conditions are usually attached controlling the types of radioactive substances which may be discharged, and the manner in which they may be discharged.

Atmospheric discharges

13.45 The main pathways by which the public are exposed to radiation as a result of atmospheric discharges are external exposure, inhalation, contact with contaminated materials, or consumption of contaminated foods. Deposition of radioactive substances on to soil and vegetation can result in contamination of food products.

13.46 Atmospheric radioactivity comes from a number of the major nuclear establishments in the UK, and includes the chemically inert gases argon-41 and krypton-85. The amount of radioactivity discharged is not necessarily the best guide to its significance, and in particular, krypton-85 makes a negligible contribution to the radiation exposure of the local population. Figure 13.5 shows recent trends in radioactive discharges of these gases from major nuclear installations. Between 1980 and 1990, the majority of discharges were from Sellafield, although these were well within the limits set under authorisation over the period.

Liquid discharges

13.47 In the case of liquid discharges from nuclear plants, consumption of seafood or direct exposure to sediments contaminated with radiation usually result in the highest dose. Authorisations setting specific numerical limits to the levels of radioactivity in liquid discharges cover all major nuclear establishments.

13.48 Sellafield has been the main source of radioactive liquid discharges between 1980 and 1990, and accounted for 59 per cent of all liquid beta discharges from nuclear sites in 1990. There were substantial reductions in the discharges of both beta and alpha radiation from Sellafield over this period. Discharges of beta radiation, excluding tritium, from the site have reduced from 4,306 TBq in 1980 to 71 TBq in 1990, against an authorised discharge limit of 500 TBq in 1990 (see Figure 13.6). Discharges of alpha radiation at Sellafield were reduced from 39 TBq in 1980 to 2 TBq in 1990, against an authorised discharge limit in 1990 of 10 TBq.

13.49 Discharges of tritium are numerically large compared with discharges of other radionuclides from nuclear installations, but

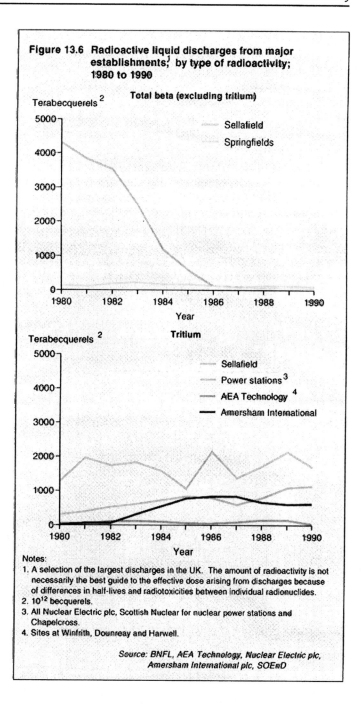

Figure 13.6 Radioactive liquid discharges from major establishments,[1] by type of radioactivity; 1980 to 1990

Notes:
1. A selection of the largest discharges in the UK. The amount of radioactivity is not necessarily the best guide to the effective dose arising from discharges because of differences in half-lives and radiotoxicities between individual radionuclides.
2. 10[12] becquerels.
3. All Nuclear Electric plc, Scottish Nuclear for nuclear power stations and Chapelcross.
4. Sites at Winfrith, Dounreay and Harwell.

Source: BNFL, AEA Technology, Nuclear Electric plc, Amersham International plc, SOEnD

their contribution to the total radiation dose is very small. The majority of liquid discharges of tritium took place at Sellafield (see Figure 13.6). The steady increase in tritium discharges from power stations reflects the increased power outputs from advanced gas cooled reactors since commissioning. The large increase in tritium discharges in 1989 is due primarily to Heysham I and II, and Hartlepool only recently achieving full power operation, and the build up of power generation from Torness which came into operation at the end of 1987.

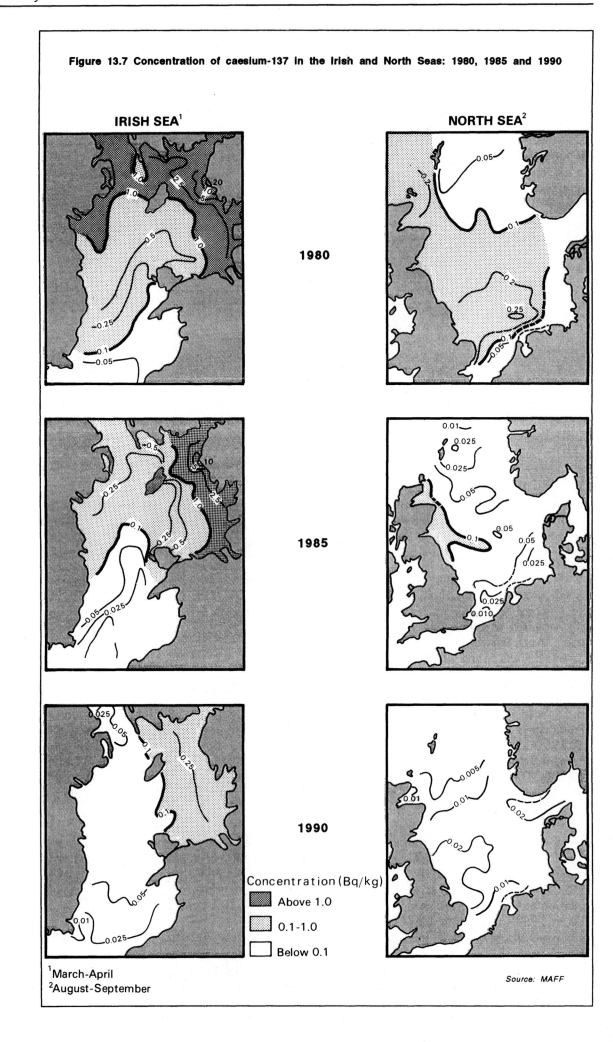

Figure 13.7 Concentration of caesium-137 in the Irish and North Seas: 1980, 1985 and 1990

IRISH SEA[1]

NORTH SEA[2]

1980

1985

1990

Concentration (Bq/kg)

Above 1.0

0.1-1.0

Below 0.1

[1]March-April
[2]August-September

Source: MAFF

Radioactivity in the marine environment

13.50 Liquid radioactive waste from nuclear sites contributes to the distribution of radioactivity in the marine environment. Figure 13.7 shows the concentration of caesium-137 in the Irish and North Seas for 1980, 1985 and 1990. The decrease in the concentrations over this period reflect the reduction in marine discharges of caesium-137 from 3,000 TBq in 1980 to 24 TBq in 1990. It takes about one year for such discharges to affect concentrations in the Irish Sea, and about three years for them to affect concentrations in the North Sea.

13.51 Discharges to the marine environment from Sellafield have given rise to the highest exposures to radioactivity and contribute to exposures near many other nuclear sites. However, discharges of radioactivity, excluding tritium, to the marine environment from Sellafield in 1990 were around 3 per cent of 1979 levels. As a result, the radiation dose to the local critical group (consumers of large quantities of local shellfish) from marine discharges has fallen. The figure in 1990 (0.16 mSv), most of which is due to past discharges, is within the current dose limit of one mSv recommended by ICRP.

13.52 Discharges to the marine environment from the BNFL sites at Sellafield and Springfields, result in sediments containing radionuclides in the muddy creeks of the Ribble estuary. The critical group assessed for external exposure in the Ribble estuary has been people living on houseboats moored

Table 13.3 External exposure to radiation [1], houseboat occupiers in the Ribble estuary 1986 to 1990	
Year	dose [2] (mSv)
1986	0.34
1987	0.24
1988	0.27
1989	0.17
1990	0.18

Notes:
1. Dose is largely due to Cs-137 from Sellafield.
2. ICRP recommended annual dose limit for members of the public is 1 mSv.

Source: NRPB

in these creeks. Table 13.3 shows the external doses of radiation that this group has received annually since 1986. In 1990 the external exposure of the group was 0.18 mSv. Details of other radioactive discharges and the associated environmental impacts for surface and coastal waters of the British Isles are published annually by MAFF[22].

Disposal of radioactive solid waste

13.53 Waste from the nuclear industry can remain radioactive for hundreds or thousands of years depending on the materials involved. Waste is categorised according to its radioactivity content. There are three broad categories of waste: high, intermediate and low level, and these are explained in Box 13.5.

Sea disposal

13.54 Disposal of drummed intermediate (ILW) and low level (LLW) waste in the Atlantic

---

Categories of radioactive waste from the nuclear industry   Box 13.5

High level (heat generating) waste (HLW)

HLW is waste in which the temperature may rise significantly as a result of its radioactivity, so this factor has to be taken into account in designing storage or disposal facilities. HLW is the concentrated and highly radioactive liquid waste produced when spent nuclear fuel is reprocessed. It contains over 95 per cent of the radioactivity in waste from the nuclear power programme, and is small in volume compared to the other waste categories. The liquid HLW is being vitrified (solidified into a glass-like form) at Sellafield. HLW will be stored for at least 50 years. Spent nuclear fuel would also be included in this category if it were to be disposed of without reprocessing.

Intermediate level waste (ILW)

ILW is not as active as high level waste but requires shielding during handling and transportation. ILW material is generally of low heat content and high bulk. It includes some of the materials used in the production of nuclear energy and the processing of radioactive materials, for example plutonium contaminated materials, fuel cladding, and liquids used to store spent fuel before reprocessing.

Low level waste (LLW)

This includes lightly contaminated waste mostly arising from the operation of nuclear facilities. LLW has low radionuclide content and does not require shielding during normal handling and transport. Examples of LLW are paper, discarded protective clothing, and laboratory tools used in work with radioactive materials.

was suspended in 1983, and in 1988 the Government announced that it would not resume sea dumping of drummed waste. The disposal at sea of high level waste (HLW) is prohibited under the global London Dumping Convention, to which the UK is a party.

Land disposal

13.55 LLW constitutes about 90 per cent by volume of all radioactive waste arising annually, and it is disposed of at approved shallow waste disposal sites. HLW is being converted into glass-like blocks (vitrified) at Sellafield, which reduces its original volume by two thirds. HLW will be stored at Sellafield and Dounreay for at least 50 years to allow it to cool and the radioactivity to decay before eventual disposal. ILW is currently stored at nuclear sites awaiting eventual disposal in the deep disposal facility which is being investigated by UK Nirex Ltd.

13.56 Table 13.4 shows the volume of the different categories of waste stocks held at Sellafield and by other types of nuclear

Table 13.4 Volume of radioactive waste stocks by waste category, 1986 to 1989 [1] UK				
				m³
	1986	1987	1988	1989
High level waste				
Sellafield	1,200	1,250	1,250	1,320
Other nuclear industry	0	0	0	0
Research, medical, industrial	151	180	213	255
Intermediate level waste				
Sellafield	28,200	29,000	32,900	30,500
Other nuclear industry	9,580	9,840	9,990	9,890
Research, medical, industrial	4,090	4,740	4,890	4,920
Low level waste				
Sellafield	120	540	523	601
Other nuclear industry	1,480	955	368	902
Research, medical, industrial	825	848	110	12,200 [2]

Notes:
1. As at 1 January for each year
2. Arises from the temporary suspension of disposal at Dounreay

Source: Electrowatt Consulting Engineers and Scientists

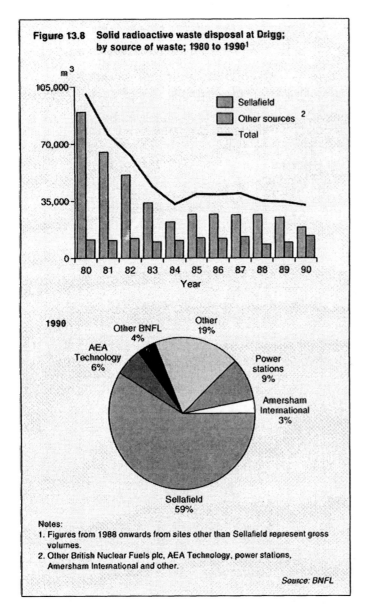

Figure 13.8 Solid radioactive waste disposal at Drigg; by source of waste; 1980 to 1990 [1]

Sellafield
Other sources [2]
Total

1990

Other BNFL 4%
AEA Technology 6%
Other 19%
Power stations 9%
Amersham International 3%
Sellafield 59%

Notes:
1. Figures from 1988 onwards from sites other than Sellafield represent gross volumes.
2. Other British Nuclear Fuels plc, AEA Technology, power stations, Amersham International and other.

Source: BNFL

establishment between 1986 and 1989. Most is ILW held at Sellafield. Most ILW (and some LLW) requires further treatment (conditioning) on the site of arising before being transported from the site for disposal. Waste volumes shown in Table 13.4 are those which still require conditioning before disposal, whereas conditioned wastes are in a form suitable for disposal.

13.57 The main disposal site for LLW is at Drigg in Cumbria. The site is operated by BNFL and is mainly used for the disposal of LLW from Sellafield, although waste is also received from other UK nuclear sites and from hospitals, research establishments and industry. Figure 13.8 shows the volume of solid wastes disposed of at Drigg since 1979, by source of waste. Most of the waste disposed of at the site is in the form of plastics, scrap metal, protective clothing, paper and excavation spoil. Waste from sites other than Sellafield is put in sealed containers and then placed in vaults.

Monitoring

13.58 Operators of sites are required to carry out their own environmental monitoring programmes[23], which are independently verified by authorising government departments. Box 13.6 covers the environmental objectives of radiation monitoring programmes.

13.59 In England and Wales, the overall responsibility for initiating and coordinating monitoring lies with DOE and MAFF, with advice from NRPB, the Radioactive Waste

Management and Advisory Committee (RWMAC)[25] and the Radioactivity Research and Environmental Monitoring Committee (RadREM). A full description of the environmental monitoring programmes carried out by site operators, government departments and other bodies in England and Wales has been published by DOE[26] and MAFF[27] . In Scotland, the responsibility for initiation and co-ordination of monitoring rests with the Scottish Office. Reports and data are published every two years[28].

13.60 The UK environmental surveillance programme jointly operated by NRPB and AEA Technology provides baseline data giving nationwide average values against which measurements made in the vicinity of nuclear installations can be compared. Results of the programme are published annually[29].

13.61 A programme to monitor radioactivity in public water supplies has been carried out by DOE, in conjunction with the water industry in England and Wales, since the 1950s. Originally aimed at detecting any contamination that might arise from atmospheric nuclear weapons tests, the programme now fulfils international obligations on environmental monitoring and provides a record of background levels of radioactivity. Thirty sources of raw drinking water, including reservoirs, rivers, groundwater and water treatment plants are sampled. Results are published annually[30]. A similar programme of monitoring of selected sources of drinking water in Scotland is undertaken by the Scottish Office.

13.62 Radioactivity in agricultural produce is monitored by the Terrestrial Radioactivity Monitoring Programme (TRAMP), run by MAFF. Samples of agricultural produce, in particular milk, are collected from the vicinity of the major nuclear sites in England and Wales. The results of this monitoring are published annually[31]. Similar monitoring is undertaken in Scotland by HMIPI in consultation with the Scottish Office[28]. A summary of the scope of programmes

monitoring radioactivity in food has also been published[32].

13.63 Cows' milk is a potentially important pathway to humans. Milk contributes a large proportion of total diet for young children, and those living near nuclear installations are a critical group. Table 13.5 shows annual average concentrations of strontium-90 and caesium-137 per litre of milk sampled from farms in the vicinity of the Sellafield, Dounreay, Harwell and Winfrith nuclear sites. Deposition from the Chernobyl reactor accident in 1986 caused a sharp increase in concentrations of caesium-137 in milk. Peak concentrations occurred in May and June 1986, with secondary peaks later in the year as a result of consumption of contaminated silage by cows during the winter months. Average levels in milk remained high for part of 1987, but fell in 1988. See also Figure 13.3.

Table 13.5 Annual average concentrations of strontium-90 and caesium-137 in milk from farms near Sellafield, Dounreay, Harwell and Winfrith

Becquerels/litre

Farms within (km):	Sellafield 0 to 3	Sellafield 3 to 6	Dounreay 0 to 2.5	Harwell 0 to 6	Winfrith 0 to 6
Strontium-90					
1981	0.4	0.2	0.4	0.05	0.1
1982	0.4	0.3	0.3	0.03	0.06
1983	0.4	0.2	0.4	0.04	0.06
1984	0.5	0.2	0.3	0.03	0.06
1985	0.4	<0.3	0.3	0.03	0.05
1986	0.5	<0.5	0.3	0.03	0.06
1987	<0.4	<0.4	0.3	0.03	0.05
1988	<0.4	<0.3	0.2	0.03	0.04
1989	0.4	<0.3	0.2	0.03	0.03
1990	<0.4	<0.3	..	0.03	0.03
Caesium-137					
1981	3.8	1.4	<1.1	<1.1	0.2
1982	3.6	0.7	<1.1	<1.1	0.2
1983	1.4	0.6	<1.1	<1.1	0.1
1984	0.9	0.3	<1.1	<1.1	-
1985	0.6	<0.4	<1.1	<1.1	-
1986 [1]	13.4	8.9	8.9	<1.1	0.8
1987 [1]	6	13	7.8	<1.1	0.4
1988	1.1	1.4	<5.8	<1.0	0.1
1989	0.8	1.2	<5.8	<1.0	0.1
1990	1.1	0.5	<1.0	<1.0	-

Note: 1. Higher levels resulting from the Chernobyl reactor accident.

Source: BNFL, AEA Technology

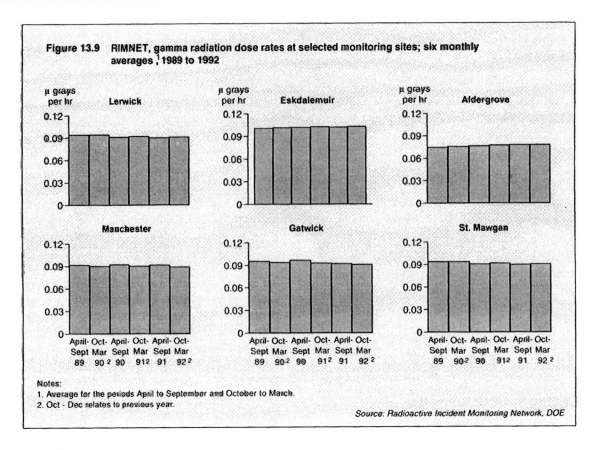

Figure 13.9   RIMNET, gamma radiation dose rates at selected monitoring sites; six monthly averages , 1989 to 1992

Notes:
1. Average for the periods April to September and October to March.
2. Oct - Dec relates to previous year.

Source: Radioactive Incident Monitoring Network, DOE

13.64 Monitoring of aquatic foods and other materials in the vicinity of the major nuclear establishments, and further afield, is undertaken by MAFF, and data are published annually [22].

13.65 HMIP monitors exposure of the public to radiation from non-food pathways which might arise from areas used for leisure purposes, for instance beaches and riverbanks. The programme consists of surveys of external dose rates and analysis of samples collected at specified locations in the vicinity of certain nuclear sites and industrial premises. Results are published annually[33].

Radioactive Incident Monitoring Network (RIMNET)

13.66 This network has been installed in two phases and will consist of 92 continuously operating monitoring stations throughout the UK. The phase I system, which has been in operation since the end of 1988, continuously measures gamma radiation dose rates at 46 monitoring stations throughout the UK[34]. Data collected at the phase I sites have been published twice a year in monthly summary form[35]. Figure 13.9 gives a summary of the six monthly averages at selected phase I sites since April 1989. Phase II of the RIMNET programme will fully automate all the gamma radiation dose rate monitoring processes.

References and further reading

1. Hughes, J. S. et al, National Radiological Protection Board, (1989). Radiation Exposure of the UK Population, 1988 Review. NRPB-R227. NRPB.

2. Shimizu, Kato and Schull, (1988). Life Span Study Report 11 Part 2: Cancer Mortality in the Years 1950-85 based on the recently Revised Doses (DS86). Hiroshima, Radiation Effects Research Foundation .

3. United Nations, UNSCEAR, (1988). Sources, Effects and Risks of Ionizing Radiation. 1988 Report to the General Assembly. UN, New York.

4. Kennedy, V. H., Horrill, A. D., Livens, F. R., Nature Conservancy Council, (1990). Radioactivity and Wildlife. Focus on Nature Conservation No 24. NCC.

5. Committee on Medical Aspects of Radiation in the Environment (COMARE), (1988). Second Report: Investigation of the possible increased incidence of Leukaemia in Young People near the Dounreay Nuclear Establishment, Caithness, Scotland. HMSO.

6. Department of Health, (1991). The Health of the Nation. Cmnd 1986. HMSO.

7. National Radiological Protection Board, (1989). Living with Radiation. NRPB.

8. Department of the Environment, (1990). The Householder's Guide to Radon (2nd edition). DOE.

9. Wrixon, A. D., et al, National Radiological Protection Board, (1988). Natural Radiation Exposure in UK Dwellings, NRPB-R190. NRPB.

10. National Radiological Protection Board, (1990). Limitation of Human Exposure to Radon in Homes. NRPB.

11. Institution of Environmental Health Officers, (1987-88, 1989, 1990). Three Reports on the IEHO Surveys on Radon in Homes. IEHO.

12. Green, B. M., et al, National Radiological Protection Board, (1989). Gamma-radiation Levels Outdoors in Great Britain, NRPB-R191. NRPB.

13. Smith-Briggs, et al, (1984). Measurement of Natural Radionuclides in UK Diet. Science of the Total Environment 35, 431.

14. Shrimpton, P. C., Jones, D. G., et al, National Radiological Protection Board, (1991). Survey of CT Practice in the UK Part 2: Dosimetric Aspects, NRPB-R249. NRPB.

15. Department of the Environment and Welsh Office, Her Majesty's Inspectorate of Pollution, (1988). Radioactive Substances Act 1960. Disposal of Radioactive Waste. List of Premises in England and Wales currently authorised under the Radioactive Substances Act 1960 to Dispose of Radioactive Waste.

16. Hughes, J. S., Lawson, G., et al, National Radiological Protection Board, (1988). Current Dose Distributions in the UK - Implications of ICRP Publication 60, NRPB-M286. NRPB.

17. Bottom, D. A., et al, (1991). Exposure to Radon in British Mines. Procedural Conference on Occupational Radiological Protection, Guernsey (papers). British Nuclear Energy Society.

18. International Commission on Radiological Protection, (1991). Publication 60, ICRP. Pergamon Press, Oxford.

19. Kendall, G. M., et al, National Radiological Protection Board, (1992). First Analysis of the National Registry for Radiation Workers. Occupational Exposure to Ionising Radiation and Mortality, NRPB-R251. HMSO.

20. International Commission on Radiological Protection, (1985). Statement from the 1985 Paris Meeting of the ICRP. Publication 15, No. 3. Pergamon Press, Oxford.

21. National Radiological Protection Board, (1986). Dose Limits for Members of the Public, NRPB-GS4. NRPB.

22. Ministry of Agriculture Fisheries and Food (Annual Reports). Radioactivity in Surface and Coastal Waters of the British Isles. Aquatic Environment Monitoring Reports. Directorate of Fisheries Research, Lowestoft.

23. British Nuclear Fuels plc. Annual Report on Radioactive Discharges and Environmental Monitoring. BNFL.

24. Department of the Environment, (1989). Sampling and Measurement of Radionuclides in the Environment. A Report by the Methodology Sub-Group to the Radioactivity Research and Environmental Monitoring Committee (RADREM). HMSO.

25. Radioactive Waste Management Advisory Committee. Annual Reports. HMSO.

26. Department of the Environment, (1988). Monitoring of Radioactivity in the UK Environment. HMSO.

27. Ministry of Agriculture, Fisheries and Food, (1992). Monitoring of Radioactivity in the UK: an Annotated Bibliography of Current Programmes. Fisheries Research Data Report. Directorate of Fisheries Research, Lowestoft.

28. Scottish Office, (biennial). Environmental Monitoring for Radioactivity in Scotland 1981 to 1985 (1987), 1983 to 1987 (1989).

29. Smith, D. M., et al, National Radiological Protection Board, (1991). Radioactivity Surveillance Programme: Results for 1990, NRPB-252. NRPB.

30. Department of the Environment, (annual). Digest of Environmental Protection and Water Statistics, Statistical Bulletin. HMSO.

31. Ministry of Agriculture Fisheries and Food, (1991). Terrestrial Radioactivity Monitoring Programme (TRAMP). Report for 1990. Radioactivity in Food and Agricultural Products in England and Wales. MAFF.

32. Ministry of Agriculture Fisheries and Food, (1990). Programmes to Monitor Radioactivity in Food. The 28th Report of the Steering Group on Food Surveillance; Working Party on Radionuclides in Food. Food Surveillance Paper 28. HMSO.

33. Her Majesty's Inspectorate of Pollution, (1992). Environmental monitoring programme. Radioactive Substances Report for 1990. HMIP.

34. Department of the Environment, (1989). The National Response Plan and Radioactive Incident Monitoring Network (RIMNET). Phase 1. HMSO.

35. Department of the Environment, Statistical Bulletins (bi-annual since April 1989). RIMNET Gamma-radiation Dose Rates at Monitoring Sites throughout the United Kingdom. HMSO.

Scottish Office, (biennial). Radioactive Waste Disposals from Nuclear Sites in Scotland.
      1980 to 1984, 1/1986
      1982 to 1987, 2(E)1988
      1984 to 1988, 1(E)1990
      1986 to 1989, 2(E)1990

Scottish Office, (1986). Chernobyl Accident, Monitoring for Radioactivity in Scotland. 1(E)1988

Electrowatt Consulting Engineers and Scientists, (annual reports since 1986). The UK Radioactive Waste Inventory. Reports prepared for UK Nirex Ltd and the Department of the Environment. Published as part of the UK Nirex Reports series.

Department of the Environment, (1988). The National Response Plan and Radioactive Incident Monitoring Network (RIMNET). A Statement of Proposals. HMSO.

Department of the Environment. Radioactive Waste Management Environment in Trust series. DOE

14 Environment and health

☐ **Lead intake from food has declined since the mid seventies by around 5 per cent per year on average (14.8). In 1987, the average UK intake was of the order of 60µg per day, well below the World Health Organisation tolerable intake levels (14.7, 14.8).**

☐ **Approximately 3 per cent of drinking water samples taken in England and Wales during 1991 contained more than 50 µg/l of lead. In Scotland, over 4 per cent of samples exceeded the standard in 1990 (14.11).**

☐ **Concentrations of lead in blood in the UK have been found to be significantly higher in men, older people, heavy smokers and drinkers, and in certain parts of the UK such as north west England (Figures 14.3 to 14.5).**

☐ **Concentrations of cadmium measured in kidneys have been found to be consistently higher in smokers than in non-smokers (14.22).**

☐ **Evidence of a link between nuclear installations and increased incidence of leukaemia is inconclusive and further research is underway (14.58 - 14.60).**

☐ **In the UK, about 3 per cent of cancer deaths may, it is estimated, result from the effects of all types of radiation (14.57). Of 39,000 UK deaths from lung cancer each year, current estimates suggest that 5 to 6 per cent may be partly attributable to radon (14.64).**

☐ **About 3½ million UK homes suffer from condensation and damp which is thought to contribute to allergic diseases (14.67).**

14.1 There are many influences on an individual's health including general lifestyle, diet and hereditary factors. Health may also be affected, in certain circumstances, by environmental conditions which may be man-made or occur naturally. For example, health may be affected if unacceptable quantities of toxic substances such as heavy metals (eg lead and cadmium) and organic compounds (eg dioxins and pesticides), are inhaled or ingested. Health may also be affected by radioactive particles and gases (eg radon) and by indoor pollution (eg smoking and damp). These and other environmental influences on health (many of which are discussed briefly elsewhere in the report) are covered here. This chapter deals only with the effects of the broader environment on health; influences which arise from people's work environment are outside the scope of this report.

14.2 In the past some kinds of gross pollution were a direct hazard to health.

Contaminated water and severe smogs as in London in the 1950s were conspicuous examples. Most of those more immediate environmental threats to health have now been dealt with in the UK. Concern now centres on the possibility of longer term effects from what are usually lower levels of environmental pollution. But it is not easy to establish precise effects here. Although it is possible to estimate exposure levels from monitoring programmes for various contaminants in air, food and water, it is often difficult or even impossible to produce definitive statistics linking these exposure levels with health effects which may take a long time to emerge. There is need for more research to pinpoint more accurately the possible linkages between the quality of the environment and health consequences; and as announced in the Health of the Nation White Paper (Command 1986) the Government is exploring the possibility of a new Institute for Environment and Health to create a new focus for work in this area.

197

Lead

14.3 Illness resulting from lead poisoning is rare in the UK. In adults, it is most likely to be caused by poorly controlled occupational exposure to lead. In a child, lead from paint is a probable source. For most people, however, environmental exposures from air, water, soil, dust and food are too low to cause classical lead poisoning. Nevertheless, low-level exposures may have adverse effects on the way children's intelligence and behaviour develops. The effects, though important, are small and difficult to detect with certainty.

14.4 The risks of adverse effects of lead increase as larger amounts are absorbed but the level at which harm starts to be done is not known at present. Blood lead concentrations are reported in µg/100ml and the current Department of Health (DH) advisory action limit in the UK is 25 µg/ 100ml of blood. If a person's blood lead concentration reaches the action level, the exposure to lead should be investigated and reduced.

14.5 Lead is absorbed in humans via four main pathways, food, air, water, and soil or dust. The sources and pathways are shown diagrammatically in Figure 14.1 [1]. Exposure from these different sources or pathways varies considerably between individuals and population groups.

Lead in food

14.6 Diet is the main source of lead exposure for most people who do not spend much time near areas of high exposure, such as lead processing works. Food can be contaminated by lead from the air, water and soil (as well as more directly from food storage containers) but there is considerable uncertainty about their relative contributions.

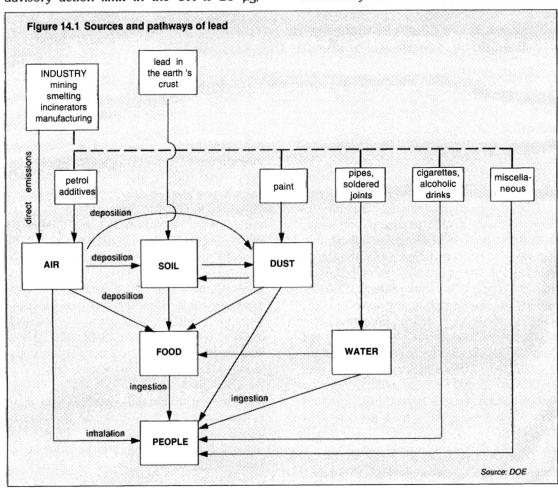

Figure 14.1 Sources and pathways of lead

Source: DOE

14.7 The maximum concentration of lead allowed in food has been reduced from a general limit of 2 mg/kg in 1961 to 1 mg/kg in 1979. There are lower limits for baby foods. The World Health Organisation (WHO) has recommended a Provisional Tolerable Weekly Intake (PTWI) of lead for adults of 50 μg/kg body weight (equivalent to 3 mg for a 60 kg adult). For infants and children the recommended PTWI is 25 μg/kg body weight.

14.8 Since the early 1970s, the intake of lead (and other contaminants) from food has been assessed using samples collected from the Total Diet Study. The National Food Survey gathers information on the purchases of food by a nationally representative sample of about 7,000 households each year, and this information is used to build up a picture of the average person's diet in the UK. The results from the Total Diet Study for lead intake from food over the period 1971 to 1987 are given in Figure 14.2 [2]. Two lines are shown which give a range of lead intakes. The upper values assume that where samples are below the limit of detection (LOD), such values are at the LOD; and the lower values assume that where samples are below the limit of detection such values are zero. The figure shows that lead intake from food has declined substantially since the mid 1970s, by around 5 per cent per year on average. In 1987 intake was of the order of 60 μg per day.

Lead in air

14.9 Lead is released into the air by emissions from, for example, leaded petrol, coal and from metal works. In recent years airborne lead concentrations have been decreasing owing to the changeover to unleaded petrol (though lead in petrol remains the largest single source of emissions into the air) and greater controls on emissions from industrial processes. However, there is little danger to most people of breathing in excessive amounts of lead. Exposure is more likely to be indirect, following deposition onto soil or plants.

Lead in water

14.10 For most people exposure to lead through water is also generally low in comparison with exposure from food. Contamination generally occurs during distribution through lead piping and may be a particular problem in areas where the water is soft and can dissolve lead. For some groups, especially bottle-fed babies in these areas, water can be the most important source of lead intake.

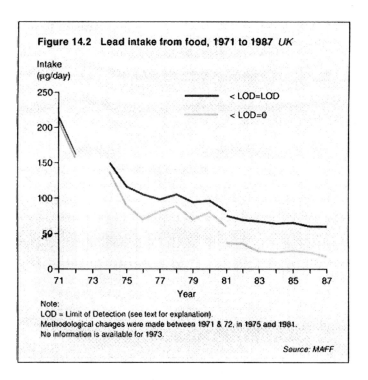

Figure 14.2 Lead intake from food, 1971 to 1987 *UK*

Note:
LOD = Limit of Detection (see text for explanation).
Methodological changes were made between 1971 & 72, in 1975 and 1981.
No information is available for 1973.

Source: MAFF

14.11 The Water Supply (Water Quality) Regulations 1989 and the Water Supply (Water Quality) (Scotland) Regulations 1990 (collectively referred to as the Water Quality Regulations), set a standard for lead of 50 μg/l at the time of supply (when the water passes from the water undertaker's pipe into the consumer's pipe). Where there is a risk of 50 μg/l being exceeded at a consumer's tap, water undertakers must treat the water to reduce the risk. In 1991 water undertakers in England and Wales took over 58,000 samples, and found that 1736 (3 per cent) exceeded 50 μg/l. In Scotland, from over 2,600 samples, 115 (4 per cent) were found to exceed the standard in 1990. Water undertakers are investigating treatment options, and where these can reduce lead concentrations they will be installed, in most cases by 1995.

Lead in soil

14.12 Lead in soil may derive from the earth's crust or it may be deposited from the air. It may also be present in sewage sludge or other waste material used as a fertiliser on agricultural land. Other sources could be coal ash, batteries and paint. Crops grown in areas with a high level of lead in soil and in areas where there is atmospheric contamination (for example near smelters or directly beside roads) may contain relatively high concentrations of lead.

14.13 Most plants however do not readily absorb lead, and contamination results from particles of lead on their surfaces. The highest concentrations of lead in plants have been found in leafy vegetables, but removing outer

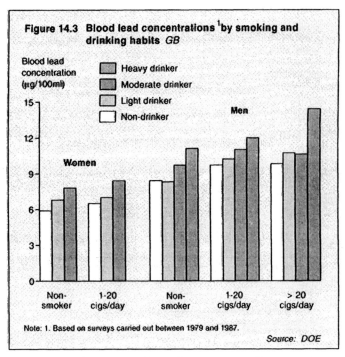

Figure 14.3 Blood lead concentrations [1] by smoking and drinking habits *GB*

Blood lead concentration (µg/100ml)

- Heavy drinker
- Moderate drinker
- Light drinker
- Non-drinker

Women   Men

Non-smoker / 1-20 cigs/day / Non-smoker / 1-20 cigs/day / > 20 cigs/day

Note: 1. Based on surveys carried out between 1979 and 1987.

Source: DOE

leaves and washing is usually enough to reduce contamination to acceptable levels.

Concentrations of lead in blood

14.14 A programme of surveys was carried out to investigate changes in concentrations of blood lead over the period 1984 to 1987 [1]. One of the aims of this programme was to assess in broad terms the effect of the reduction of lead in petrol (see also the sections on unleaded petrol in Chapter 2). A survey of lead in blood was also carried out on a representative sample of adults in GB between 1986 and 1987 to provide information on diet and nutrition [3].

14.15 Results from these and earlier EC surveys [4] showed that average blood concentrations were generally low and only a small proportion of people had raised

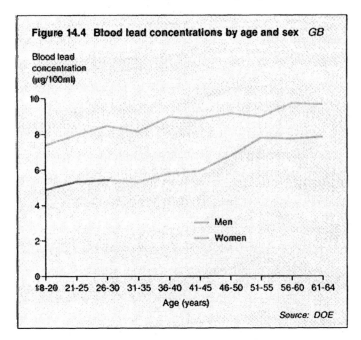

Figure 14.4 Blood lead concentrations by age and sex *GB*

Blood lead concentration (µg/100ml)

Men
Women

18-20 21-25 26-30 31-35 36-40 41-45 46-50 51-55 56-60 61-64
Age (years)

Source: DOE

levels. In the diet survey 15 of the 1,855 samples had blood lead concentrations above 25 µg/100ml, the level above which DH advise that the person's environment should be investigated for sources of lead and steps taken to reduce exposure [5]. Five of the elevated concentrations were above 35 µg/100ml but these are thought to have arisen from occupational exposure or DIY activities involving stripping paint with a high lead content.

14.16 Results also show that blood lead concentrations were significantly higher in men, older people, heavier smokers and drinkers, and in certain parts of the UK such as North West England. These points are illustrated in Figures 14.3 to 14.5 which show blood lead concentrations by smoking and drinking habits; by age and sex; and by region and country.

Cadmium

14.17 Cadmium is distributed widely throughout the natural environment and is also a by-product of zinc refining. Ninety five per cent of primary cadmium is obtained from zinc production, and the level of production of zinc, not the demand for cadmium, governs supply. The use of cadmium has increased greatly since the 1940s and UK consumption fluctuated in the late 1980s between 1,100 and 1,400 tonnes per annum [6].

14.18 For the general population exposure to cadmium is derived mainly from food although relatively minor amounts can be taken up from drinking and through inhalation: smokers are exposed to and can take up higher levels. The main sources of cadmium in the diet are from atmospheric deposition, and the use of fertilisers and sewage sludge on agricultural land. The average weekly intake of cadmium from food is estimated to be less than 140 µg/person [7]. This is less than the PTWI for cadmium of 420 µg for a 60 kg adult [8]. People will take in more cadmium if they regularly consume kidney and brown crab meat, both of which store cadmium.

14.19 The accumulation of cadmium in the body can be measured by concentrations in the kidneys. A survey carried out in 1986 [9] aimed to determine the concentration of cadmium in human kidneys and to assess variations by age, sex, smoking habits and cause of death. The study was the first of its kind in the UK and provided benchmark data on cadmium concentrations based on laboratory analyses of almost 1,000 kidneys. The results of the study showed that there is a high variability in cadmium concentrations, related to factors such as age and smoking habits.

14.20 Accumulation of cadmium in the kidney above a critical concentration can affect the reabsorption of amino acids, glucose and minerals, which are vital compounds the body needs to remain healthy. Some cadmium compounds are known to cause cancers when inhaled by laboratory animals; human data are limited, but the International Agency for Research on Cancer has classified cadmium as probably carcinogenic in humans.

14.21 Figure 14.6 shows how cadmium concentrations in kidneys were found to vary with age. Concentrations (in the cortex of the kidney) were found to be successively higher in age groups up to 50-59 years after which they were successively lower.

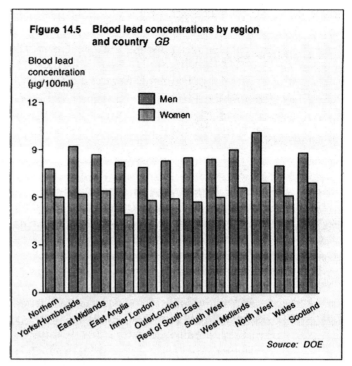

Figure 14.5   Blood lead concentrations by region and country GB

Source: DOE

Table 14.1   Mean [1]cadmium concentrations in kidneys by smoking habits UK

µg/g

Smoking category	Cortex [2]	Medulla [2]
Heavy smoker	19.1	9.7
Light smoker	16.2	8.3
Non-smoker	13.9	7.0
Not known	14.9	7.6
Overall	15.3	7.8

Notes :
1. Geometric mean
2. Parts of the kidney

Source : Aughey et al (1986)

14.22 Concentrations of cadmium in kidneys were also found to be consistently higher in people who smoked compared with non-smokers, across all age groups. Non-smokers had the lowest average cadmium concentrations, light smokers (including pipe and cigar smokers) had concentrations that were some 15 to 20 per cent higher; heavy smokers (over 25 cigarettes smoked a day) had concentrations that were 15 to 20 per cent above those of light smokers. Table 14.1 shows cadmium concentrations in the

cortex and medulla parts of kidneys, by smoking habits.

14.23 Cadmium can be present in small amounts in some materials used in plumbing systems. The Water Regulations set a standard for cadmium of 5 µg/l at the time of supply. In 1990, water undertakers took approximately 4,000 samples from consumers taps and found that only 16 (0.04 per cent) exceeded 5 µg/l.

Other metals

14.24 Trace metal concentrations have been measured in air at a number of sites in the UK (eg zinc and iron) and some examples of results from monitoring of trace elements

Table 14.2   Metals: estimated dietary intake

	mg/person/day
Mercury	0.003
Aluminium	6
Antimony	0.002-0.29
Chromium	0.08-0.10
Cobalt	0.03
Indium	0-0.023
Nickel	0.22-0.23
Thallium	0.005-0.06

Source DOE

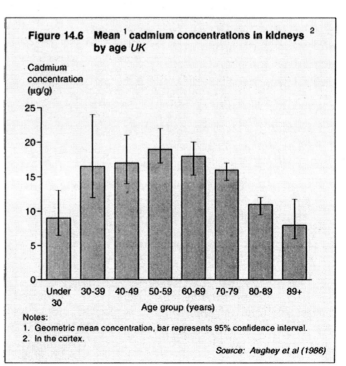

Figure 14.6   Mean [1] cadmium concentrations in kidneys [2] by age UK

Notes:
1. Geometric mean concentration, bar represents 95% confidence interval.
2. In the cortex.

Source: Aughey et al (1986)

are given in Chapter 2 on air quality and pollution. Values for estimated dietary intakes for various other metals are given in Table 14.2. The dietary intake of mercury is estimated at 0.003 mg/day for a 60 kg adult which is much less than the WHO PTWI of 0.3 mg/person/week when taken over the same period [10].

Dioxins

14.25 Dioxins form a group of 210 closely related chemicals, most of which are considered to pose no health hazard at the levels commonly found. However there are 17 of these compounds, especially 2,3,7,8 - tetrachlorodibenzo-p-dioxin (2,3,7,8-TCDD), which are of rather more concern because of their toxicity to certain species of laboratory animals. Dioxins have no known use and are not produced intentionally, except in very small quantities for analytical purposes. They are formed as by-products in some chemical processes and, in very small quantities, in various combustion processes such as waste incineration and coal burning. They are all around us in very low concentrations, typically of the order of one nanogram per kg. They are also present in effluent from sewage treatment works and in sewage sludge. They are stable solids, almost insoluble in water and are strongly absorbed by soil and organic matter, where they may persist for many years.

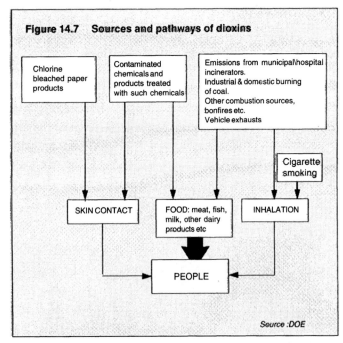

Figure 14.7 Sources and pathways of dioxins

Source :DOE

14.26 The main sources and pathways of dioxins to humans are shown (very approximately) in Figure 14.7. The main pathway to humans is through food and they are most likely to enter the food chain when grazing animals ingest contaminated plants and soil. They are absorbed via the

digestive tract and accumulate in fatty tissues. People may also become exposed to dioxins by breathing them in and by skin contact although for most people these pathways are of little significance.

14.27 Dioxins have been found in meat, cows' milk, fats and oils, milk products, fish, and other foods; also, in the air, and in consumer products particularly where chlorine bleaching processes have been used in their production (eg in manufacture of sanitary products and disposable nappies).

14.28 The health effects of dioxins are difficult to identify, because in studies which have been carried out, people have not been exposed to pure dioxins, but to mixtures of substances in which these compounds were contaminants. The most reliable indicator of extreme acute exposure to dioxins is thought to be the skin condition chloracne, although damage to the nervous system and transient liver damage have also been observed.

14.29 The UK Committee on the Toxicity of Chemicals in Food, Consumer Products and the Environment (COT) [11] concluded that no deaths have been attributed conclusively to any dioxin in the UK or elsewhere. Nor did it find any convincing evidence of a link between exposure to dioxins and cancer. No major adverse health effects, other than chloracne, have so far been identified in people exposed to much higher levels of dioxins than the general population.

14.30 Although knowledge of past levels of dioxins in the environment is very limited, current evidence is that the amount added to the environment is decreasing [12], but amounts already in the environment will take decades to disappear. Measures are being taken to control emissions (ie controls on combustion conditions and emissions from all incinerators). It is also expected that the phasing out of leaded petrol will substantially reduce dioxin emissions from vehicles. Dioxin emissions are also likely to be reduced when stubble burning is banned in England and Wales after the 1992 harvest.

14.31 A survey of dioxins in the British diet found that levels of dioxins in UK food are very low and average dietary intake is estimated to be 0.125 ng TEQ/day for a 60 kg adult [15], which is less than a quarter of the WHO Tolerable Daily Intake. There is little likelihood of a risk to the general population from these chemicals in food.

14.32 Recently, surveys of dioxins in cows' milk have identified one locality in the UK (in Bolsover) where dioxin concentrations are relatively high (above 0.7 ng TEQ/kg -

---

Guidelines for emissions and daily intake of dioxins   *Box 14.2*

In the UK, emissions of dioxins from certain processes are controlled under the Environmental Protection Act 1990. Process Guidance Notes for these processes have been recently published and specify an emission limit of 1 ng Toxic Equivalents/m^3 (ng TEQ/m^3) and a guide value of 0.1 ng TEQ/m^3. The TEQ takes account of the toxicity of each dioxin, relative to 2,3,7,8-TCDD, and weighting factors are used to give the equivalent concentration of 2,3,7,8-TCDD.

A WHO Europe expert group meeting at Bilthoven in December 1990 set a Tolerable Daily Intake of 0.01 ng/kg body weight/day for 2,3,7,8-TCDD, the most toxic dioxin [13]. This has been endorsed in the UK by the independent Committee on the Toxicity of Chemicals in Food, Consumer Products and the Environment (COT) and used to derive a "Maximum Tolerable Concentration" in milk of 0.7 ng TEQ/kg on a whole product basis [14].

---

see Box 14.2) and milk from two farms has been withdrawn from supply [14]. Investigations currently being carried out include work to determine the source of contamination in this area, and the rate at which the levels of dioxins in milk fall when cattle are fed a dioxin-free diet.

PCBs

14.33 Polychlorinated biphenyls (PCBs) have been used extensively in electrical components such as transformers and capacitors. From 1971, PCBs produced in the UK were limited to use in products where the PCB was contained, and were finally phased out in 1977 (although production continued in some other European countries until 1984). As a result of earlier use, it is estimated that there are still significant quantities of PCBs contained in electrical equipment in use in the UK.

14.34 PCBs from obsolete equipment are currently disposed of by high temperature incineration. At 1,200 to 1,400 degrees Celsius, PCBs can be destroyed with efficiencies of 99.9999 per cent and the flue gas treated to prevent dioxin formation. Limited and diminishing quantities of PCBs in commercial and household wastes are for the most part disposed of at landfill sites. In March 1990, it was agreed at the Third International Conference on the Protection of the North Sea that member states would take measures to phase out and destroy all identifiable PCBs, substantially by 1995 and completely by 1999 at the latest.

14.35 Concentrations of PCBs in drinking water are negligible. There are no specific standards in the Water Quality Regulations, but PCBs are included in the standard for pesticides and related products. Fatty foods have been analysed for PCBs and it is estimated that the dietary intake by the average consumer is less than 10 μg/person/day [16]. PCBs, however, are found at higher levels in oily fish, such as mackerel or herring; the estimated dietary intake for consumers of large quantities of oily fish is up to 38 μg/person/day.

Nitrate

14.36 The Water Quality Regulations set a maximum nitrate concentration in drinking water of 50 mg/l. Higher concentrations may cause infantile methaemoglobinaemia (a condition in which the capacity of the blood to carry oxygen is reduced) and health authorities monitor for the condition where they are informed of nitrate levels above 50 mg/l. However, the condition is very rare in the UK. Only 14 cases have been reported in the last 42 years which were attributable to nitrate in drinking water. The last reported confirmed case in the UK was in 1972. Water undertakers have undertaken to take steps to secure compliance with the 50 mg/l standard by 1995 or earlier, with one exception, for the few supplies that do not comply at present.

14.37 The latest results of monitoring show that the estimated average dietary intake of nitrate by consumers in the UK is 54 mg/person/day [17]. For vegetarians, however, the dietary intake is estimated to be in the range 185-194 mg/person/day due to the relatively high concentrations of nitrate in vegetables. These estimates are within the EC Scientific Committee for Food's recommended Acceptable Daily Intake of 0-219 mg/kg for a 60 kg adult [18]. See also Chapters 4,5,7 and 15 on soil, land, water quality and pressures on the environment.

Pesticides

14.38 There are approximately 450 individual substances used in the UK as pesticides. They are used mainly in agriculture but they also have non-agricultural applications such as usage in wood preservatives and antifouling paints.

14.39 Most pesticide intake is through food consumption. The Working Party on Pesticide Residues (WPPR) reports annually the results of extensive monitoring of the food supply. The latest reports covering the period 1988 to 1990 show that from a sample of 7,000 foodstuffs, 69 per cent were found not to contain detectable pesticide residues.

14.40 Pesticides may also occur in drinking water. The Water Quality Regulations incorporate EC non-health related standards for pesticides in drinking water of 0.1 µg/l for individual substances and 0.5 µg/l for the total of the detected concentrations of individual substances. Some pesticides have been detected in excess of the 0.1 µg/l standard in some water supplies. Water undertakers in England and Wales have undertaken to take steps to secure or facilitate compliance with the pesticides standard, in most cases by 1995. The concentrations of pesticides detected during 1989 to 1991 however, were far smaller than amounts which are known to be harmful or are likely to damage public health. See also Chapters 4,7,8,10 and 15 on soil, water quality, the marine environment, wildlife and pressures on the environment.

Toxic algae

14.41 Toxic algal blooms (blue-green in freshwater, mainly dino-flagellate in marine water) have always occurred from time to time. Nutrient enrichment of waters, particularly by phosphate and nitrate compounds is regarded as a possible contributory factor to enhanced algal growth. See also the references to algal blooms and eutrophication in Chapters 7 and 8.

14.42 Blue-green toxic algal blooms can kill fish and other aquatic life, or domestic animals which drink contaminated water from ponds, lakes or reservoirs. Problems are caused by the ingestion of algal suspensions. These however, are removed in the course of treating drinking water and no incidents of illness in humans have been recorded which could be attributed to the presence of algae or algal toxins in treated drinking water supplies.

14.43 There are numerous species of marine algae that can produce toxins which kill fish. People can also become ill through eating bivalve shellfish such as oysters, mussels and clams, which have accumulated algal toxins in the course of their normal feeding. Two important types of poisoning, Paralytic Shellfish Poisoning (PSP) and Diarrhetic Shellfish Poisoning (DSP) can result. There have been no recorded cases of DSP in shellfish in UK waters, but PSP in shellfish occurs seasonally along the north east coasts of England and south east Scotland. Improved monitoring of shellfish stocks since an incident in 1968 when 78 people became ill from PSP, has meant that there have been no reported cases of illness since then. See also Chapters 7 and 8 on water quality and the marine environment.

Cryptosporidium

14.44 Cryptosporidium is a parasite found worldwide. It was first recognised as a cause of human disease in the mid 1970s. In the UK it is the sixth commonest cause of diarrhoea. Cryptosporidiosis may be contracted from a number of sources including animals and person to person contact. In recent years, there have been a few outbreaks in which water was the route of infection; one particular outbreak occurred in Oxfordshire and Swindon in early 1989, following which an expert group was set up to advise on the significance of crypto-sporidium in water supplies. The group's report [19] and the Government's response [20] were published in 1990.

14.45 A national survey in 1990 revealed that the oocysts (which are the environmentally resistant transmissible form of cryptosporidium excreted in the faeces of an infected host) of cryptosporidium were absent or present at only very low levels in water supplies. There were two outbreaks of cryptosporidiosis during 1990 in which water supplies may have been the route of infection.

14.46 As a result of recommendations from the expert group, the Drinking Water Inspectorate (DWI) initiated a national programme of research which includes improving the monitoring of and assessing the viability of cryptosporidium in water supplies, and determining the minimum infective dose for man. A progress report has been published in 1992 [21].

Disinfection by-products

14.47 Various disinfectants are used in water treatment. During this process the disinfectant may react with organic substances many of which are naturally present in source waters. The chemicals formed as a result of these reactions are called disinfection by-products and there are health concerns over some of these.

14.48 Chlorine for example, is widely used as a disinfectant and produces chlorination by-products, including trihalomethanes (THM) such as chloroform. Some individual disinfection by-products have been found to cause cancers in laboratory animals when large doses have been administered over long periods, but there is no strong evidence of any health risk from the low concentrations of chlorination by-products in water supplies. Nevertheless, concentrations should be kept as low as is compatible with efficient disinfection.

PAHs

14.49 Polycyclic aromatic hydrocarbons (PAHs) are emitted by motor vehicles and other processes involving incomplete combustion. Some PAHs, such as benzo-(a)pyrene, are known to be carcinogenic to animals.

14.50 PAHs may occur in water supplies where water is distributed in mains laid before the mid 1970s; these were given an internal anti-corrosion coating of coal-tar pitch before being laid in the ground and coal tar pitch can contain up to 50 per cent PAHs.

14.51 Surveys during the 1980s of the occurrence of coal-tar pitch particles and PAHs in water supplies showed that PAHs were not detected in the majority of samples taken, but concentrations occasionally exceed the standards now set in the Water Quality Regulations of 0.01 µg/l as an annual average for benzo(a)pyrene and 0.2 µg/l for the sum of six specified PAHs which includes benzo(a)pyrene. In 1991, water undertakers in England and Wales made over 17,000 determinations for PAHs; 2 per cent of which exceeded 0.2 µg/l. The requirements for benzo(a)pyrene were met in nearly all water supply zones. Water undertakers have given undertakings to take steps to secure or facilitate compliance with the standards for PAHs.

14.52 Low levels of PAHs also occur in air as a result of combustion processes such as coal burning, waste incineration and vehicle emissions. See also Chapter 7 on water quality.

Marine waters and sewage

14.53 Marine waters polluted with sewage effluent are potentially harmful to health though risks of serious illness are minimal. A four year research programme is being carried out to quantify the risks of minor illnesses from bathing in sewage contaminated sea water, with final results available in 1993.

14.54 Bivalve molluscan shellfish such as oysters, clams, mussels and cockles are farmed and fished at many inshore marine sites for subsequent sale and human consumption in the UK. Such shellfish filter seawater in the course of their normal feeding, and this results in microbial pathogens derived from sewage becoming concentrated and retained in their flesh. Since these species are traditionally consumed raw, or only lightly cooked, precautions (such as ensuring that farms and fishing areas are sited away from sewage outfalls) have to be taken to deal with the public health risks from the discharge of sewage into the sea.

14.55 The diseases most commonly associated with shellfish consumption are viral gastroenteritis and viral hepatitis (hepatitis A) [22]. Incidences are recorded by the Public Health Laboratory Service. One hundred and eighty three outbreaks were recorded in England and Wales between 1941 and 1990. Of these, 107 outbreaks, involving over 3,500 cases, occurred in the last decade, reflecting not only growth in the popularity of shellfish as a food, but also improved surveillance and awareness of shellfish-related illness.

14.56 Extensive monitoring programmes were started in late 1991 to classify all known commercial shellfish harvesting areas in the UK into categories according to the degree of microbiological contamination in the shellfish. From January 1993, these categories will determine whether shellfish can be eaten direct from waters or whether prior treatment is required. Results of the programme will be published and will provide an indication of whether measures to improve sewage treatment under the EC's Urban Waste Water Directive are proving successful. See references to this Directive in Chapter 7, and Chapter 8 on the marine environment.

Radioactivity

14.57 The main concerns about environmental radiation and health centre on the possible effects of nuclear establishments (in particular the investigation of leukaemia clusters), ultraviolet (UV) radiation and radon in homes (radon is a radioactive gas, which is discussed in the section on indoor pollution, in this chapter). In the UK, about 3 per cent of cancer deaths may, it is estimated, result from the effects of all types of radiation.

Radioactivity and leukaemia clusters

14.58 In 1984 the Black Advisory Group investigated the possible increased incidence of cancer in west Cumbria, around the Sellafield nuclear reprocessing plant. Initial studies found that there was a higher incidence of leukaemia in young people in the vicinity of the plant. Further investigation was undertaken by the Committee on Medical Aspects of Radiation in the Environment (COMARE). This research concluded that there were also statistically significant excesses of leukaemia in young people near the Dounreay reprocessing plant in Scotland, and of leukaemia and other childhood cancers near the atomic weapons research establishments at Aldermaston and Burghfield in England.

14.59 However, after further, more detailed analyses of the data, COMARE judged that radioactive discharges from the nuclear establishments in question were most unlikely to explain the findings of the studies undertaken. The theoretical link remains, and findings in this area have been controversial.

14.60 Further research work is underway into other possible mechanisms of exposure to radiation, for example, a study is underway into the health of children of people working at plants where they come into contact with radioactivity. See also Chapter 13 on radioactivity.

Ultraviolet radiation

14.61 The ozone layer shields the earth from lethal ultraviolet-C radiation from the sun, and substantially reduces the amount of harmful ultraviolet-B radiation from reaching the earth's surface. High doses of UV radiation can cause skin cancers and cataracts and reduce immunity to disease.

14.62 Depletion of the ozone layer is expected to lead to an increase in the amount of ultraviolet radiation reaching the surface of the earth. Measurements of UV radiation in GB began in 1989 but it may take some years before any trends can be detected, because of natural variations such as cloud cover. See also Chapter 3 on the global atmosphere.

Indoor pollution

14.63 The presence of toxic and other substances indoors may be a direct or indirect cause of ill health or discomfort. People spend on average 90 per cent of their time indoors and 75 per cent of that time in the home. Exposure to indoor pollutants can form a significant proportion of an individual's total exposure and pollutants resulting from indoor sources can contribute to wider environmental pollution. The following sections discuss some of the main concerns.

Radon

14.64 Radon is a gas formed from the radioactive decay of uranium which is naturally present in small quantities in soils and rocks throughout the UK. Radon gas can accumulate in buildings, and inhalation of the gas or its decay products, known as radon daughters, increases the risk of lung cancer. Of 39,000 UK deaths from lung cancer each year, current estimates suggest that 5 to 6 per cent may be partly attributable to radon. Householders have been recommended to take precautions to reduce radon concentration when it gets above a prescribed action level. See also the section on radon in Chapter 13.

Environmental tobacco smoke

14.65 Tobacco smoke is a major source of many pollutants including organic compounds, particulates, carbon monoxide and nitrogen dioxide. In addition to irritating the eyes, nose and throat, exposure to environmental tobacco smoke (passive smoking) can cause respiratory illness in vulnerable groups like children, and can increase the risk of lung cancer in non-smokers.

14.66 In December 1991 a code of practice was issued on smoking in public places. This states that non smoking should be the norm in buildings frequented by the public, for example waiting rooms, shops and restaurants, with special provision for smoking where appropriate.

The Small Area Health Statistics Unit (SAHSU)   Box 14.3

In the course of its enquiry into childhood cancers around the nuclear processing plant at Sellafield in Cumbria in 1984, the Black Advisory Group recommended that health statistics for small areas should be monitored near other potentially hazardous installations. Consequently the SAHSU, an independent unit, was established to investigate concern about alleged health risks and effects at sites of environmental pollution such as industrial works. The SAHSU holds data on UK health "events", including death and cancer registrations, identified by geographical location, and can produce event rates and other statistical analyses relating to small areas.

The unit became fully operational in July 1990 with the commencement of the first study into cancer incidence in the vicinity of waste solvent and oil incinerators. The results of this study were published in April 1992 [24].

Condensation/damp

14.67 The condition of a building itself can also lead to increased levels of exposure to indoor air pollutants. Wet surfaces, whether caused by condensation or damp, encourage the formation of moulds (and possibly the proliferation of mites, although the link between mites and damp is not proven) which may lead to allergic illnesses. About three and a half million homes are affected by damp [23].

14.68 Spores which in the right circumstances are responsible for the proliferation of mould, are present in the air in all dwellings. Typical average winter levels are about 100 to 450 spores per cubic metre of air. To develop into mould indoors, spores require four things: oxygen, a reasonable temperature, nutrients and moisture. Under normal living conditions, the first three of these are usually present inside buildings, but there will usually only be enough moisture if the building surfaces suffer from condensation or other dampness. When conditions are right for them, typical spore levels may increase tenfold, and levels up to 450,000 spores per cubic metre have been detected in homes suffering from severe damp and mould.

Domestic activities

14.69 Cooking and heating with gas may lead to increased levels of carbon monoxide and nitrogen dioxide in indoor air. If the supply of fresh air or ventilation is insufficient, lethal concentrations of carbon monoxide can result and about a hundred deaths a year may be attributable to carbon monoxide poisoning resulting from faulty appliances and inadequate ventilation. The risk however is declining with improvements in appliances and ventilation. There is some evidence that nitrogen dioxide from gas cookers may increase the risk of respiratory illness in children, but the results of epidemiological studies have been inconclusive.

14.70 The use of consumer products such as cleaning materials, aerosols and perfumes, and the removal of old paint and application of new paint, can also lead to increased levels of indoor air pollutants known as volatile organic compounds. There is little evidence however, about the possible effects of long term exposure to low levels of indoor pollutants resulting from activities of this kind.

Building materials and furnishings

14.71 Pollutants are also found in some building materials. For example, asbestos has been used, in association with other materials, for a number of purposes, including sprayed insulation, lagging suspended ceilings and partitioning. Asbestos, when damaged, is known to be a carcinogen and can cause lung cancer. Organic pollutants, in particular formaldehyde, are given off by building materials such as urea formaldehyde foamed insulation and pressed wood products such as particle board. Organic pollutants may also come from furnishings and fittings.

Noise

14.72 Noise may damage people's health, particularly their hearing. Prolonged exposure to high noise levels may cause permanent loss of hearing, but this is commonest among people who work in noisy environments, such as factory production lines and construction sites. Exposure to high noise levels can cause ringing in the ears (tinnitus). There are other possible adverse effects to human health at extreme levels. Continuous noise can lead to stress-related health problems such as raised blood pressure and increased risk of heart and blood supply disorders; exposure to noise during sleep may alter normal sleep patterns. However, there is no conclusive evidence of any more general adverse effects of noise on health beyond noise induced hearing loss and tinnitus. See Chapter 12 for more detailed information on noise.

References and further reading

1. Department of the Environment, (1990). UK Blood Lead Monitoring Programme 1984 to 1987 - Results for 1987. Pollution Report No 28. HMSO.

2. Ministry of Agriculture, Fisheries and Food: The Working Party on Inorganic Contaminants in Food, (1989). Survey of Lead in Food, Progress Report. Third Supplementary Report. The twenty-seventh Report of the Steering Group on Food Surveillance. HMSO.

3. Gregory, J., Foster, K., Taylor, H. and Wiseman, M., (1990). The Dietary and Nutritional Survey of British Adults. HMSO.

4a. Department of the Environment, (1981). Pollution Report No 10. HMSO.

4b. Department of the Environment, (1983). Pollution Report No 18. HMSO.

5. Department of Environment and Welsh Office, (1982). Lead in the Environment. Circular 22/82. HMSO.

6. March Consultants, (1990). Impact of Cadmium and Cadmium Compounds Directive. Report to Department of Trade and Industry.

7. Ministry of Agriculture, Fisheries and Food, (1983). Survey of Cadmium in Food, First Supplementary Report. Food Surveillance Paper No 12. HMSO.

8. World Health Organisation, (1972). Evaluation of certain Food Additives. Sixteenth Report of Joint Food and Agriculture Organisation, World Health Organisation Expert Committee on Food Additives. WHO.

9. Aughey, E., Fell, G.S., Scott, R., Quinn, M.J., (1986). Cadmium Concentrations in Human Kidneys from the UK. Human Toxicology, 5.

10. Ministry of Agriculture, Fisheries and Food, (1987). Survey of Mercury in Food, Second Supplementary Report. Food Surveillance Paper No 17. HMSO.

11. Committee on the Toxicity of Chemicals in Food, Consumer Products and the Environment, (1989). The Human Health Hazards of PCDDs and PCDFs; summarised in Dioxins in the Environment, Pollution Paper No 27. HMSO.

12. Kjeller, L-O, Jones, K.C., Johnston, A.E. and Rappe, C., (1991). Increases in the Polychlorinated Dibenzo-p-dioxin and Furan (PCDD/PCDFs). Environmental Science Technology 25, pp1619-1627.

13. WHO Regional Office for Europe, (1991). Consultation on Tolerable Daily Intake from food of PCDDs and PCDFs. Summary report EUR/ICP/PCS 030(S) 0369n. WHO Regional Office for Europe, Copenhagen.

14. Ministry of Agriculture, Fisheries and Food, (1991). Results of testing for Dioxins in Milk. News release FSD 43/91. MAFF.

15. Ministry of Agriculture, Fisheries and Food, (1992). Survey of Dioxins in Food. Food Surveillance Paper No 31. HMSO.

16. Ministry of Agriculture, Fisheries and Food, (1983). Polychlorinated biphenyl (PCB) Residues in Food and Human Tissues. Food Surveillance Paper No 13. HMSO.

17. Ministry of Agriculture, Fisheries and Food, (1992). Nitrate, Nitrite and N-Nitroso Compounds in Food, Second Report. Food Surveillance Paper No 32. HMSO.

18. Commission of the European Communities Scientific Committee for Food, (1992). Report of the Scientific Committee for Food on Nitrate and Nitrite, 26th Series. CEC.

19. Departments of Environment and Health, (1990). Cryptosporidium in Water Supplies. Report of the Group of Experts. HMSO.

20. Departments of Environment and Health, (1990). Government's response to Report on Cryptosporidium in Water Supplies by the Group of Experts. HMSO.

21. Department of the Environment, (1992). Cryptosporidium in Drinking Water - Research Progress. Report of the Cryptosporidium Research Steering Committee. DOE.

22. Scoging, A.C., (1991). Illness associated with Seafood. Communicable Disease Report, 1 (11), pp 117-125. Public Health Laboratory Service, Communicable Disease Surveillance Centre.

23. Department of the Environment, (1990). Memorandum from the DOE to the Commons Environment Committee, Enquiry into Indoor Pollution.

24. Elliot, P., Hills, M., Beresford, J., Kleinschmidt, I., Jolley, D., Pattenden, S., Rodrigues, L., Westlake, A. & Rose, G., (1991). Incidence of cancers of the larynx and lung near incinerators of waste solvents and oils in GB. SAHSU.

Department of Health, (1991). The Health of the Nation. A Consultative Document for Health in England. HMSO

Department of Health, (1992). The Health of the Nation. A Strategy for Health in England. Cmnd 1986. HMSO

Department of Environment, (1986). Nitrate in Water. Pollution Paper No 26. HMSO.

Department of Environment, (1986). Organotin in Antifouling Paints, Environmental Considerations. Pollution Paper No 25. HMSO.

Department of Environment, (1989). Dioxins in the Environment. Pollution Paper No 27. HMSO.

United Nations Environment Programme, World Health Organisation, (1992). Global Environment Monitoring System. Human Exposure to Pollutants. Report on the Pilot Phase of the Human Exposure Assessment Locations Programme. UNEP/WHO.

15 Pressures on the environment

☐ The area of land used for growing crops in the UK has remained broadly constant between 1980 and 1990. Cereals covered 73 per cent of the land used for crops in 1990, compared with 79 per cent in 1980. The area used for growing oilseed rape increased fourfold in this period, and accounted for 8 per cent of the crop area by 1990 (Table 15.1). The number of cattle decreased by 10 per cent over the decade, while the number of sheep and lambs increased by 39 per cent over the same period (Figure 15.2).

☐ Between 1984 and 1990, 52,000 km of hedgerows in GB were removed while 26,400 km of new hedgerows appeared. This and other habitat changes have led to decreases in populations of plants, wild animals and other species reliant on traditional farming techniques and landscapes (15.11).

☐ Between 1980 and 1990, the total tonnage of pesticide active ingredient used decreased by 20 per cent (Figure 15.4). In the same period, the total area treated by pesticide increased by 9 per cent while the area used for growing crops and grass fell by 2 per cent (15.17).

☐ Total primary energy consumption in the UK in 1991 was equivalent to 207.7 million tonnes of oil, of which 187.2 million tonnes were fossil fuels (Figure 15.5). In 1991, electricity generated from non-fossil fuel resources (nuclear, natural flow hydro and wind power) amounted to 16.6 mtoe, while net imports of electricity accounted for 3.9 mtoe (15.23).

☐ Energy consumption by final users, ie direct use of primary fuels such as coal and natural gas, or after conversion to secondary fuels, amounted to 60 billion therms, or 142.2 mtoe in 1991, an increase of 20 per cent since 1960. The difference between primary energy consumption and consumption by final users, at 65.5 mtoe in 1991, was accounted for by the fuel used by energy producers, mainly to generate and transmit electricity.

☐ Since 1979, there has been a gradual decline in energy consumption by industry; in 1991 it accounted for 25 per cent of total final consumption, compared with 42 per cent in 1973 and 1960. The greatest area of growth was the transport sector; in 1991 it accounted for 31 per cent of total final consumption compared with 17 per cent in 1960 (Figure 15.6).

☐ It is estimated that improved insulation and heating appliance efficiency resulted in a reduction of delivered energy consumption in the domestic sector of about 32 per cent in 1989, compared with the energy which would have been required if insulation standards and heating appliance efficiency had remained at 1970 levels (Figure 15.8).

☐ Nuclear power stations accounted for 21 per cent of electricity supplied by UK generating companies in 1991, compared with 10 per cent in 1975 (15.54). In 1991, about 2 per cent of the fuel input for electricity generation came from hydropower (15.56).

☐ The volume of traffic on the roads has increased sixfold since the early 1950s (Figure 15.12). Road transport accounted for 93 per cent of passenger travel and 81 per cent of freight moved in 1990. The largest growth since the 1950s was in car use which increased more than tenfold, to over 330 billion vehicle km in 1990 (15.61).

☐ Between 1980 and 1990, the length of motorway increased by 20 per cent to 3,100 km, but the volume of traffic on motorways more than doubled. In the same period, the overall length of road network (including motorways) increased by about 5 per cent but traffic volume increased by 50 per cent (15.68).

☐ The total volume of freight traffic has increased by over 60 per cent since 1955 and the proportion carried by road has increased from 75 per cent to almost 81 per cent (15.71). The proportion of freight traffic carried by rail has declined from 21 per cent in 1955 to 6 per cent in 1990 (Figure 15.14).

15.1 Earlier chapters of this report concentrated on the condition of various environmental resources - air, water, land and wildlife. They identified and discussed various areas where man's activities have created pressure on the environment. This chapter looks at a few of the areas of most concern or interest, and describes what has been happening.

15.2 The chapter considers in particular, the pace of change and its impact on the environment in agriculture, the energy industries, and transport. It does not attempt to be comprehensive and where appropriate, refers to other chapters where issues are discussed in more detail. Nor does it look at the policies put in place to modify environmental impacts of many of the activities in the above sectors, except where these have had a significant effect on the trends described. Policy responses are the subject of the White Paper "This Common Inheritance" and its annual follow up Reports (the first published in September 1991, the second published in October 1992).

Agriculture

15.3 Agriculture now produces nearly 75 per cent of the types of food which can be produced in the UK, compared with about 60 per cent in 1970. Agricultural land management is essential for the maintenance of the quality of the environment. Centuries of cultivation have shaped the appearance of the countryside and much of the wildlife depends on the mix of habitats which has been created by agriculture. About 77 per cent of the UK land area is used for agriculture and some 630,000 people are engaged in the industry, which contributes about 1½ per cent of GDP.

15.4 Some farming practices however, can have an adverse effect on the quality of the

Table 15.1	Agricultural land use and crop areas 1980, 1985, 1990 UK		
			Thousand hectares
	1980	1985	1990
Total agricultural land	18,953	18,703	18,542
Total crop areas	4,972	5,224	5,013
of which			
Cereals	3,938	4,015	3,657
Other arable *	559	806	971
Potatoes	205	191	177
Horticulture	270	212	208
* of which			
Oil seed rape	92	296	390

Source: MAFF

environment. Using fertilisers, draining land, and changing the type of crops grown can have an effect on soil and water resources. Using pesticides (which include herbicides, insecticides and fungicides) to control weeds, insect pests and disease can reduce the numbers of some species of wild plants, insects, birds and mammals. Farm animals, particularly cattle and sheep, produce methane, one of the greenhouse gases, and emit ammonia and other odours. Waste products, silage effluent and slurry, if handled inappropriately, can cause contamination of watercourses or harm wildlife. For further discussion on many of these issues, see Chapters 2 to 7 and 10 on air quality, the global atmosphere, soil, land, water resources and quality, and wildlife.

15.5 More intensive farming in recent years has caused some of these problems, but others have been present for many years, unrecognised or only partly understood. Modern research and better measurement have given them new significance. This section looks at the basic trends in agriculture which may exert pressures on the environment.

Basic trends

15.6 The main trends in agriculture in the decade up to 1990 have been the decline in the area of land used for growing cereals, potatoes and horticultural crops, the increase in land used for growing other arable crops, in particular oilseed rape, the decline in the dairy herd and the increase in the number of sheep.

15.7 In 1990 there were 18.5 million hectares of land in agricultural use in the UK compared with 19.0 million hectares in use in 1980 (see Table 15.1). Of these, 5 million hectares were used for growing crops in 1990. The crop area increased from under 5 million hectares in 1980 to 5.3 million in 1987, but has since declined, partly as a result of the introduction of the Set Aside Scheme, which was introduced in 1988 by the European Community (EC) to contain the costs of the cereals regime.

15.8 There has been a significant change in the pattern of crops grown since 1980 (see Figure 15.1). Cereals covered 79 per cent of land used for crops in 1980, but by 1990 this had decreased to 73 per cent. The decline was mainly in barley, which was grown on 1.5 million hectares of land in 1990, compared with over 2.3 million in 1980. Land used for growing wheat increased from 1.4 million to just over 2 million hectares over the same period. Non-cereal crops were grown on 1.4 million hectares in 1990, compared with 1.0 million hectares in 1980. The largest increase was in the area used for growing oilseed rape, which increased fourfold in this period to 0.4 million hectares and accounted for 8 per cent of the total crop area in 1990.

15.9 In 1990 there were over 12 million cattle in the UK, a decrease of 10 per cent compared with 1980. In this period the number of beef cattle increased by 8 per cent, while the number of dairy cows decreased by 12 per cent. There were 44 million sheep and lambs in the UK in 1990, an increase of 39 per cent since 1980. The number of pigs decreased by 5 per cent in the same period to 7.4 million (see Figure 15.2).

15.10 These changes in patterns of farming result partly from improvements in farming productivity, partly from movements in market prices, and partly from measures taken under the EC's Common Agricultural Policy to encourage or discourage particular products. The environmental effects which give most concern are, as far as arable farming is concerned, the loss of habitat for

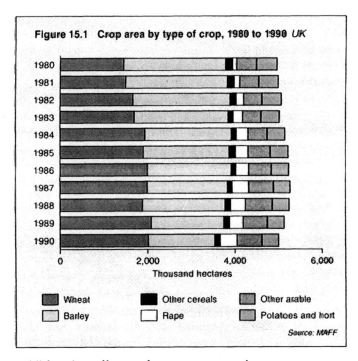

Figure 15.1 Crop area by type of crop, 1980 to 1990 *UK*

Thousand hectares

Wheat   Other cereals   Other arable

Barley   Rape   Potatoes and hort

Source: MAFF

wildlife; the effects of new crops such as oilseed rape and linseed on wildlife and landscape; and water pollution. There are also concerns about the changed appearance of the countryside, with fewer hedges and bigger fields, and unfamiliar crop colours. It has been suggested that the increased acreage of oilseed rape may be responsible for increases in the incidence of hay fever and other allergic reactions, though there is no direct evidence for this. As far as livestock farming is concerned, the main potential problems are overgrazing, air and water pollution. The large increase in sheep has had a damaging effect on heather in some uplands, with detrimental consequences for other wildlife. Intensive livestock units can give rise to farm waste disposal problems. See also Chapters 5 to 7, 10 and 11 on land, water resources and quality, wildlife and waste.

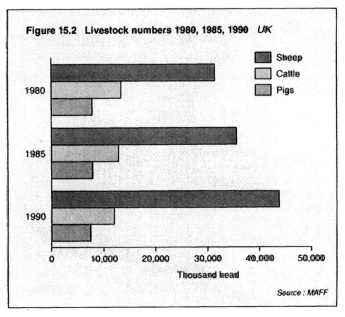

Figure 15.2 Livestock numbers 1980, 1985, 1990 *UK*

Sheep
Cattle
Pigs

Thousand head

Source : MAFF

15.11 Changes in agricultural practices have led to the loss of wetlands, heaths and other habitats. Between 1984 and 1990, 52,000 km of hedgerows in GB were removed while 26,400 km of new hedges appeared. Government grants encouraging the planting and management of hedgerows have now been supplemented with a new Countryside Commission scheme to encourage improved hedgerow management. The overall reduction in length of hedgerow and other habitat changes have led to decreases in populations of plants, wild animals and other species reliant on traditional farming techniques and landscapes. There are now a range of Government schemes such as Environmentally Sensitive Areas, Countryside Stewardship and in respect of SSSIs, which are specifically aimed at securing sympathetic management of valued habitats. See also Chapter 5 on land and Chapter 10 on wildlife.

15.12 The quality of the soil is an important feature of arable farming. Crops need the nutrients in soil for growth. Where crops are repeatedly harvested, these nutrients need to be replaced so that the soil does not become infertile. Without fertilisers, productivity would decline markedly and the UK would need to import more food. Some 80 million tonnes of animal manure and slurry from housed livestock are recycled back to the land each year as fertiliser, as are considerable quantities of sewage sludge. See also Chapters 4 and 11 on soil and waste. In addition, in 1991, some 1,525,000 tonnes of manufactured nitrogen, 380,000 tonnes of phosphate and 465,000 tonnes of potash were applied.

15.13 Problems however, can arise if farmers use more of these materials than plants need. For example, too much nitrogen, whether as inorganic fertilisers or as manures, can increase the amount of nitrate leaching from the soil into groundwater and surface water. The Anglian and Severn Trent water regions have been among the areas worst affected by this problem. The Government have established Nitrate Sensitive Areas to investigate ways of reducing leaching. See also Chapters 4 to 7 on soil, land, water resources and quality.

15.14 Manure, slurry and sewage sludge may also contain impurities which can be harmful. Pig slurry may contain concentrations of copper which is sometimes used as an additive in commercial pigfeed, and sewage sludge can be contaminated by high levels of heavy metals. The continuous use of manure, slurry and sewage sludge on land can lead to the accumulation of metals in the soil. Spreading farm wastes and silage effluent on land can also cause serious pollution of rivers and lakes. The number of reported water pollution incidents involving farm pollution, rose steadily during the 1980s reaching a peak of 4,141 in 1988 before falling back sharply to 2,889 in 1989. Figures for the following two years show a slight rise again with 3,325 incidents in 1991 although expressed as a percentage of all incidents it is the same as in 1989 (11 per cent) and lower than in 1981 (19 per cent). This relative improvement in recent years reflects a number of factors, including over £162 million spent by farmers on pollution control equipment since 1989. Recent Government initiatives include Regulations on the construction of farm waste facilities and a Code of Practice on Water Protection for farmers. See also Chapters 4 to 7 on soil, land, water resources and quality.

15.15 In Scotland, 521 recorded water pollution incidents arose from agricultural activities in 1990, more than double the number recorded in 1982 (263), but below the peak of 763 incidents recorded in 1985. In Northern Ireland, agricultural pollution accounted for 30 per cent of the 2,000 water pollution incidents reported in 1991; in 1988, the proportion due to agricultural pollution was 50 per cent. See also Chapter 7 on water quality.

15.16 Pesticides and substances licensed as veterinary medicines such as sheep dip are widely used for the control of weeds, fungi and insects. However, these substances can be dangerous if misused, or if used or disposed of in a careless manner. Some sheep dip chemicals can also be poisonous to birds (including geese and hens), fish and other water life, and can cause serious damage to

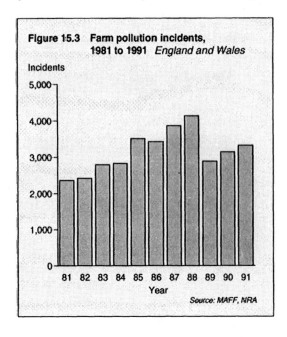

Figure 15.3 Farm pollution incidents, 1981 to 1991 England and Wales

Incidents

Year

Source: MAFF, NRA

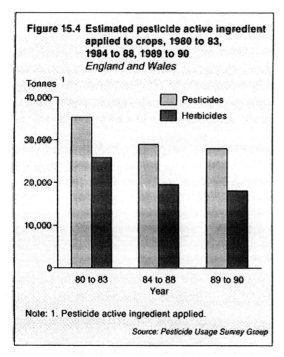

Figure 15.4 Estimated pesticide active ingredient applied to crops, 1980 to 83, 1984 to 88, 1989 to 90
England and Wales

Note: 1. Pesticide active ingredient applied.

Source: Pesticide Usage Survey Group

wildlife if they are allowed to enter a watercourse. In 1991, pesticides were found to be involved in 193 cases of poisoning of mammals. Most were cases of misuse or abuse of pesticides, often associated with attempts to kill predators of game or livestock. See also Chapter 10 on wildlife.

15.17 Between 1980 and 1990, the total tonnage of pesticide active ingredient used decreased by 20 per cent (see Figure 15.4). In the same period the total area treated by pesticides increased by 9 per cent while the area used for growing crops and grass fell by 2 per cent.

15.18 The burning of crop residues can produce smoke and ash, which can be a major public nuisance. It can also cause fire damage to field margins, hedgerows and trees. The area of cereals residues burnt has fallen by 50 per cent since it was decided to prohibit this practice, and burning will be banned in England and Wales before the 1993 harvest. See also Chapter 2 on air quality.

Energy

15.19 The production and distribution of energy is a major industry in its own right and also provides an essential input to other economic activities. The energy production and supply industries account for about 10 per cent of GDP and employ over 300,000 workers.

15.20 The energy industry exerts a number of pressures on the environment, which differ according to the type of fuel involved, and which manifest themselves at different stages in the energy cycle.

15.21 Most energy is extracted from underground sources in the form of fossil fuels (coal, oil and gas). Some of this energy can be used directly by the consumer (eg natural gas), but much of it has to be converted into a usable form. Crude oil is refined into a range of petroleum products such as motor spirit (petrol), fuel oil, and diesel fuel. Some fuels are used to generate other forms of energy; for example coal, fuel oil, nuclear energy and gas are all used in power stations to generate electricity.

15.22 Fuels in their original form, whether used directly or as inputs to convert them to other forms of energy, are called primary fuels, while those produced as the result of a conversion process are called secondary fuels. The impact of a secondary fuel on the environment is not limited to its own use, but also includes the environmental impacts of the conversion processes used to produce the fuel. Most notably, using electricity creates a demand for the primary fuels to be used in generating that electricity. Though electricity itself is a clean fuel, its generation in coal, oil or gas fired power stations has a considerable environmental impact. See also Chapters 2 and 3 on air quality and the global atmosphere.

Primary energy consumption

15.23 In 1991 consumption of primary fuels and equivalents for energy use in the UK amounted to 207.7 million tonnes of oil or oil equivalent (mtoe), 31 per cent higher than in 1960. Of the total, fossil fuels accounted for 187.2 mtoe, having increased by 19 per cent since 1960. In 1991, electricity generated from non-fossil fuels sources (nuclear power, hydro and wind power) amounted to 16.6 mtoe, while net imports of electricity accounted for 3.9 mtoe.

15.24 Overall consumption for energy use (ie excluding consumption of fuels used as raw materials in chemical processes) increased steadily from 1960 to 1973, at an average rate of about 2 per cent per year. The 1973 peak of over 200 mtoe was reached again in 1979, but after allowing for climatic effects 1973 is normally regarded as the year of higher energy consumption, with a temperature corrected level of 208.3 mtoe. In the years 1982 to 1984, energy consumption fell back but since then it has grown again, reaching 1973 levels again by 1990.

15.25 Figure 15.5 shows that within this overall pattern the consumption of natural gas grew rapidly in the late 1960s and early 1970s. Throughout the period from 1960

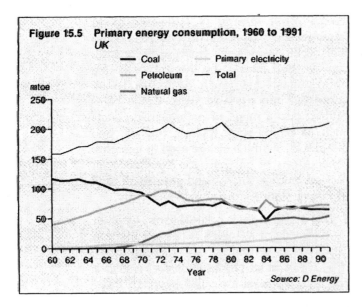

Figure 15.5   Primary energy consumption, 1960 to 1991 *UK*

Source: D Energy

by industry has steadily declined and now stands at 25 per cent of total consumption. Industry's share is now less than that of the domestic sector which has almost the same share as in 1960. The greatest growth has been in the transport sector which had a share of 17 per cent in 1960, but by 1991, accounted for 31 per cent. Other final users (commerce, public administration, agriculture etc) accounted for 12 per cent of total consumption in 1960 and 14 per cent in 1991.

to 1991, there was a general decline in coal consumption. Petroleum consumption peaked in 1973, when fuel prices rose rapidly following the oil supply crisis, and has since declined.

Energy consumption by final user

15.26 Energy consumption by final users, ie through direct use of primary fuels such as coal and natural gas, or after any conversion to secondary fuels, amounted to 60 billion therms, or 142.2 mtoe in 1991, an increase of 20 per cent since 1960. The trends in consumption by final users are shown in Figure 15.6. The difference between primary energy consumption and consumption by final users, 65.5 mtoe in 1991, was accounted for by fuel used by energy producers, mainly to generate and transmit electricity, and is not attributed to final users. From 1960 to 1973, industry (including iron and steel) was the sector which used the most energy, maintaining a level of about 42 per cent of total final consumption over the period. However, since 1973 consumption

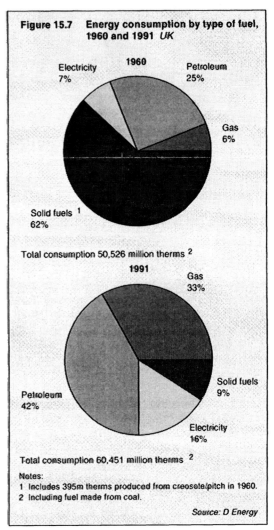

Figure 15.7   Energy consumption by type of fuel, 1960 and 1991 *UK*

Notes:
1 Includes 395m therms produced from creosote/pitch in 1960.
2 Including fuel made from coal.

Source: D Energy

15.27 Trends in final energy consumption for individual fuels have also varied (see Figure 15.7). In 1960, coal and other solid fuels accounted for over 60 per cent, but solid fuels have declined steadily in importance, first as the level of oil consumption increased but also since 1973 as natural gas usage has squeezed out both solid fuel and oil consumption. Electricity consumption has risen steadily over the last thirty years, from 7 per cent of the total in 1960 to 16 per cent in 1991.

Energy Efficiency

15.28 There has been growing interest in energy efficiency over recent years. In the

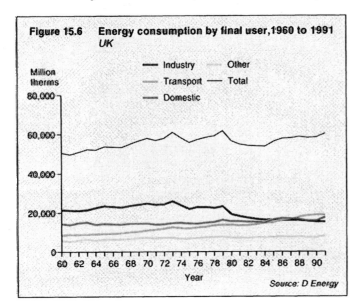

Figure 15.6   Energy consumption by final user, 1960 to 1991 *UK*

Source: D Energy

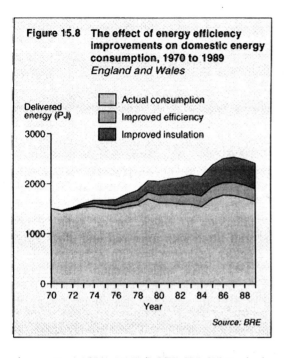

Figure 15.8 The effect of energy efficiency improvements on domestic energy consumption, 1970 to 1989
England and Wales

Delivered energy (PJ)

Actual consumption
Improved efficiency
Improved insulation

Year

Source: BRE

domestic sector insulation is particularly important because space heating accounts for 57 per cent of energy consumption. Between 1974 and 1989, loft insulation increased from 42 per cent to 89 per cent of homes, and cavity wall insulation increased from 2 per cent to 20 per cent of homes in which it can be installed. Estimates have been made of the effect of energy efficiency improvements on energy consumption for domestic heating between 1970 and 1989. These estimates compare delivered energy consumption for each year, corrected for 1989 temperatures, with the energy which would have been required if insulation standards and heating appliance efficiency had remained at 1970 levels. It is estimated that in 1989, improved insulation and heating appliance efficiency have resulted in a reduction of delivered energy consumption of about 32 per cent in the domestic sector (see Figure 15.8). Improvements in insulation standards account for two thirds of this.

15.29 Combined Heat and Power (CHP) technology produces usable heat as well as electricity and achieves overall fuel efficiencies as high as 80 to 90 per cent, which makes it a significantly more efficient technology than conventional energy options. CHP is currently operated at over 600 sites in the UK and accounts for 3 per cent of electricity generated.

Coal

15.30 Coal production in 1991 was 96 million tonnes, less than half the 200 million tonnes production level of the early 1960s. Production of deep mined coal has declined by more than 60 per cent since the early 1960s while open cast coal production has increased from eight million tonnes in 1960 to the current level of 19 million tonnes, and now represents 19 per cent of total production.

15.31 Coal is used in the UK in four main market sectors: power generation, the manufacture of coke and smokeless fuels, industry and the domestic sector. The largest sector is power generation which in 1991 accounted for 78 per cent of UK consumption. Nine per cent is used for coke manufacture for blast furnaces and around 1 per cent for the production of smokeless fuels. The remaining 11 per cent is used in the domestic sector for heating and by industry.

15.32 Coal combustion is a major source of atmospheric emissions and greenhouse gases. In 1990, coal use in the UK accounted for 75 per cent of sulphur dioxide (SO_2) emissions, 29 per cent of emissions of oxides of nitrogen (NO_x), 37 per cent of black smoke emissions and 40 per cent of carbon dioxide (CO_2) emissions. Coal mines still account for some 19 per cent of total methane emissions, though their contribution is declining, and is currently estimated at about 850,000 tonnes per year. See also Chapters 2 and 3 on air quality and the global atmosphere.

15.33 Spoil tips present problems for the local environment; they often look unsightly and can contaminate water resources with acid run-off. In addition, over 5 million tonnes of minestone spoil is disposed of each year on beaches and at sea. British Coal has been asked to identify land-based alternatives to disposal on beaches by 1995 and at sea by 1997. Annual spoil production has declined from an estimated 62 million tonnes a year in 1981-82 to 45 million tonnes a year in 1990-91. This is mainly because of a decrease in the number of operational deep mines.

15.34 The 1988 Survey of Land for Mineral Workings in England recorded 8,420 hectares of land with permission for opencast coal working compared with 7,760 hectares in 1982. There were also 11,120 hectares with permission for deep mined coal spoil in 1988, down from 12,370 hectares in 1982. In 1988 there were 4,700 hectares of derelict land in England and 965 hectares in Scotland as a result of coal mining operations, though some 2,300 hectares in England had been reclaimed between 1982 and 1988. See also Chapter 5 on land.

15.35 Mining subsidence causes a problem in some areas. During 1990-91, there were 8,315 claims against British Coal for mining

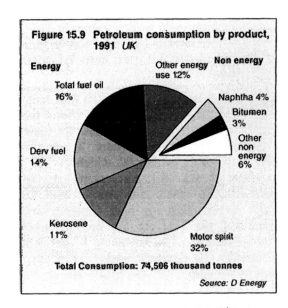

Figure 15.9 Petroleum consumption by product, 1991 UK

Energy

Other energy use 12%

Non energy

Total fuel oil 16%

Naphtha 4%

Bitumen 3%

Other non energy 6%

Derv fuel 14%

Kerosene 11%

Motor spirit 32%

Total Consumption: 74,506 thousand tonnes

Source: D Energy

subsidence damage. Just over 16,500 claims remained outstanding at the end of the year, including claims received in earlier years.

Oil

15.36 At the end of 1991 there were 19 onshore and 46 offshore UK oil fields in production, with a total output of 91.3 million tonnes of crude oil and natural gas liquids.

15.37 Oil refineries in the UK process about 90 million tonnes of crude oil each year, producing a variety of petroleum products. Over 85 per cent of these products are used for energy in the form of gases, motor spirit, diesel, gas oil, fuel oil and kerosene (see Figure 15.9). Products intended for non energy use include naphtha, bitumen, lubricating oils, propane, butane and white spirit.

15.38 Most petroleum products are used as a direct energy source in the transport, industry and domestic sectors, but in 1991, 11 per cent of the petroleum products consumed for energy purposes were used for the generation of electricity.

15.39 Potential sources of environmental pollution from offshore production include spillages of oil; the discharge of chemicals; the discharge of oil-contaminated drilling cuttings; and the discharge of water produced from the well with the oil, which it is difficult to separate completely. In addition, flaring produces partially burnt oil and oily waste, and decommissioning of oil and gas installations needs to be handled in an environmentally sensitive manner. Figures showing the quantities of oil discharged to the environment from these sources are published annually. See also Chapter 8 on the marine environment.

15.40 The average oil content of discharged water in 1991 was 0.0036 per cent (4,689 tonnes of oil in 129 million tonnes of water). This was higher than the previous year (0.0026 per cent, 3,187 tonnes in 121 million tonnes of water) because of problems with treatment equipment on several installations. The quantity of oil discharged on drill cuttings also fell, from 12,310 tonnes in 1990 to 9,380 tonnes in 1991.

15.41 Since 1986 the UK has carried out surveillance flights over offshore (North Sea) installations to check for oil spills, using aircraft fitted with infra red and ultra violet detectors and side looking airborne radar. In 1991, 234 spills totalling 192 tonnes were reported, a substantial reduction on the previous year when there were 345 spills totalling 899 tonnes.

15.42 Oil and petroleum products can pollute land and water if they are not stored properly or if accidents occur during transport or storage. Volatile organic compounds (VOCs) can evaporate from petrol during its production, storage and distribution and pollute the air. See also Chapter 2 on air quality.

15.43 In 1991 there were 5,288 reported water pollution incidents in England and Wales caused by oil and petroleum products, accounting for 24 per cent of reported incidents. Over 1 per cent of reported oil pollution incidents were classified as major, and such incidents accounted for 18 per cent of major water pollution incidents from all categories. (These figures do not include oil spillages from farms or industrial premises as these are assigned to the appropriate source). See also Chapter 7 on water quality.

15.44 Burning oil and petroleum products is a major source of air pollution and greenhouse gases. Transport is the main user of oil and petroleum products and accounts for most of the pollutant effects of petroleum. See also Chapters 2 and 3 on air quality and the global atmosphere.

15.45 Industry is the major consumer of fuel oil, butane and propane. Fuel oil accounted for 19 per cent of SO_2 emissions in 1990, 7 per cent of CO_2 emissions, 6 per cent of NO_x emissions and 3 per cent of black smoke emissions.

Gas

15.46 In 1991, gross gas production from the UK continental shelf was 55 billion cubic metres, up from 50 billion cubic metres in 1990 and 45 billion cubic metres in 1989. These increases resulted from higher capacity through the introduction of new fields, and increased domestic demand. At the end of 1991, there were 34 offshore gas fields linked by pipeline to 10 operational gas terminals in the UK. Eighty per cent of gas production in the UK in 1991 was of natural gas from the North Sea; other sources of gas include petroleum gases, blast furnace gas, coke oven gas and colliery methane.

15.47 Gas production affects the environment at the local level as there is some disturbance to habitats around gas fields and pipelines. Onshore pipelines are however usually underground which decreases their impact.

15.48 Waste products from offshore gas production include liquid hydrocarbons recovered for use in chemical production. During 1991 an average of 6.5 million cubic metres of gas a day was flared at offshore facilities, about 10 per cent less than in 1990. A major part of the reduction is accounted for by fields returning to normal operations after the installation of emergency shutdown valves to comply with new regulations introduced after the Piper Alpha accident.

15.49 Methane is a major constituent of natural gas and one of the greenhouse gases. Gas leakage from transmission along the distribution system accounted for an estimated 8 per cent of UK methane emissions in 1990 and offshore oil and gas production for a further 21 per cent. Leakage from the gas supply system is more common from the old cast-iron low-pressure systems which are being replaced by high-pressure pipelines made from polyethylene or welded steel.

15.50 Gas combustion also contributes to NO_x and CO_2 emissions, accounting for 6 per cent of NO_x emissions and 19 per cent of CO_2 emissions in 1990. See also Chapters 2 and 3 on air quality and the global atmosphere.

Electricity

15.51 In 1960, coal provided 80 per cent of the fuel for electricity generation, with oil accounting for most of the remainder. Coal still provides the major input in 1991, but its share has fallen to 68 per cent (see Figure 15.10). Between 1968 and 1972, oil use increased, rising to 28 per cent of fuel input in 1972, but after the oil supply crisis in the following year, its use declined except for a temporary increase during the 1984-85 miners' dispute. In 1991, the use of oil for electricity generation had fallen to 8 per cent. Electricity generation using nuclear

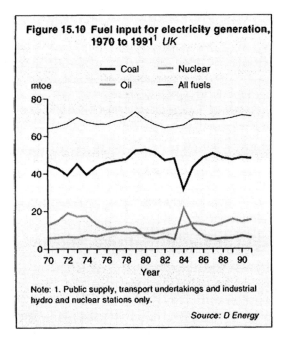

Figure 15.10 Fuel input for electricity generation, 1970 to 1991[1] UK

Note: 1. Public supply, transport undertakings and industrial hydro and nuclear stations only.

Source: D Energy

fuels, has grown steadily from virtually zero in 1960 to 21 per cent in 1991.

15.52 Various renewable energy sources are used for the production of electricity and heat, including passive and active solar power, photovoltaics, onshore wind power, hydropower, geothermal aquifers and biofuels. Biofuels include the combustion of landfill gas, biogas produced from industrial, commercial, domestic, agricultural and

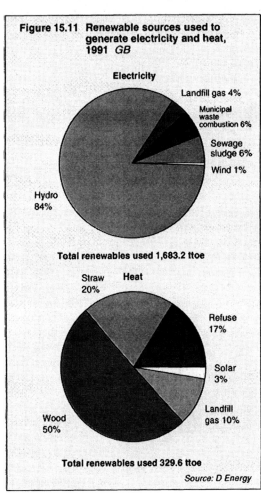

Figure 15.11 Renewable sources used to generate electricity and heat, 1991 GB

Electricity

Total renewables used 1,683.2 ttoe

Heat

Total renewables used 329.6 ttoe

Source: D Energy

217

forestry wastes, and other sources such as energy plantations. Most renewable energy sources are used for the generation of electricity, but biofuels can produce heat or electricity or can be used as fuel for transport. In 1991, biofuels accounted for 16 per cent of the electricity derived from renewable sources and 97 per cent of the heat generated from renewable sources (see Figure 15.11).

15.53 The main environmental pressures from electricity relate to power generation, but pylons used for transmission can spoil the landscape.

Nuclear power

15.54 Nuclear power stations accounted for 21 per cent of electricity supplied by UK generating companies in 1991, compared with 10 per cent in 1975. Although they produce much less greenhouse gases than conventional power stations, nuclear power stations present potential risks to the environment in three main areas: routine emissions of radioactivity, waste disposal and radioactive contamination accidents.

15.55 In general, discharges by the nuclear power industry account for about 0.1 per cent of the total annual average radiation dose of the UK population, though certain critical groups, such as those in the vicinity of the Sellafield reprocessing plant, will receive higher doses. It is estimated that approximately 760,000 cubic metres of radioactive, mainly Low Level, waste will arise from the activities of nuclear power generators, including the reprocessing of spent nuclear fuel, up to the year 2030. See also Chapter 13 on radioactivity.

Hydropower

15.56 In 1991, about 2 per cent of the fuel input for electricity generation in the UK came from hydropower. In general, large scale schemes are sited in mountainous areas such as northern Scotland and Wales and need big dams and reservoirs. These schemes remove one habitat by flooding, and have a major visual impact, but also create a new aquatic habitat. Small scale hydro plants use the water flow from rivers, reservoirs and waterfalls.

15.57 Run-of-river schemes have the disadvantage for power generation of less storage capacity. However, they provide certain environmental benefits such as water oxygenation by turbines, and they avoid the visual impact of large dams.

Transport

15.58 The transport industry accounts for about 4 per cent of GDP. Over 900,000 workers are directly employed in transport, with a similar number in transport related industries. Transport is both a major component of consumers' expenditure (around 15 per cent of the average household budget) and a substantial element within business costs.

15.59 However, most forms of transport can have adverse effects on the environment, both locally and globally. Vehicle emissions from exhausts and fuel evaporation are major sources of greenhouse gases and air pollutants including CO_2, NO_x, SO_2, lead, particulates, carbon monoxide and volatile organic compounds. See also Chapters 2 and 3 on air quality and the global atmosphere. In addition, some of these primary pollutants react together in warm sunny weather to produce photochemical smog, including ground level ozone. A reduction in permitted lead levels in petrol in 1985, followed by the introduction of unleaded petrol, has resulted in a considerable fall in airborne lead levels. However, in some places, particularly urban areas and beside major roads, the soil retains traces of lead from vehicle exhausts which may be introduced into the food chains of people, birds and animals. Traffic noise from roads, rail and air is a major nuisance which causes stress and annoyance for those people affected. See Chapters 4, 12 and 14 on soil, noise and health.

15.60 The extent of many of the environmental problems associated with transport are of recent origin, reflecting the growth of passenger and freight road transport: the volume of traffic on the roads has increased sixfold since the early 1950s.

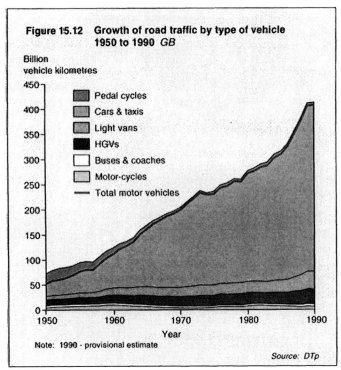

Figure 15.12 Growth of road traffic by type of vehicle 1950 to 1990 GB

Billion vehicle kilometres

Legend:
- Pedal cycles
- Cars & taxis
- Light vans
- HGVs
- Buses & coaches
- Motor-cycles
- Total motor vehicles

Note: 1990 - provisional estimate

Source: DTp

The growth in car usage and road traffic has increased air pollution, noise and the production of greenhouse gases.

15.61 Road transport is the principal means of transport in GB, accounting for 93 per cent of passenger travel and 81 per cent of freight moved in 1990. As can be seen in Figure 15.12, the largest growth since the early 1950s has been in the use of cars which has increased more than tenfold, to over 330 billion vehicle km in 1990.

15.62 The number of cars on the road has risen from about 2.5 million in 1951 to about 20 million in 1990 and cars now comprise about 80 per cent of road vehicles. Figure 15.13 shows forecasts made by the Department of Transport which predict a steady increase in both cars and goods vehicles over the next thirty five years.

15.63 NO_x emissions from road transport in the UK increased by 42 per cent over the period 1985 to 1990. Road transport was the source of 19 per cent of CO_2 emissions in 1990, producing an estimated 30 million tonnes of CO_2 (carbon equivalent) compared with an estimated 24 million tonnes in 1985. VOCs emissions from motor traffic increased by 16 per cent between 1985 and 1990 and accounted for 41 per cent of total VOCs emissions.

15.64 In 1991, changes were made to the MOT test to ensure that all vehicles meet specific emissions standards. In addition, for petrol engined cars, fitting three way catalytic converters to new cars from the end of 1992 should result in a significant reduction in harmful emissions from each vehicle, though it will not affect the level of cold start emissions.

15.65 Emissions of lead from cars have been falling since 1985, when legislation reduced permitted levels of lead in petrol. Surveys of lead concentrations in air showed a fall of around 50 per cent between early 1985 and early 1986. The relative reduction of fuel duty on unleaded petrol compared with leaded petrol since 1987 has resulted in consumption of unleaded petrol rising from under 5 per cent of petrol sales at the beginning of 1989 to over 45 per cent in March 1992. Total lead emissions fell by about 30 per cent between 1988 and 1990.

15.66 The UK now uses over 24 million tonnes of petrol per year, compared with 19 million tonnes in 1980. Lighter materials and improvements in engine technology led to an increase of about 20 per cent in the fuel efficiency of new cars between 1978 and 1985. However as fuel efficiency has

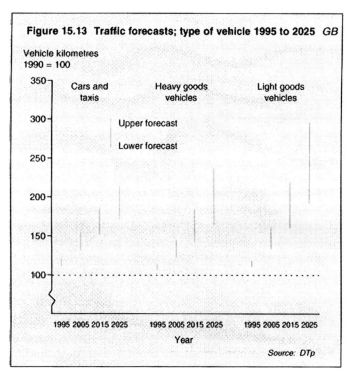

Figure 15.13 Traffic forecasts; type of vehicle 1995 to 2025 *GB*

Vehicle kilometres 1990 = 100

Cars and taxis   Heavy goods vehicles   Light goods vehicles

Upper forecast

Lower forecast

1995 2005 2015 2025   1995 2005 2015 2025   1995 2005 2015 2025

Year

Source: DTp

increased, consumers have tended to move towards models with larger engines and higher performance.

15.67 The growth in the volume of road traffic has meant that traffic noise has become an increasing problem, particularly in urban areas and close to roads with high levels of traffic. Complaints about traffic noise have steadily increased over the last decade. Tyre

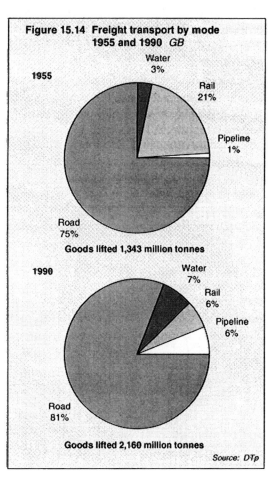

Figure 15.14 Freight transport by mode 1955 and 1990 *GB*

1955

Water 3%
Rail 21%
Pipeline 1%
Road 75%

Goods lifted 1,343 million tonnes

1990

Water 7%
Rail 6%
Pipeline 6%
Road 81%

Goods lifted 2,160 million tonnes

Source: DTp

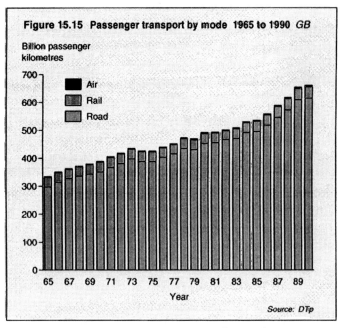

Figure 15.15 Passenger transport by mode 1965 to 1990 *GB*

Billion passenger kilometres

Legend:
- Air
- Rail
- Road

Source: DTp

noise is a major source of complaint in some areas, especially in the vicinity of motorways.

15.68 Congested roads heighten the impact of environmental problems, particularly local air pollution and noise. Between 1980 and 1990, the length of motorway increased by 20 per cent to a total of 3,100 km, but the volume of road traffic on motorways more than doubled. In the same period, the overall length of road network (including motorways) increased by about 5 per cent while there was a 50 per cent increase in traffic volume. Increased traffic volume meant that on comparable networks of central London roads, 1986 average speeds (about 12 mph) were roughly 3 mph faster than in 1912, and almost identical to those of 1936.

15.69 An estimated 14,070 hectares of land in GB was developed for highways and road transport between 1985 and 1990. About 50 per cent was land previously used for

agriculture, around 16 per cent was semi-natural or uncultivated land or forest, and about 33 per cent land previously in urban use, mainly as roads and previously developed vacant land.

15.70 Pedal cycles comprise about 1 per cent of road traffic compared with about 10 per cent in 1960. Levels of use vary in different parts of the country, owing in part to differing terrains, and because some local authorities provide special facilities such as cycle lanes and cycle tracks to encourage the use of pedal cycles. During the last twenty years, usage of pedal cycles has fluctuated about the 5 billion vehicle km level.

Freight Transport

15.71 The total volume of freight traffic in GB has increased by over 60 per cent since 1955 and the proportion carried by road has increased from 75 per cent to some 81 per cent (see Figure 15.14). Road freight carried by heavy goods vehicles is forecast to treble by 2025. These changes have a global impact in terms of air pollution and greenhouse gases, and at a local level there can be an impact from air pollution, noise vibration and traffic congestion, particularly in urban areas and near major roads.

15.72 Exhaust fumes from heavy diesel vehicles are the major source of urban smoke in the UK, as well as a significant source of NO_x.

15.73 The proportion of freight traffic carried by rail has declined from 21 per cent in 1955 to 6 per cent in 1990 (see Figure 15.14). Freight trains, like heavy goods vehicles, can present a noise problem. The opening of the channel tunnel link and the introduction of high speed freight trains and tracks is likely to increase such problems.

Passenger transport

15.74 Road passenger transport more than doubled between 1965 and 1990 (see Figure 15.15), mainly because of the increased use of cars and taxis. Travel by cars and taxis is estimated to have increased from 224 billion passenger km to 561 billion passenger km over this period. As a result the proportion of passenger transport accounted for by cars and taxis has increased from 67 per cent to 85 per cent.

15.75 Conversely the proportion of passenger travel by bus and coach has declined steadily from 18 per cent in 1965 to 6 per cent in 1990. Over this period, the number of passenger km accounted for by bus and coach travel has fallen from 58 billion in 1965 to 41 billion in 1990. In

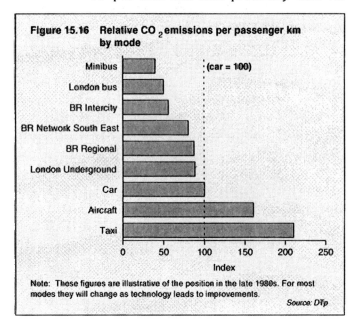

Figure 15.16 Relative CO_2 emissions per passenger km by mode

(car = 100)

- Minibus
- London bus
- BR Intercity
- BR Network South East
- BR Regional
- London Underground
- Car
- Aircraft
- Taxi

Index

Note: These figures are illustrative of the position in the late 1980s. For most modes they will change as technology leads to improvements.

Source: DTp

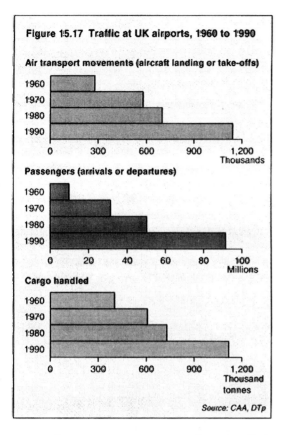

Figure 15.17 Traffic at UK airports, 1960 to 1990

Source: CAA, DTp

terms of emissions, a passenger km travelled by bus emits less CO_2, on average, than a km travelled in a car, train, taxi or plane (see Figure 15.16).

15.76 Rail passenger travel increased from 35 billion passenger km to 41 billion passenger km between 1960 and 1990. The proportion of passenger transport by rail has however fallen from 10 per cent to 6 per cent over the same period.

15.77 The only sector of passenger transport which has shown a larger increase than cars and taxis is domestic air transport. Air travel in the UK increased more than sixfold from 0.8 billion passenger km in 1960 to 5.1 billion passenger km in 1990, but it still accounts for less than 1 per cent of all domestic passenger travel. Both domestic and international air travel are forecast to more than double by 2025.

15.78 Figure 15.17 shows the growth in traffic at UK airports. Aircraft noise around airports has been a concern for many years, though technological advances in engine design and improvements in aircraft construction have led to considerable progress in the reduction of noise levels, with the result that modern jet aircraft are up to 25 decibels quieter than their predecessors. See also Chapter 12 on noise.

15.79 Emissions from aircraft are small in relation to transport generally, although CO_2 emissions per passenger km are more than

double those for bus and rail. Total aircraft emissions worldwide are estimated to form between 2 and 3 per cent of total man-made emissions, and 10 to 15 per cent of total worldwide transport emissions. Emissions data relating to the UK cover the impact of aircraft on local air quality up to a height of one km. Aircraft are a major source of emissions, particularly NO_x, in the atmosphere above one km and a major international research project is in hand to investigate this.

15.80 Fuel costs are a substantial component of aviation running costs and the high price of fuel has contributed to the continued development of more fuel efficient engines and larger aircraft. The fuel efficiency of aircraft per seat mile has roughly doubled during the past twenty years.

Waterborne transport

15.81 Waterborne transport includes international marine traffic and domestic waterborne traffic (ie goods transferred between ports in the UK, traffic between ports and UK offshore installations, and traffic on inland waterways and estuaries). Goods moved on rivers and inland waterways comprise only 4 per cent of domestic waterborne traffic, but the facilities are very important for leisure and recreational purposes. These uses can pollute the watercourses with chemicals, fuel and sewage.

15.82 Figure 15.18 shows port traffic at all ports in GB. Almost all international trade is seaborne: 97 per cent of export tonnage is seaborne trade and 98 per cent of imports (excluding oil and gas in pipelines). Foreign trade comprises about two thirds of freight traffic through the ports of GB, with bulk fuel transfers accounting for about half of this trade. Combined transport is becoming

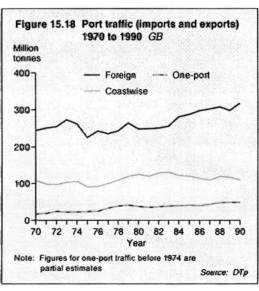

Figure 15.18 Port traffic (imports and exports) 1970 to 1990 GB

Note: Figures for one-port traffic before 1974 are partial estimates

Source: DTp

increasingly important for non-bulk traffic on the short sea routes. The number of road goods vehicles (including unaccompanied trailers) carried on roll-on/roll-off ferry services nearly doubled between 1980 and 1990, to reach a level of 3.4 million vehicles.

15.83 Pollution at sea, whether by oil, chemicals or garbage, can affect the marine environment, the coast and wildlife over a wide area. Discharges from ships of noxious liquids are either banned or controlled, according to their potential pollution hazard. In 1988, it was estimated that almost 2,800 tonnes of controlled substances were discharged from ships into the North Sea. See also Chapter 8 on the marine environment.

References and further reading

Ministry of Agriculture, Fisheries and Food, (1991). Agriculture in the United Kingdom 1990. HMSO.

Ministry of Agriculture, Fisheries and Food, (1992). Agriculture in the United Kingdom 1991. HMSO.

Ministry of Agriculture, Fisheries and Food, (1992). Farm Incomes in the United Kingdom. HMSO.

Central Statistical Office, (1992). Annual Abstract of Statistics 1992. HMSO.

Fertiliser Manufacturers Association, (1992). Fertiliser Review, 1992. FMA.

National Rivers Authority, (1991). Water Pollution Incidents in England and Wales 1990. NRA.

Department of Trade and Industry, (1992). Digest of United Kingdom Energy Statistics 1992. HMSO.

Department of Trade and Industry, (1992). Development of the Oil and Gas Resources of the United Kingdom 1992. HMSO.

Building Research Establishment, (1992). Domestic Energy Fact File. Report No. 220.

Department of the Environment, (1991). Survey of Derelict Land in England 1988. HMSO.

Department of the Environment, (1991). Survey of Land for Mineral Workings in England 1988. HMSO.

Department of Transport, (1991). Transport Statistics Great Britain 1991. HMSO.

Department of Transport, (1991). Transport and the Environment, 1991. HMSO.

Department of Transport, (1989). Transport Statistics Report. Traffic speed on roads in central London. HMSO.

16 Public attitudes

☐ **In the second half of the 1980s interest in environmental issues increased in the media and among the general public, but from a low level of interest at the start of the decade (16.3, 16.5).**

☐ **The main environmental concerns in England and Wales in 1989 were chemicals put into rivers and the sea, sewage contamination of beaches and bathing water, and radioactive waste. In Scotland, the main concerns in 1991 were pollution of rivers, lochs and seas, and raw sewage in the sea (Figure 16.3). In Northern Ireland, sewage on beaches, water quality and nuclear waste were identified as the main concerns in 1991 (16.11).**

☐ **Two main local problems were identified in England and Wales in 1989: fouling by dogs, and litter and rubbish. Over two thirds of respondents identified these as problems in their area (16.18).**

☐ **People were asked about personal action taken to limit damage to the environment. In England and Wales, the most common action which people had already adopted was use of ozone-friendly aerosol sprays. Half said they picked up other people's litter and 40 per cent said that they took bottles to a bottle bank (Figure 16.7). In Scotland, over 40 per cent often chose green products when shopping, and a third said they pick up other people's litter (Figure 16.8). In Northern Ireland, putting litter in bins, safe disposal of household chemicals and buying ozone friendly aerosols were the most frequent activities (16.33).**

☐ **During the 1980s people have expressed an increased willingness to accept the economic costs of measures to protect the environment (Table 16.2).**

☐ **The membership of environmental groups in the UK has increased significantly since 1981 (Table 16.4).**

☐ **Around nine out of ten people say they are interested in the environment, and eight out of ten say they want to know more. Almost half feel they do not fully understand environmental issues (16.46). Most people get their information about the environment from television, and to a much lesser extent, from newspapers (16.48).**

16.1 Environmental concern is not new. In the second half of the 1980s however, there has been an increased interest in environmental issues. Even when the environment is not people's main concern they continue to show environmental awareness, and producers and retailers are taking increasing account of these environmental concerns in marketing.

16.2 Data on public attitudes to the environment in the UK are collected by survey research undertaken for various reasons. Although some surveys assess public opinion about the environment regularly, most are occasional. Responses to questions give "snapshots" of public attitudes to the environment and environmental issues at specific points in time, but some trend data are available where the questions asked have been repeated in later surveys. The main surveys undertaken are given in Box 16.1.

Public concern about the environment relative to other issues

16.3 Opinion polls can give an indication of the level of concern about the environment relative to other current issues. Market and Opinion Research International (MORI) carry out a regular monthly survey[19] which includes

Surveys on public attitudes to the environment   *Box 16.1*

Many surveys about public attitudes to the environment are undertaken and published by organisations such as MORI, Gallup, NOP, and Social and Community Planning Research, in its British Social Attitudes survey. An extension of the British Social Attitudes survey has been carried out in Northern Ireland since 1989. Government also funds research into public attitudes about the environment.

Department of the Environment Public Attitude Surveys

The Department of the Environment (DOE) has funded two surveys, carried out by NOP, in September 1986 and May 1989. The surveys covered England and Wales[1,2], and people were asked a range of questions about their attitudes on a number of environmental issues.

DOE has also undertaken studies on attitudes to energy conservation, and these have included some coverage of environmental issues generally[3,4].

Scottish Office Environment Department

A survey was commissioned by the Scottish Office Environment Department in late 1990[5] to examine public awareness of and attitudes towards environmental issues.

Department of the Environment (Northern Ireland)

A 1991 survey carried out by Ulster Marketing Surveys [6] examined 11 listed environmental issues, similar to those in the DOE and Scottish Office surveys.

The British Social Attitudes Survey

The British Social Attitudes Survey has been carried out by Social and Community Planning Research (SCPR) each year since 1983 [7,8]. The survey includes a section on attitudes to the environment and countryside.

The Northern Ireland Social Attitudes Survey

The Northern Ireland Social Attitudes Survey[9], introduced in 1989, is carried out by the Northern Ireland Policy, Planning and Research Unit (PPRU) under the direction of SCPR. In 1990, the survey included a section on attitudes to the environment and countryside.

MORI surveys

MORI carry out an Omnibus survey every month which includes questions on the environment.

A survey covering environmental issues was conducted in January 1983[10]. It probed attitudes towards environmental problems in Britain and the world in general, with particular emphasis on measures to save natural resources.

MORI also conducted two surveys in March 1987, for the Nature Conservancy Council[11], and for Friends Of the Earth and the World Wildlife Fund[12].

A survey conducted on behalf of the World Wide Fund for Nature was undertaken in March 1990[13]. Environmental issues people said they were most concerned about were river and sea pollution, and the greenhouse effect.

Gallup surveys

A question in the Gallup political index over a period of years has asked what people see as "the most urgent problem facing the country".

A 1987 survey[14] was undertaken on behalf of the Royal Society for Nature Conservation and a 1988 survey[15] for the Daily Telegraph covered concern about a wide range of environmental issues.

Eurobarometer surveys

The European Community (EC) Eurobarometer series covers public attitudes in individual countries in the EC on a comparable basis. Questions on the environment are included periodically, in 1982, 1986, 1988 and 1992 (results from the latter survey are expected in late 1992) [16,17,18].

the questions "what would you say is the most important issue facing Britain today?" and "what do you see as other important issues facing Britain today?". Figure 16.1 shows the percentage of respondents who named pollution/the environment in answer to these questions (the combined figure gives the proportion of people saying each issue is *either* the most important *or* one of the most important - ie they are not counted twice). Before November 1988 the environment was not considered significant enough to be ranked separately from other concerns. In surveys in the 1970s which gathered similar information, the proportion mentioning environment or pollution as the major issue fluctuated between 1 per cent and 7 per cent [20].

16.4 People's perception of the importance of the environment relative to other major issues is strongly influenced by media coverage. As mentions of other issues rise, mentions of the environment tend to fall. In each year, mentions of the environment peak in July. The peak of concern in July 1989 coincided with the European elections, when some 35 per cent of respondents said the environment was the most, or one of the most important issues. In June 1992, 15 per cent of responses mentioned the environment, well below the July 1989 peak, but the highest figure recorded for almost a year.

16.5 In the Department of the Environment (DOE) surveys, in 1986 and 1989, respondents were asked (before any mention of environmental issues) what were the most important problems the Government should be dealing with. Figure 16.2 summarises the responses. In 1989, 30 per cent included environmental or pollution issues in responses, compared with 8 per cent in 1986. In the 1989 survey, concern about the environment was second only to concern about the National Health Service and other social services.

16.6 During 1988, Eurobarometer carried out a survey on attitudes to the environment in the European Community (EC)[18]. Participants from all member countries were asked if they thought that pollution and environmental damage were: mainly an urgent and immediate problem; more a problem for the future; or not really a problem at all. In the UK, 67 per cent thought that pollution and environmental damage was an urgent and immediate problem, compared with 74 per cent for the EC as a whole. Twenty five per cent of respondents in the UK thought that environmental pollution was

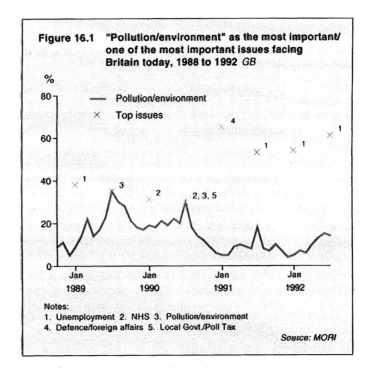

Figure 16.1 "Pollution/environment" as the most important/ one of the most important issues facing Britain today, 1988 to 1992 *GB*

Notes:
1. Unemployment 2. NHS 3. Pollution/environment
4. Defence/foreign affairs 5. Local Govt./Poll Tax

Source: MORI

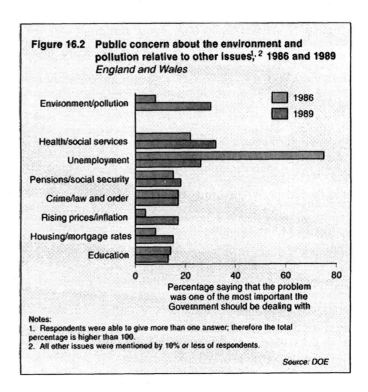

Figure 16.2 Public concern about the environment and pollution relative to other issues[1, 2] 1986 and 1989 *England and Wales*

Notes:
1. Respondents were able to give more than one answer; therefore the total percentage is higher than 100.
2. All other issues were mentioned by 10% or less of respondents.

Source: DOE

more a problem for the future, compared with 20 per cent for the EC.

The degree of concern about specific environmental issues

16.7 In many surveys, people are asked to indicate their level of concern, ranging from "very worried/serious" to "not at all worried/ serious", against a list of specific environmental issues which gives an indication of relative concerns about one issue compared with another. This is discussed further below.

Table 16.1 Changes in public concern about selected environmental issues [1] 1971-89 UK				
			percent saying 'very serious'	Changes
	1971	1972	1989	1971-1989
Pollution of seashore and beaches	46	44	77	+31
Water pollution	50	56	76	+26
Air pollution	36	36	55	+19

Note: 1. "How serious do you think the problem of .... is in Britain ?"

Source: MORI

Changes in concern about specific issues since 1971

16.8 Table 16.1 shows that twenty years ago, half the respondents in Britain believed that water pollution was "very serious" and almost as many thought that pollution of the seashore and beaches was "very serious". Just over one third thought that air pollution was "very serious". There has been a substantial increase since then in the percentage of people who thought that these problems were "very serious".

Issues raised in more recent surveys

16.9 A comparison of the 1989 DOE survey and the 1991 Scottish survey shows a number of similar concerns in England and Wales, and in Scotland. Pollution of the seas, and sewage head the list of concerns in both surveys while derelict land and access to parks

were of much less concern. Figure 16.3 compares the results for those who were "very worried" about specific environmental issues.

16.10 Pollution of rivers, lochs and seas was the single environmental problem of most concern in Scotland. Over half of those interviewed regarded it as a "very serious" problem. This and raw sewage in the sea, are the issues which worried people most. In England and Wales almost two thirds of respondents were "very worried" about chemicals put into rivers and the sea and almost as many are very worried about sewage. Comparison of results from the two DOE surveys shows an increase in concern about almost all environmental problems.

16.11 In the 1991 survey in Northern Ireland, sewage on beaches, water quality, nuclear waste, traffic fumes, litter and depletion of the ozone layer were the issues which people most frequently rated as "very important".

16.12 The extent to which people are concerned about environmental issues depends on their age, sex, social class and income. Generally speaking, both in England and Wales and in Scotland, women were slightly more concerned about environmental issues than men. In England and Wales, retired people were more concerned about local problems such as fouling by dogs, litter and noise. Those in a higher social class tend to be more concerned about global problems and pollution.

16.13 A 1991 DOE study[3] found that although environmental concern appeared to be widespread and growing, the understanding of many issues was poor. For example, when respondents were asked what is the largest single cause of global warming, about a quarter chose (incorrectly) the destruction of the ozone layer; less than a quarter of respondents correctly chose carbon dioxide emissions.

16.14 The results of the British Social Attitudes Survey showed that similar issues to those identified in the DOE surveys were seen as the most important. In general, more people said they thought specific problems were a threat to the environment in 1990 than in 1983. Figure 16.4 shows changes in concern about some specific environmental problems included in the survey. In 1990, 75 per cent thought the problem of industrial wastes discharged to the rivers and the sea was "very serious" compared with 62 per cent in 1983. Fifty eight per cent of people thought that the problem of industrial fumes

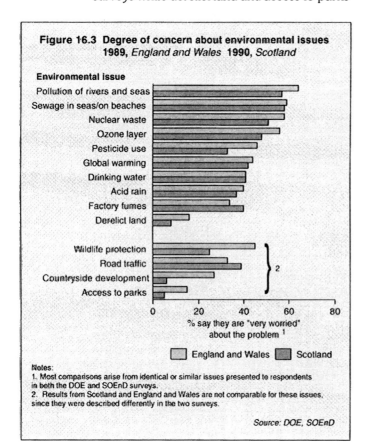

Figure 16.3 Degree of concern about environmental issues 1989, England and Wales 1990, Scotland

Environmental issue

% say they are "very worried" about the problem [1]

☐ England and Wales ▓ Scotland

Notes:
1. Most comparisons arise from identical or similar issues presented to respondents in both the DOE and SOEnD surveys.
2. Results from Scotland and England and Wales are not comparable for these issues, since they were described differently in the two surveys.

Source: DOE, SOEnD

in the air was a "very serious" threat to the environment compared with 41 per cent in 1983. Twenty eight per cent of respondents in 1990 considered noise and dirt from traffic a threat, compared with 23 per cent in 1983.

16.15 In 1990, the Northern Ireland Social Attitudes Survey (see Box 16.1) included the same topics on the environment and countryside as the GB survey. When asked how serious an effect selected threats had on the environment, people in NI and GB ranked the threats in a broadly similar order. However, the proportion of respondents in NI viewing problems as "very serious" tended to be less than in GB. For example, 61 per cent of NI respondents thought that the problem of industrial wastes discharged to the rivers and seas was "very serious" compared with 75 per cent in the GB survey.

Relative concern about specific environmental issues

16.16 A survey looking at public attitudes to the environment in Britain was undertaken for the World Wide Fund for Nature in March 1990[13]. Fifty five per cent of respondents said that river and sea pollution was the environmental problem that they were "most concerned" about and 54 per cent said the greenhouse effect. Around half were most concerned about air pollution from exhaust fumes (49 per cent) and the destruction of tropical rain forests (48 per cent).

Relative concern at different geographic levels

16.17 People express degrees of concern over different issues at local, national and international level. These are discussed below.

Local environmental problems

16.18 Results from the 1989 DOE survey show that fouling by dogs, and litter and rubbish, are the two main local problems. Over two thirds of respondents identified these as problems in their area and 14 per cent identified each of them as the most important problem in their area. The quality of drinking water was thought to be the most important local environmental problem by 10 per cent of respondents.

16.19 Over half thought that problems were worse in 1989 in the area in which they lived than five years previously. In both surveys about four fifths thought problems in their area were the same or better than elsewhere in Britain; about 15 per cent thought they were worse.

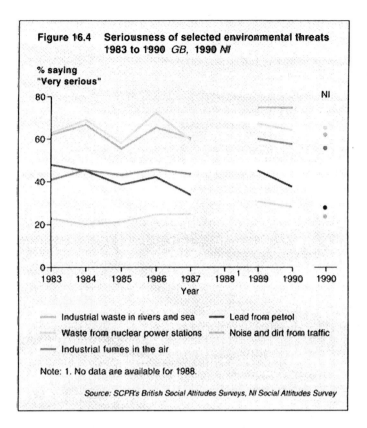

Figure 16.4 Seriousness of selected environmental threats 1983 to 1990 *GB*, 1990 *NI*

% saying "Very serious"

Industrial waste in rivers and sea — Lead from petrol
Waste from nuclear power stations — Noise and dirt from traffic
Industrial fumes in the air

Note: 1. No data are available for 1988.

Source: SCPR's British Social Attitudes Surveys, NI Social Attitudes Survey

16.20 Local problems were not specifically examined in the Scottish survey, although people were asked if they were personally affected a great deal by the problems. Litter, and fouling by dogs, were not included in the list of problems in the Scottish survey, but three of the four next most important local issues in the DOE survey were those which "personally affected people a great deal" ie road traffic, the quality of drinking water, and pollution of rivers and the sea.

Environmental problems at national level

16.21 In both DOE surveys, people were asked to select what they considered to be "the most important problem in Britain". In both 1986 and 1989, radioactive waste was selected as the single most important problem (by 52 per cent of respondents in 1986 and 18 per cent in 1989). The other four problems selected in 1989 were: chemicals put into rivers and the sea (10 per cent); destruction of the ozone layer (10 per cent); decay of inner cities (9 per cent); and sewage contamination of beaches and bathing water (8 per cent). Respondents were also asked to select the three most important problems in Britain; chemicals put into rivers and the sea, and radioactive waste, were both included by a third in their top three in 1989.

16.22 In the Scottish survey, respondents were asked to choose what they thought was "the one issue to improve". Global warming was chosen most (selected by 12

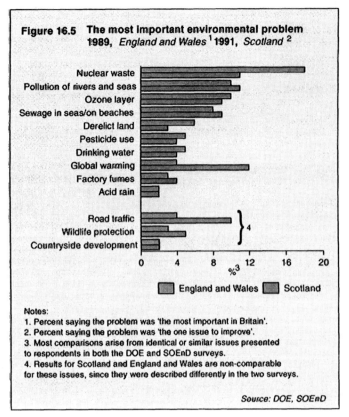

Figure 16.5 The most important environmental problem 1989, *England and Wales* [1] 1991, *Scotland* [2]

- Nuclear waste
- Pollution of rivers and seas
- Ozone layer
- Sewage in seas/on beaches
- Derelict land
- Pesticide use
- Drinking water
- Global warming
- Factory fumes
- Acid rain

- Road traffic
- Wildlife protection
- Countryside development

} 4

0   4   8   12   16   20
%[3]

■ England and Wales   ■ Scotland

Notes:
1. Percent saying the problem was 'the most important in Britain'.
2. Percent saying the problem was 'the one issue to improve'.
3. Most comparisons arise from identical or similar issues presented to respondents in both the DOE and SOEnD surveys.
4. Results for Scotland and England and Wales are non-comparable for these issues, since they were described differently in the two surveys.

Source: DOE, SOEnD

per cent). Nuclear waste, pollution of rivers lochs and seas, and road traffic were each chosen by 10 per cent or more. Nuclear waste, pollution of the seas, the ozone layer, and sewage, were selected by significant minorities in both DOE and Scottish surveys. Figure 16.5 compares results for England and Wales, in 1989, and Scotland, in 1991.

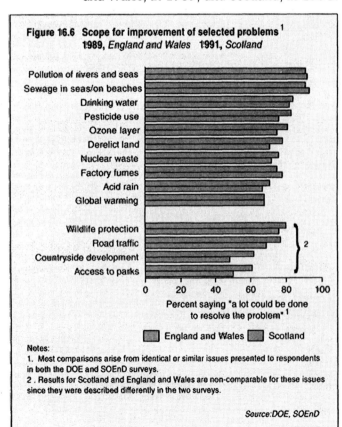

Figure 16.6 Scope for improvement of selected problems [1] 1989, *England and Wales* 1991, *Scotland*

- Pollution of rivers and seas
- Sewage in seas/on beaches
- Drinking water
- Pesticide use
- Ozone layer
- Derelict land
- Nuclear waste
- Factory fumes
- Acid rain
- Global warming

- Wildlife protection
- Road traffic
- Countryside development
- Access to parks

} 2

0   20   40   60   80   100
Percent saying "a lot could be done to resolve the problem" [1]

■ England and Wales   ■ Scotland

Notes:
1. Most comparisons arise from identical or similar issues presented to respondents in both the DOE and SOEnD surveys.
2. Results for Scotland and England and Wales are non-comparable for these issues since they were described differently in the two surveys.

Source: DOE, SOEnD

Most important problems in the world

16.23 In the 1989 DOE survey, four problems were identified as the most important environmental problems in the world ie destruction of the ozone layer (22 per cent of respondents); radioactive waste (20 per cent); destruction of tropical forests (17 per cent); warming of the atmosphere because of the greenhouse effect (14 per cent).

Scope for improvement and allocation of responsibility for action

Scope for improvement

16.24 In the 1989 DOE survey, at least half felt "quite a lot could be done about" all the problems listed. Over 90 per cent thought that quite a lot could be done about chemicals put into rivers and the sea, and sewage contamination of beaches and bathing water. Those problems where more people thought some action could be taken, were also those causing most concern.

16.25 In Scotland, respondents were also asked to select which issues "a lot could be done about". The relative importance of issues were very similar in both the DOE and Scottish surveys. The proportions of respondents who believed quite a lot could be done about specific issues are shown in Figure 16.6.

Allocation of responsibility

16.26 Respondents were asked to consider who ought to be doing something about environmental problems. In England and Wales, the majority of respondents allocated responsibility mainly according to the geographical extent of the problem; for example 57 per cent thought that local councils should be doing something about litter and rubbish and 71 per cent thought that international bodies should be doing something about the destruction of the rainforest. In the 1989 survey, respondents were also asked who they felt ought to make sure that something was actually done about each problem. For two thirds of the problems listed, the majority of respondents thought that the British Government was responsible for making sure something was done.

16.27 Results from the Scottish survey indicate that the majority of people look to central government to resolve, and pay for, many environmental problems. The main exceptions where responsibility was allocated elsewhere were drinking water, which was seen to be the responsibility of local authorities, and global warming, where people allocated a shared responsibility.

Surveys about specific environmental issues

16.28 A number of surveys and studies cover public attitudes on specific environmental issues, such as the countryside, wildlife, transport and pollution, electricity generation and nuclear power, in more depth. Many surveys are also undertaken on attitudes of particular groups of people such as farmers and businessmen. Some of these studies are listed at the end of the chapter.

Personal action and paying to protect the environment

Personal action and green consumers

16.29 In the 1989 DOE survey, people were asked which actions they personally already took to limit damage to the environment, and which actions they were prepared to take (see Figure 16.7). The most common action already taken was the use of ozone-friendly aerosol sprays. At the time the survey was carried out, 64 per cent of respondents were using these. Half the people in the survey said that they picked up other people's litter, and 40 per cent said that they took bottles to a bottle bank. A quarter of people in the survey said that they used recycled paper and a further 60 per cent said they would consider doing so. The question did not cover the frequency of activity however, and respondents may not always do what they say they do.

16.30 Analysis by demographic group showed that the groups most likely to be taking action already were those living in rural areas, those engaged in professional, senior managerial or administrative work, and those in the middle age group (45-64). Young people in the age range 18-24 were least likely to be taking action.

16.31 Further research work, covering England, undertaken in October 1991 on behalf of DOE[21], complemented these earlier results. Respondents were asked to identify from a list, what they had done over the previous year as a result of concern about the environment. The most common action was buying ozone friendly aerosols (or buying no aerosols) and three quarters of respondents had done this. Next most common actions were cutting down on the use of electricity (65 per cent), buying products sold in recycled packaging (56 per cent), buying products made from recycled materials (53 per cent), and regular use of a bottle bank (45 per cent). Generally, people appear to concentrate on a few issues where simple and clear action is possible.

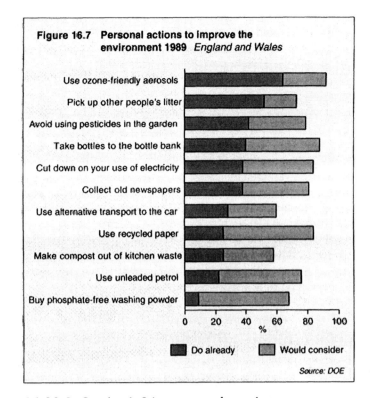

Figure 16.7 Personal actions to improve the environment 1989 *England and Wales*

Source: DOE

16.32 In Scotland, 94 per cent of people in the 1990 survey believed that they could protect the environment by changes to their own behaviour. Figure 16.8 shows environmental actions taken in the twelve months prior to the survey. Over 40 per cent of people claimed that they often chose green products when shopping. A third said they pick up other people's litter. Between a quarter and a third claimed that they often took bottles to a bottle bank, took waste

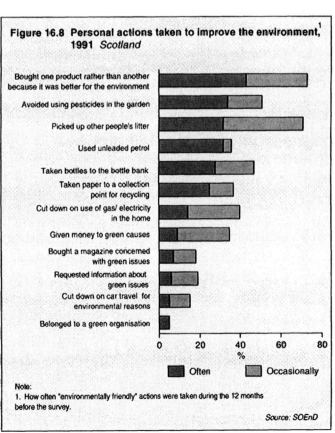

Figure 16.8 Personal actions taken to improve the environment,[1] 1991 *Scotland*

Note:
1. How often 'environmentally friendly' actions were taken during the 12 months before the survey.

Source: SOEnD

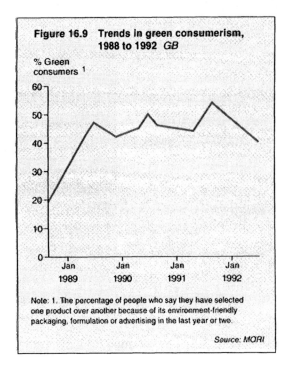

Figure 16.9 Trends in green consumerism, 1988 to 1992 *GB*

% Green consumers [1]

Note: 1. The percentage of people who say they have selected one product over another because of its environment-friendly packaging, formulation or advertising in the last year or two.

Source: MORI

paper to a collection point or avoided using pesticides in the garden. As in England and Wales in 1989, young people (18-24) were less likely than other age groups to take action, and in particular, less likely to recycle materials. However, over half the young people in the survey said they often used environmentally friendly products.

16.33 In Northern Ireland, the 1991 survey showed a very similar pattern between men and women claiming to participate in environment friendly activities. Putting litter in bins, disposing of chemicals safely and buying ozone-friendly aerosols were the activities which people most frequently claimed to do. People aged 35-49 were more aware of environmental issues than those in other age groups and were most likely to take newspapers and glass for recycling, and buy phosphate-free washing powder or organic food. Affluent households were more

likely to take newspapers and glass for recycling.

16.34 The increase in publicity about the environment in early 1989 led to a change in consumer behaviour. Figure 16.9 indicates that in Britain, the proportion of "green" consumers (people who choose one product over another because of its environment-friendly packaging, formulation or advertising) more than doubled between September 1988 and July 1989. Since mid 1989, the proportion of consumers buying environment-friendly products has continued at a high level, which suggests that selective purchasing on environmental grounds is now established, despite the less clear pattern of interest in the importance of the environment against other major issues shown in Figure 16.1.

Environmental protection and economic growth

16.35 Through the 1980s the evidence from various surveys suggests an increased willingness to accept the economic costs of measures to protect the environment. One particular question about the perceived "trade-off" between protecting the environment and economic growth has been repeated in various surveys[1,15,16]. Table 16.2 shows the percentage of respondents in these surveys who favoured priority being given to environmental protection, or economic growth.

16.36 A similar "trade-off" question was put to people participating in the 1986 and 1988 Eurobarometer survey (covering the UK) and the 1989 DOE survey (covering England and Wales), with an additional judgemental option about economic development and environmental protection. Table 16.3 shows the results. In all three surveys about half the respondents thought that protecting the environment was necessary to ensure economic development. About a third thought that it was sometimes necessary to make a judgment between the environment and the economy but people were generally unenthusiastic about development of the economy taking priority over the environment.

Paying for environmental improvements

16.37 Various surveys have attempted to address the question of whether people would be willing to pay for environmental improvements, either through higher prices for environment-friendly goods, or through higher taxes to fund government spending on environmental protection. However, the

Table 16.2 Attitudes to the 'trade-off' between protection of the environment and economic growth, 1982 to 1988				
	Percentage of respondents selecting their priority			
	Oct 1982 (UK)	May 1985 (GB)	Sept 1986 (E&W)	Oct/Nov 1988 (GB)
Priority should be given to:				
The environment [1]	50	57	54	70
Economic growth [2]	36	32	29	17
Neither/don't know	14	11	17	12

Notes:
1. "Protection of the environment should be given priority, even at the risk of holding back economic growth."
2. "Economic growth should be given priority, even if the environment suffers to some extent."

*Sources: 1982, Eurobarometer
1985 + 1988, Gallup
1986, DOE*

results of this kind of question have to be treated with considerable caution, since what people say they do and what they actually do in practice may be quite different. This means that it is not possible to identify "willingness to pay" from general surveys. For example, in the 1989 DOE survey, 71 per cent of respondents said that they thought it was "a good idea" to increase the price of petrol by 17 pence per gallon so that less harm was done to the air, but it was not until the Government reduced the duty on unleaded petrol to make it cheaper than leaded, that a significant proportion of customers began to buy unleaded fuel. Nevertheless, questions about paying for environmental improvements give some indication of the degree of concern people feel about environmental issues.

16.38 In the DOE surveys, several questions were included on willingness to pay. About 2 per cent felt that the country could not afford to spend anything on environmental problems (1989). Thirty five per cent felt that "the fairest way" to find money needed to solve environmental problems would be for government and local councils to cut back on other areas of public spending. There were however, different responses according to socioeconomic status; people in lower socioeconomic groups were much more inclined to favour spending cuts in other areas of public spending to pay for environmental improvements than those in higher income groups. Those in higher groups were inclined to favour higher taxes, not cuts in other areas.

16.39 As an alternative to spending less in other areas, about 30 per cent of respondents in both the 1986 and 1989 DOE surveys thought that "the fairest way" of finding the money needed to solve environmental problems would be for industry to charge higher prices for goods which cause pollution when they are made. Those in higher socioeconomic groups were more in favour of industry charging higher prices than those in the lowest group.

16.40 Five per cent of respondents in the 1991 Scottish survey felt that nothing should be spent on environmental protection because "we cannot afford it". Two thirds agreed that money to protect the environment should be found by paying more for environment- friendly products. Half the respondents agreed that money to protect the environment could be found from additional taxation but there was a strong socioeconomic bias: wealthier, better educated people were much more prepared

Table 16.3 Attitudes to protection of the environment and economic growth 1986, 1988 and 1989
UK, England and Wales[1]

Percentage of respondents selecting their priority			
	1986 (UK)	1988 (UK)	1989 (E&W)
Protecting the environment and preserving natural resources are necessary conditions to ensure economic development	48	51	46
Sometimes it is necessary to make a judgement between economic development and protection of the environment	32	34	40
Development of the economy should take priority over questions of the environment	11	9	6
Neither/don't know	9	6	7

Note: 1. 1986 and 1988 results are from Eurobarometer surveys, and cover UK. The 1989 DOE survey covered England and Wales only.

Source: DOE

to pay higher taxes than pensioners, single parent families and tenants. A less strong, though similar, bias applied to buying environment-friendly goods. In general, the degree to which people were environmentally aware or "green" in their everyday actions determined their attitudes towards payment and responsibility. People who were better informed about the environment supported in particular, controlling industry, paying more for environment-friendly products and paying higher taxes to protect the environment. Most people felt that resolving global problems such as global warming and

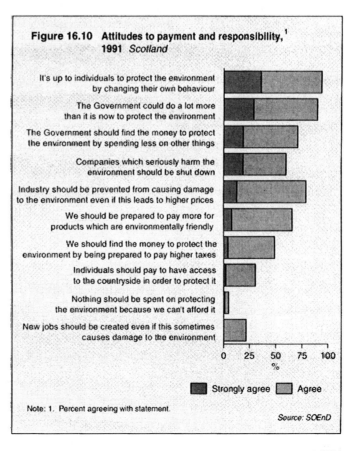

Figure 16.10 Attitudes to payment and responsibility,[1] 1991 *Scotland*

It's up to individuals to protect the environment by changing their own behaviour

The Government could do a lot more than it is now to protect the environment

The Government should find the money to protect the environment by spending less on other things

Companies which seriously harm the environment should be shut down

Industry should be prevented from causing damage to the environment even if this leads to higher prices

We should be prepared to pay more for products which are environmentally friendly

We should find the money to protect the environment by being prepared to pay higher taxes

Individuals should pay to have access to the countryside in order to protect it

Nothing should be spent on protecting the environment because we can't afford it

New jobs should be created even if this sometimes causes damage to the environment

0 25 50 75 100
%

■ Strongly agree ■ Agree

Note: 1. Percent agreeing with statement.

Source: SOEnD

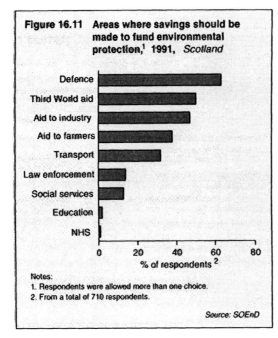

Figure 16.11 Areas where savings should be made to fund environmental protection,[1] 1991, *Scotland*

% of respondents [2]

Notes:
1. Respondents were allowed more than one choice.
2. From a total of 710 respondents.

Source: SOEnD

ozone depletion, should be paid for by everybody. A large majority (79 per cent) thought that industry should be prevented from causing damage to the environment even if this led to higher prices. Figure 16.10 summarises the responses to a number of statements about allocation of responsibility and methods of payment in Scotland.

16.41 Also in Scotland, 71 per cent of respondents felt that the Government should find the money to protect the environment by spending less on other policy areas. Of these, 63 per cent said that defence spending should be cut, half thought that aid to the Third World should be cut, and just less than half (47 per cent) thought aid to industry should be cut to pay for environmental improvements. Figure 16.11 shows the areas

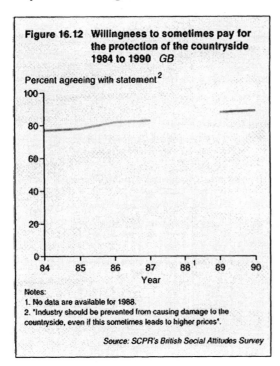

Figure 16.12 Willingness to sometimes pay for the protection of the countryside 1984 to 1990 *GB*

Percent agreeing with statement [2]

Year

Notes:
1. No data are available for 1988.
2. 'Industry should be prevented from causing damage to the countryside, even if this sometimes leads to higher prices'.

Source: SCPR's British Social Attitudes Survey

where people thought savings should be made to fund environmental protection.

16.42 In SCPR's British Social Attitudes Survey, people were asked if they agreed with the statement "Industry should be prevented from causing damage to the countryside, even if this sometimes leads to higher prices". Figure 16.12 shows that the proportion agreeing with this statement rose from 77 per cent in 1984 to 89 per cent in 1990. The equivalent figure in Northern Ireland in 1990 was 82 per cent. The questions however, did not say how much higher prices might be.

16.43 In a March 1990 survey conducted on behalf of the World Wide Fund for Nature[13], almost half (46 per cent) of respondents claimed they would be willing to pay a one penny increase in income tax for measures to protect the environment and conserve natural resources. Forty one per cent said they would pay a small increase (of about 5 per cent) in their food bill to pay for food guaranteed to be free of pesticides. A similar proportion (44 per cent) said they would be prepared to pay an extra two pence per gallon of petrol to reduce air pollution and acid rain from that source. Nineteen per cent of respondents said they would not be willing to pay for any of the environmental improvements listed in the survey.

Membership of environmental organisations

16.44 Membership of the main environmental organisations in the UK can be used as an indicator of committed concern about the environment. Table 16.4[22] shows the membership of main environmental groups since 1971. Many people may be members of more than one organisation. Most groups reported a significant upturn in membership applications during 1988.

16.45 In Scotland, 13 per cent of those in the 1991 survey said that they were members of at least one organisation concerned with the environment, compared with some 18 per cent in England (from a survey carried out in October 1991[21]). The organisations mentioned the most were the National Trust, Royal Society for the Protection of Birds, World Wide Fund for Nature, Greenpeace, and Friends of the Earth.

Sources of information and education

16.46 In a survey carried out in October 1991[21], nine out of ten people said they were interested in the environment, and eight out of ten said they want to know more about it. Almost half feel they do not fully understand environmental issues. In spring 1991, DOE published a leaflet "Wake up to what you can do for the Environment" aimed at raising public awareness about environmental issues, and informing people about how they can help protect the environment. Five million leaflets were printed and around 3.4 million were distributed. In research conducted some 6 months later [21], 7 per cent of the public covered in the research claimed to have seen the leaflet (suggesting nationally around 3 million people). About half the people who saw the leaflet claimed to have read all or most of it. One in ten said they had not looked at it.

16.47 DOE are currently conducting a three year energy efficiency publicity campaign which began in November 1991. The main aim is to educate people about the link between energy use in the home and global warming. The first phase of advertising appeared in newspapers between 5 November 1991 and 1 December 1991 and a report has been produced[3] which evaluated the effectiveness of the adverts and people's awareness about, and knowledge of, global warming. This showed that there is confusion over the causes of global warming and what individuals can do to help. Nevertheless, over half the respondents realised that saving energy in the home helps to reduce global warming.

16.48 A survey carried out in Britain for the Nature Conservancy Council[11] in 1987 asked respondents where they currently obtained most information about wildlife and countryside conservation. Seventy three per cent said that television was an important source of information. Twenty six per cent said newspapers, 10 per cent said magazines and 6 per cent said the radio. Only 3 per cent said that schools, universities, societies or groups were of importance in getting environmental information.

16.49 Results from the 1991 Scottish survey show that television is the principal source of information on environmental issues for most people (72 per cent), and then newspapers (46 per cent). Over half the people felt that too little information was available, and wanted to know more about environmental issues. For this latter group, the most trustworthy sources were thought to be environmental organisations (57 per cent) and scientists (23 per cent). Those considered least trustworthy were central government (43 per cent) and industry (19 per cent). Half the parents in the survey wanted their children

Table 16.4 Membership of selected voluntary environmental organisations 1971, 1981, 1990			
			thousand
	1971	1981	1990
National Trust	278	1,046	2,032
National Trust for Scotland	37	110	218
Royal Society for the Protection of Birds [1]	98	441	844
Greenpeace UK [2]	..	30	372
Civic Trust [3]	214	..	302
Royal Society for Nature Conservation [4]	64	143	250
World Wide Fund for Nature	12	60	247
Friends of the Earth [5]	1	18	110
Ramblers Association	22	37	81
Woodland Trust	..	20	66
Council for the Protection of Rural England	21	29	44

Notes:
1. Includes the Young Ornithologists' Club.
2. Data shown for 1981 refers to 1985 membership. Data for earlier years are not available.
3. Members of local amenity societies registered with the Civic Trust.
4. Does not include junior organisations, WATCH, or clubs affiliated to WATCH.
5. England and Wales only. Friends of the Earth (Scotland) is a separate organisation founded in 1978.

Source: Social Trends, organisations concerned

to be taught more about environmental issues in school.

16.50 A survey of the attitudes of 11-16 year olds carried out in 1987[23] showed, as might be expected, that schools and then television were the most common sources of information about the environment for children. The results also showed that there was a greater awareness of environmental matters among 15-16 year olds who were being formally taught about environmental matters at school than amongst those pupils of the same age where such courses were not available.

16.51 The Environment Council publishes a Directory of Environmental Courses as part of its information programme[24] which provides a guide to academic, professional and vocational courses related to the environment.

References and further reading

1. Department of the Environment, (1986). Digest of Environmental Protection and Water Statistics No 9. HMSO.

2. Department of the Environment, (1989). Digest of Environmental Protection and Water Statistics No 12. HMSO.

3. BMRB, (1992). Energy Efficiency Campaign October 1991 to January 1992. Research Report prepared for the COI Research Unit on behalf of the Departments of the Environment and Energy.

4. Department of the Environment, (1991). Attitudes to Energy Conservation in the Home, Report on a Qualitative Study "The Hedges Report". HMSO, London.

5. Scottish Office, Central Research Unit, (1991). Public Attitudes to the Environment in Scotland. Scottish Office.

6. Ulster Marketing Surveys, (1991). The Public Attitudes and Awareness Survey (Northern Ireland). A Survey carried out for the Department of the Environment Northern Ireland Environment Service. Ulster Marketing Surveys, Belfast.

7. Social and Community Planning Research, (1992). British Social Attitudes, Cumulative Sourcebook: the First Six Surveys. Gower Publishing.

8. Jowell, R., Brook, L., & Taylor, B., (eds), (1991). British Social Attitudes: The 8th Report (1990 results). Dartmouth, Aldershot.

9. Stringer, P. & Robinson, G., (eds), (1992). Social Attitudes in Northern Ireland. Blackstaff Press.

10. MORI, (1983). Earth's Survival. Report of a Survey for the Programme Organising Committee of the Conservation and Development Programme for the UK, January 1983. MORI, London.

11. MORI, (1987). Public Attitudes towards Nature Conservation. Report of a survey for the Nature Conservancy Council, March 1987. MORI, London.

12. MORI, (1987). Public Attitudes to the Environment. Report of a Survey for Friends of the Earth and the World Wildlife Fund, April 1987. MORI, London.

13. MORI, (1990). Tropical Rainforests and the Environment. A survey of Public Attitudes. World Wide Fund For Nature March 1990. MORI, London.

14. Gallup, (1987). Attitudes towards Nature Conservation. Report of a Study for the Royal Society for Nature Conservation, July 1987. Social Surveys (Gallup Poll) Ltd.

15. Gallup, (1988). Report of a Survey for the Daily Telegraph, November 1988. Social Surveys (Gallup Poll) Ltd.

16. Riffault, H., (ed), (1982). The Europeans and their Environment. Commission of the European Communities, Brussels.

17. Tchernia, J.F., Gattolin, A., (eds), (1986). The Europeans and their Environment in 1986. Commission of the European Communities, Brussels.

18. Riffault, H., Tchernia, J.F., (eds), (1988). The Europeans and their Environment in 1988. Commission of the European Communities, Brussels.

19. The MORI Omnibus Survey - a regular survey carried out every month which contains some questions on the Environment, asked for by clients of MORI.

20. Opinion Research Centre, (1976 and 1978). Corporate Strategy Guide. ORC, London.

21. MORI, (1992). Measuring the Effectiveness of the Wake Up leaflet. Research Survey conducted for the Department of the Environment, October 1991.

22. Department of the Environment, (1992). Digest of Environmental Protection and Water Statistics No 14. HMSO.

23. Gayford, C.G., (1987). Environmental Education, Experiences and Attitudes. Council for Environmental Education. Report submitted to the Department of Environment.

24. The Environment Council, (1992). Directory of Environmental Courses 1992-1993. A Guide to Academic, Professional and Vocational Courses related to the Environment. The Environment Council, London.

Jacobson, H.K., Price, M.F., (1990). A Framework for Research on the Human Dimensions of Global Environmental Change. For the International Science Social Council Standing Committee on the Human Dimensions of Global Change, Series 3. ISSC/UNESCO.

Worcester, R.M., Barnes, S.H., (1991). Dynamics of Societal Learning about Global Environmental Change, for the International Science Social Council Standing Committee on the Human Dimensions of Global Environmental Change, Series 4. ISSC/UNESCO.

Cuttle, S.P., (1989). Public Perceptions of Agriculture and the Environment. A Review of Survey Information. A Report prepared for the Ministry of Agriculture, Fisheries and Food. Welsh Plant Breeding Station, Aberystwyth.

British Market Research Bureau, (1983). Countryside Survey, conducted for the Country Landowners Association, March 1983. CLA Press Release 1247, April 1983.

Ministry of Agriculture Fisheries and Food, (1991). Public Attitude Survey of Environmentally Sensitive Areas. Carried out for MAFF by KPMG in association with Mass Observation Ltd.

Ministry of Agriculture, Fisheries and Food, (1985). Survey of Environmental Topics on Farms. England and Wales, 1985. Stats 244/85. MAFF.

Miller, F.A. & Tranter, R.B., (eds), (1988). Public Perception of the Countryside. CAS Paper 18. Centre for Agricultural Strategy, Reading.

MORI, (1987). Farmers' Attitudes towards Nature Conservation. Report of a qualitative survey for the Nature Conservancy Council, March 1987. MORI, London.

MORI, (1986). Childrens' Attitudes to the Environment. A Survey for the World Wildlife Fund. MORI, London.

Research Surveys of Great Britain, (1988). Public Attitudes towards Farmers. Report of a survey for the National Farmers Union, March/April 1988. Research Surveys of Great Britain Ltd.

System Three Scotland, (1987). Survey on Attitudes Towards Conservation in the Countryside. Report for the Countryside Commission for Scotland. System Three Scotland, Edinburgh.

Brand New Product Development Ltd and Diagnostics Market Research Ltd, (1989). Green, Greener, Greenest? The Green Consumer in the UK, Netherlands and Germany. Michael Peters Group plc, London.

MORI, (1989). The Greening Consumer: Attitudes to the Environment among the General Public and Senior Managers. Research Study conducted for Colman RSCG. MORI, London.

MORI, (1990). Public Attitudes Towards Transport and Pollution. Research Study conducted for the World Wide Fund for Nature. MORI, London.

MORI, (1991). Survey of Britain's Environment Journalists. Report to Participants. MORI, London.

Corrado, M. & Ross, M., (1990). Green Issues in Britain and the Value of Green Research Data. Paper prepared for the ESOMAR Congress, Monte carlo, September, 1990.

MORI, (1991). Business and the Environment. A Survey of Public Attitudes and Behaviour. Report to Environment Journalists. MORI, London.

MORI, (1991). Tropical Rainforests and the Environment. A Survey of Public Attitudes. World Wide Fund for Nature. April 1991, MORI, London.

MORI, (1991). Environmental Activism. An Eight Country Survey of Public Attitudes. Conducted for WWF International. September- November 1991.

MORI, (1991). Concern about Marine Wildlife. A Survey of Public Attitudes. Conducted for RSPB. August 1991.

17 Expenditure on the environment

☐ **Accurate figures on the amount spent on protecting the environment do not exist, either in the UK or in other countries (17.2). Definition of environmental expenditure is difficult, and the estimates given in the chapter should be regarded as broad orders of magnitude only (17.9). They relate to expenditure directly concerned with improving or maintaining the quality of the environment which can be clearly identified or estimated (17.6).**

☐ **Separately identifiable direct environmental expenditure in the UK in 1990-91 is estimated to have been around £14 billion, about 2½ per cent of GDP. Expenditure on water accounts for almost half of identified total expenditure, and expenditure on waste accounts for about one fifth of the total (Table 17.1 and Figure 17.2).**

☐ **About 60 per cent of total identified expenditure, around £8.8 billion, relates to pollution abatement. Of this, about 36 per cent is spent on water pollution abatement, 29 per cent on waste collection, treatment and disposal, and 26 per cent on air pollution control (Figure 17.1, Table 17.1 and Figure 17.3).**

☐ **The remaining expenditure is on natural resource management (24 per cent), improvement of amenities (8 per cent), and general administration, environmental conservation, research and development, education and training (together accounting for 7 per cent) (Table 17.1 and Figure 17.1).**

☐ **Government (including government departments, local authorities, and various public sector organisations) and enterprises incur around 94 per cent of total expenditure. The remainder consists of expenditure by households and voluntary or other non-profit organisations (Table 17.2).**

☐ **Of the estimated total £14 billion spent in 1990-91, capital expenditure is estimated to be around 35 per cent (about £4.9 billion), and current (operating) expenditure is estimated to be around 65 per cent (about £9.1 billion) (17.19).**

17.1 This chapter presents the available information about expenditure on protecting or maintaining the quality of the environment. Such expenditure also has positive benefits, both economic and in respect of health and the quality of life, but these effects are very hard to quantify and no estimates of them are given here. It should therefore be borne in mind that the figures presented give only a partial picture, and do not allow for the economic benefits of expenditure on environmental protection.

17.2 Even expenditure on protecting the environment itself is difficult to quantify, and accurate figures on the amount spent do not exist, either in the UK or in other countries. This is partly because much of the data has not been collected, but mainly because of the difficulty in defining what is meant by environmental expenditure, which in itself leads to difficulties in collecting the necessary data.

17.3 Environmental expenditure is often an integral and indistinguishable part of expenditure on goods and services. For example, cars and industrial plant are designed to meet certain environmental standards (such as maximum noise levels) and maintenance of these standards will be reflected in the cost of producing the goods and the price paid by the consumer.

17.4 As environmental standards become established, some are taken into account in product design and in the processes used in the manufacture of goods. This is often

---

Definition of environmental expenditure   *Box 17.1*

Environmental expenditure is defined in this chapter as the expenditure which can be estimated or identified as incurred in pursuit of environmental objectives. It includes both current (operating) and capital expenditure.

Where an amount has been identified, it has been attributed by environmental activity, and where possible by spending group. Both "activity" and "spending group" are explained in the section on "breakdown of expenditure". Where it has not been possible clearly to identify environmental expenditure, estimates have been made. In many cases both the identified data and to a greater extent the estimates give only an indication of the level of activity undertaken for environmental purposes.

The *capital expenditure* figures included are the estimated or identified annual values of investment expenditure on fixed assets attributable to environmental management and protection.

The *current expenditure* figures included represent annual expenditure on operating costs, and maintenance, and charges such as fees paid for waste collection and disposal. They also include the costs of servicing capital debt where data are available.

Estimates are made only of expenditure incurred; income associated with this outlay is not taken into account. The estimates are therefore of gross rather than net expenditure. Where this expenditure has included expenditure on direct or indirect taxation these elements have been included in the estimates.

No attempt has been made to include hidden costs not resulting in separately identifiable expenditure, even where these may have environmental objectives. For example, there is an increased operational cost to airlines required to comply with restrictions on night-time flights to reduce exposure to noise around airports. This type of cost has not been included in the estimates.

Most values relate to expenditure in the UK in 1990-91 (in millions of pounds) at 1990 prices.

---

referred to as using "clean technology". Other environmental standards have to be met by adding pollution control and management systems to the production process, for example to reduce emissions of harmful substances to the air. This additional item of pollution control and environmental management is often referred to as "end of pipe" technology. Expenditure on end of pipe pollution control is generally the only environmental expenditure which can be separately identified, although at present the majority of expenditure is thought to be of this type.

17.5 As new standards are introduced so new investment in pollution control equipment or new products is required. But investment decisions may be made for a number of reasons as well as environmental ones and in many cases it is not possible to identify that part of the investment expenditure which has environmental objectives.

17.6 This chapter therefore attempts to quantify only expenditure which is directly concerned with improving or maintaining the quality of the environment, and which can clearly be identified or estimated.

17.7 The broad definition of environmental expenditure used in this chapter, and an explanation of terms used are given in Box 17.1.

17.8 The figures presented have been derived from an interim report commissioned to investigate the feasibility of collecting environmental expenditure for the UK, and from a variety of internal sources. For government, local authorities and regulatory agencies, figures have generally been taken from published sources. For the private sector, some figures are from published sources, whilst others are based on results from research [1]. Estimates of household expenditure are based on market research of expenditure on environmental products. Spending by voluntary and interest groups (non-profit organisations) has mostly been identified from published accounts.

17.9 The estimates should be regarded as indicating broad orders of magnitude only. Apart from the substantial definitional difficulties described above, figures have been estimated in a number of areas because data are not available.

Breakdown of expenditure

17.10 Although environmental expenditure is often thought of only as the costs incurred by industry for purposes of pollution abatement, monitoring and control, this chapter also includes other elements of environmental expenditure.

17.11 Expenditure has been categorised on the basis of a classification system which is being developed by the European Community[2], and covers the following activities:

- Pollution abatement
- Environmental conservation
- Research and development
- Education and training
- General administration
- Management of natural resources
- Improvement of amenities.

It is also broken down by medium (waste, air, water, noise, land).

17.12 The conventions used in constructing pollution abatement expenditure are also consistent with those being used by OECD countries.

17.13 Expenditure has been attributed to organisations or groups incurring the expenditure. The spending groups identified are:

- Government
- Enterprises
- Households
- Non-profit organisations (NPOs).

17.14 Expenditure estimates are attributed to the groups which are responsible for spending, not to the ultimate source of finance. For example, if expenditure by a voluntary conservation organisation is partly financed by a government department, the expenditure is attributed to the voluntary organisation and not to government. However, expenditure by quasi-governmental

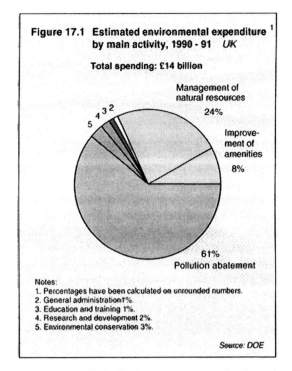

Figure 17.1 Estimated environmental expenditure [1] by main activity, 1990 - 91 *UK*

Total spending: £14 billion

Management of natural resources 24%

Improvement of amenities 8%

61% Pollution abatement

Notes:
1. Percentages have been calculated on unrounded numbers.
2. General administration 1%.
3. Education and training 1%.
4. Research and development 2%.
5. Environmental conservation 3%.

Source: DOE

agencies on behalf of government is attributed to government. The government spending group also includes other public sector organisations, such as local authorities and regulatory or enforcement agencies and nationalised industries. Expenditure by enterprises includes spending by manufacturing and process industry and also expenditure by other non-manufacturing enterprises, such as waste disposal contractors.

Direct environmental expenditure

17.15 Separately identifiable direct environmental expenditure in the UK in 1990-91 is estimated to be around £14 billion

Table 17.1   A summary of estimated environmental expenditure [1] by activity and medium 1990-91 *UK*

£ million	Medium							
Activity	Waste	Air	Water	Noise	Land	Unattrib.	Total [2]	%[3]
Pollution abatement	2,600	2,300	3,200	530	240	12	8,800	61
Environmental conservation	..	..	..	..	..	450	450	3
Research and development	..	110	45	..	55	40	250	2
Education and training	..	..	..	..	..	150	150	1
General administration	..	..	18	..	..	85	100	1
Sub-total	2,600	2,400	3,200	530	300	730	9,800	68
Management of natural resources	..	..	3,400	..	..	..	3,400	24
Improvement of amenities	380	..	86	..	690	50	1,200	8
Total [2]	3,000	2,400	6,700	530	990	780	14,000	100
%[3]	21	16	47	4	7	5	100	

Notes:
1. The numbers in the table have been rounded to two significant figures.
2. Totals are the rounded sum of unrounded numbers.
3. Percentages have been calculated on unrounded numbers.

Source: DOE

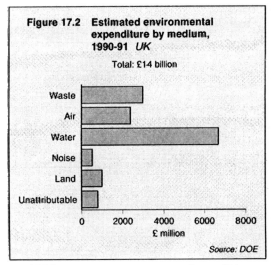

Figure 17.2 Estimated environmental expenditure by medium, 1990-91 *UK*

Total: £14 billion

Source: DOE

for almost half of the identified total and that on waste accounts for about one fifth.

17.18 Estimates of expenditure on each activity by spending group are shown in Table 17.2. Government and enterprises together account for 94 per cent of total expenditure identified.

17.19 The estimates given are of both current and capital expenditure (see Box 17.1). Of the total, capital expenditure is estimated to account for around 35 per cent (£4.9 billion), and current expenditure for about 65 per cent (around £9.1 billion). Estimates of current and capital expenditure by activity have not been made.

at 1990 prices, representing about 2½ per cent of GDP. A summary of identified estimated environmental expenditure by activity and by medium is shown in Table 17.1.

17.16 Figure 17.1 shows environmental expenditure by activity. The five core environmental activities - pollution abatement, conservation, research and development, education and training, and general environmental administration - account for almost 70 per cent of identified total environmental expenditure on the broader definition, which includes the management of natural resources and amenity maintenance and improvement. The largest proportion of the total, just over 60 per cent, relates to pollution abatement.

17.17 Estimated expenditure by medium (waste, air, water, noise, land) is shown in Figure 17.2. Expenditure on water accounts

Pollution abatement expenditure

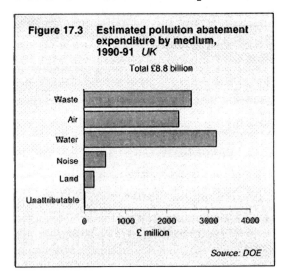

Figure 17.3 Estimated pollution abatement expenditure by medium, 1990-91 *UK*

Total £8.8 billion

Source: DOE

17.20 Direct expenditure on pollution abatement is estimated to be of the order of £8.8 billion for 1990-91. This is mostly incurred in relation to water (36 per cent), waste collection, treatment and disposal (29 per cent), and air pollution control (26 per cent). Estimated UK pollution abatement expenditure by medium is shown in Figure 17.3.

17.21 Expenditure on pollution abatement is mainly incurred by enterprises (£5.9 billion, 67 per cent) and by government, including local authorities (£2.2 billion, 25 per cent). The amount spent by government includes administrative expenditure which is clearly identifiable as spending on pollution abatement. An estimated 8 per cent is spent by households.

Waste Management

17.22 Pollution abatement expenditure on waste management covers spending on different types of wastes (ie household, commercial and industrial) and includes the cost of collection and transport of waste materials, and their disposal by different

£ million	Spending group				
Activity	Govern-ment	Enter-prise	House-holds	NPOs	Total [2]
Pollution abatement	2,200	5,900	680	..	8,800
Environmental conservation	290	..	-	160	450
Research and development	250	..	-	..	250
Education and training	90	60	-	..	150
General administration	100	..	-	..	100
Sub-total	**3,000**	**6,000**	**680**	**160**	**9,800**
Management of natural resources	630	2,800	-	..	3,400
Amenity improvement	1,200	..	-	..	1,200
Total [2]	**4,800**	**8,700**	**680**	**160**	**14,000**
%[3]	*33*	*61*	*5*	*1*	*100*

Table 17.2 A summary of estimated environmental expenditure [1] by activity and spending group 1990-91 *UK*

Notes:
1. The numbers in the table have been rounded to two significant figures.
2. Totals are the rounded sum of unrounded numbers.
3. Percentages have been calculated on unrounded numbers.

Source: DOE

methods (usually landfill or incineration - see Chapter 11 on waste). Around 87 per cent of expenditure on waste management services is incurred on pollution abatement (see Table 17.1).

Air pollution abatement

17.33 Environmental expenditure on air pollution abatement is mainly incurred by industry, usually following legal requirements on emissions. There are no published data on abatement costs, and no data about direct expenditure by enterprises. Estimates by the enterprise spending group can, however, be made by assessing the activity in the UK market for air pollution control equipment by individual sectors of industry (for example fuel processing, electricity supply, waste incineration etc). The estimate also includes expenditure by local authority environmental health departments on air pollution abatement.

17.24 For 1990-91, the estimates suggest identifiable expenditure in the UK on air pollution abatement of around £2.3 billion.

Water pollution abatement

17.25 Estimated expenditure on water pollution abatement is around £3.2 billion for 1990-91. Expenditure includes current and capital expenditure on sewerage systems and sewage treatment plant, payments for trade effluent consents and in-house treatment by industry and agriculture. It also includes expenditure on pollution abatement activities by the National Rivers Authority, River Purification Authorities in Scotland, and the Department of the Environment, Northern Ireland (DOE(NI)). The water industry in England and Wales will complete between 1990 and 1999 investment programmes worth £28 billion (at 1989-90 prices) to improve the quality of drinking water, rivers, estuaries and coastal waters.

Noise abatement

17.26 Identification of expenditure on noise abatement is one of the most difficult areas, since noise abatement is now an intrinsic part of the design process in many cases, eg in the design of new cars and aircraft, and the estimate of noise abatement expenditure by industry is therefore particularly tentative. Government expenditure figures include enforcement of compliance with noise regulations, compensation for increased highway and aircraft noise and financing of certain sound insulation.

Reclamation of previously developed land

17.27 In GB in 1990-91, a total of around £240 million was spent on reclamation of previously developed land and mineral workings restoration.

Environmental expenditure other than pollution abatement

Environmental conservation

17.28 Around £450 million was spent by the major environmental conservation organisations in the UK in 1990-91. It is not possible to break this down by medium. The total includes expenditure on environmental conservation by central government, its conservation agencies and non-departmental public bodies (NDPBs), together with spending by voluntary and interest groups, which has been attributed to non-profit organisations (NPOs). Of the total amount, 64 per cent was spent by government; the remainder was spent by NPOs.

Research and development

17.29 In 1990-91, an estimated £250 million was spent on research and development in the environmental field. There is also substantial R&D expenditure on clean technology by industry, including research into the potential effects that products might have on the environment. It has not however been possible to identify or estimate these amounts, and so they are not included in the total estimate for R&D.

Education and training

17.30 Total expenditure in this area is estimated to be of the order of £150 million. Estimates of educational expenditure are based on costs of higher education for students taking environmental science or physical geography courses. They exclude expenditure on schools and on further education, though many of the courses run by these institutions will include an environmental element. Estimates of environmental industrial training expenditure and training by the regulatory agencies have been included.

General administration

17.31 Expenditure on general administration includes the expenditure incurred by bodies, such as the Department of the Environment (DOE), which cannot be identified as attributable to a specific activity elsewhere in the framework. Expenditure by the regulatory agencies, such as local authority environmental health departments and HMIP, which can clearly be identified as directed at, say, pollution abatement, has been included under the relevant activity. On this basis, the estimated expenditure on general environmental administration amounts to around £100 million for 1990-91. This expenditure can mostly be identified from the published accounts of the various bodies.

Management of natural resources

17.32 Around £3.4 billion per year is spent on managing natural resources. Expenditure identified mainly covers that by the water companies in England and Wales, the River Purification Authorities in Scotland and DOE(NI) Water Service in Northern Ireland. It also includes expenditure on flood defences, and on the management of fish stocks. Just over 80 per cent of the total is spent by enterprises, and the remainder identified is spent by government. Estimates do not cover other media, or the costs of energy conservation.

Improvement of amenities

17.33 The total expenditure on maintenance and improvement of amenities is estimated to be £1.2 billion. Local authorities incur expenditure on services aimed at maintaining and improving local amenities, for example on street cleaning and maintenance of parks and open spaces. Central government expenditure on improvement of the urban environment is also included, together with expenditure by the NRA and British Waterways Board on maintaining the inland waterways network.

References and further reading

1. ECOTEC Research & Consulting Ltd, (1989). Industry Costs of Pollution Control. A Report Prepared for the Department of the Environment. ECOTEC, Birmingham.

2. European System for the Collection of Economic Information on the Environment (SERIEE).

Department of the Environment, (1991 and 1992). Annual reports. HMSO, London.

Department of the Environment, (1990). This Common Inheritance, Britain's Environmental Strategy. HMSO.

Department of the Environment, (1991). Survey of Derelict Land in England 1988, vols 1 and 2. HMSO.

Department of the Environment, (1991). Survey of Land for Mineral Workings in England 1988, vols 1-3. HMSO.

Welsh Office, (1991). Survey of Land for Mineral Workings in Wales 1988.

The Scottish Office, (1991). Serving Scotland's Needs. HMSO.

Water Services Association, (1991). Waterfacts 91. WSA, Sheffield.

CIPFA, (unpublished). Waste Disposal Statistics 1988-89.

Centre for the Study of Regulated Industries (CRI), (1992). Water Services and Costs 1990-91. Public Finance Foundation, London.

GLOSSARY

Absorbed dose - the amount of radiation absorbed, (eg in body tissues) expressed in terms of the energy imparted by that radiation to a unit mass of matter. The unit of absorbed dose is the Gray.

Abstraction - removal of water from surface waters (lakes, reservoirs, rivers) and groundwater (rocks) for domestic, commercial and industrial use.

Acid rain - rain, snow, fog and mist contaminated by sulphur and nitrogen.

Acid deposition - the removal from the atmosphere by trees, plants, and the earth's surface of sulphur and nitrogen containing compounds.

Adsorb - the taking up of a substance onto the surface of a solid.

Agrochemicals - chemical substances used in agricultural production including fertilisers, herbicides, fungicides and insecticides.

Algal blooms - rapid growth of phytoplankton in marine and freshwaters which may colour the water and may accumulate on the surface as a green scum. Decomposing dead cells consume large quantities of oxygen in the water which may result in the waters becoming anaerobic. Some blooms (such as certain species of blue-green algae) may produce poisons.

Alpha particles - the nuclei of helium atoms, emitted as ionising radiation by certain radionuclides.

Ambient - relating to the immediate surroundings eg ambient concentrations of pollutants, ambient temperature.

Ammonia - a colourless, highly soluble gas which is pungent and toxic at high concentration. The main source of the gas is agricultural waste products. Dissolved ammonia is an important constituent of sewage.

Anaerobic - absence of oxygen: conditions suitable only for organisms which do not require free oxygen or air for respiration.

Anticyclone - a body of moving air of higher pressure than the surrounding air.

Aquifer - a porous water-bearing underground formation of permeable rock, sand or gravel capable of yielding significant quantities of water.

Argon-41 - radioactive isotope of argon which has a very short half-life (1.8 hours) which emits beta and gamma rays. It is mainly produced from the activation of argon present naturally in the air used to cool the outside surfaces of reactor vessels and their shields.

Atom - the smallest quantity of an element that can take part in a chemical reaction.

Becquerel - a standard unit of radioactivity, equal to one radioactive disintegration per second. Measurements of radioactivity in air, water, rain and food are usually made in becquerels.

Benthic - adjective of the noun benthos meaning the flora and fauna of the sea bottom.

Benzene - a carcinogenic organic compound found in petrol and emitted mainly from car exhausts.

Beta particle - electrons emitted as ionising radiation by the nucleus of a radionuclide.

Biochemical oxygen demand - the amount of oxygen used by micro-organisms per unit volume of water at a given temperature, for a given time. It is used as a measure of water pollution by organic materials.

Biodegradable - capable of being decomposed by bacteria or other biological means.

Biodiversity - the total range of the variety of life on earth or any given part of it.

Biome - major ecological community of organisms occupying a large area (eg rain forests).

Biosphere - that part of the earth and atmosphere in which organisms live.

Blue-green algae - see algal blooms.

Brown soils - generally free-draining brownish or reddish soils overlying permeable materials.

Bryophyte - a group of primitive, non-vascular plants, eg mosses and liverworts, generally confined to damp locations.

Caesium-137 - radioactive isotope of caesium produced in nuclear reactions with a half-life of 30 years. It is chemically similar to potassium; when absorbed in the body, it spreads throughout.

Carbon monoxide - a toxic colourless gas produced by the incomplete combustion of carbon, mainly from combustion of petrol. Cigarette smoking is another important source.

Carbon dioxide - gas present in the atmosphere and formed during respiration, the decomposition and combustion of organic compounds (eg fossil fuels, wood, etc). A greenhouse gas.

Carbon cycle - natural circulation of carbon which is exchanged between the large carbon reservoirs in the land and ocean biospheres and the atmosphere.

Catalytic converter - a device fitted to the exhausts of motor cars which converts carbon monoxide and nitric oxide to carbon dioxide and nitrogen respectively, and organic compounds to carbon dioxide and water.

Catchment - area from which river systems, lakes and reservoirs collect water.

Cetacean - marine mammals belonging to the order Cetacea, including whales, dolphins and porpoises.

CFCs - chlorofluorocarbons; volatile but inert (ie without active chemical or other properties) compounds of carbon and (mainly) chlorine and fluorine. Important greenhouse gases and ozone layer depletors.

Chlorine loading - the total amount of chlorine in the atmosphere which is a measure of the potential damage to the ozone layer.

Coliforms - a group of bacteria which may be faecal or environmental in origin, and used as indicators in water of the possible presence of disease-causing organisms.

Collective dose equivalent - a measure of the total radiation dose to a group of people or a whole population. It is also called the collective effective dose equivalent and is expressed in man sievert.

Compound - two or more elements chemically bonded.

Contaminant - a substance that is present in higher concentrations than natural background levels.

Controlled waste - industrial, household and commercial waste, as defined in UK legislation. Controlled waste specifically excludes mine and quarry waste, wastes from premises used for agriculture, some sewage sludge and radioactive waste.

Cortex - the outer part of an organ, eg the kidney.

Cosmic rays - high energy ionising radiations from outer space.

Critical load - a quantitative estimate of exposure to pollutants below which no significant harmful environmental effects result. Critical load maps can be compared with current deposition maps to show areas where environmental damage occurs.

Crustaceans - belonging to the class Crustacea and typically having a thick hard shell including lobsters, crabs, shrimps and barnacles.

Cryptosporidium - a microscopic parasite which can cause disease in humans.

Cyclone - see depression.

DDT - an organochlorine insecticide widely used in the 1940s to 1960s.

Decibel (dB) - the unit of measurement of sound intensity. 0 dB corresponds to the quietest audible sound perceptible to the human ear. It is measured on a logarithmic scale.

Depression - region of low atmospheric pressure.

Derogations - modifications of requirements, in particular to the application of EC Directives, eg relaxing of ambient concentrations of atmospheric pollutants.

Derv - fuel for diesel engined road vehicles.

Determinand - any substance or characteristic of water that is measured or determined, eg by a physical, chemical or microbiological procedure.

Dewpoint - temperature at which cooled air becomes saturated with water vapour and condensation results, forming dew. The dewpoint varies with the relative humidity and temperature of the air.

Dioxins - a group of 210 closely related chemicals which can be formed as by-products in some chemical processes and in various combustion processes such as waste incineration and coal burning.

Disinfection by-products - disinfectants used in water treatment may react with organic substances (many of which are naturally present) and produce chemicals called disinfection by-products.

Dose equivalent - a measurement of the energy released by radiation when deposited in living tissue. It is obtained by multiplying the absorbed dose by a factor to allow for the varying effectiveness of different types of radiation in causing harm to tissue. It is expressed in sievert.

Ecology - the study of the relationships between living systems and their natural environment.

Ecosystem - a community of interdependent organisms and the environment they inhabit eg ponds and pond life.

EC Directive - a European Community legal instrument binding on all member states but leaving the method of implementation to national governments, and which must therefore, generally, be transposed into national legislation.

EC Regulation - European Community legislation having legal force in all member states.

Effective dose equivalent - a measurement which expresses the variety of dose equivalents for different organs in the body as a single number. It is commonly referred to as "dose" and measured in sievert. The value gives an indication of the risk to health from any exposure to radiation.

Effluent - liquid waste from industrial, agricultural or sewage plant outlets.

Epidemiology - the branch of medical science concerned with the study of environmental, personal and other factors that determine the incidence of disease.

Eutrophication - the nutrient enrichment of water (especially by compounds of nitrogen and/or phosphorus), causing an accelerated growth of algae and higher forms of plant life, producing an undesirable disturbance in the balance of organisms present and reducing water quality.

Evapotranspiration - loss of water resulting from transpiration from plants (loss of water vapour from plants through stomata) and evaporation from surface water and soil.

Fauna - all animal life.

Ferric/Ferrous - of or containing iron compounds.

Fertiliser - any material added to soil to supply nutrients for plant growth.

Field capacity - maximum amount of water that can be retained in the soil following natural drainage.

Flora - all plant life.

Fungicide - a substance which kills fungi, the cause of most plant diseases.

Gamma ray - type of electromagnetic energy (ionising radiation) emitted by some radionuclides.

Global warming - the increase in the average temperature of the earth, thought to be caused by the build up of greenhouse gases.

Gray - the unit of absorbed dose of ionising radiation.

Green belt - any zone of countryside immediately adjacent to a town or city, defined for the purpose of restricting outward expansion of the urban area.

Greenhouse effect - process by which certain gases in the atmosphere behave in effect, like glass in a greenhouse; glass allows solar radiation in, which heats the interior, but reduces the outward emission of heat radiation.

Greenhouse gases - naturally occurring gases, such as carbon dioxide, nitrous oxide, methane and ozone, and man made gases like chlorofluorocarbons.

Groundwater gley soils - soils with prominently mottled or grey subsoils caused by periodic waterlogging by a high groundwater table, and resulting in anaerobic conditions.

Ground level ozone - see ozone.

Groundwater - water held in water bearing rocks, in pores and fissures underground.

Gulf Stream - see North Atlantic Drift.

Gullies - see runoff.

Habitat - the customary dwelling place of a species or community, having particular characteristics, eg sea shore.

Half-life - the time taken for the activity of a radionuclide to lose half its value by decay.

Halogenated hydrocarbons - formed when hydrogen in hydrocarbon molecules such as methane is replaced by halogens (fluorine, chlorine, bromine and iodine). Their breakdown in the stratosphere releases chlorine and bromine which actively take part in the destruction of stratospheric ozone. The best known group of halogenated hydrocarbons is the chlorofluorocarbons (CFCs). The brominated compounds are referred to as halons.

Halons - see halogenated hydrocarbons.

HCFCs - hydrochlorofluorocarbons; used as replacements to CFCs in refrigeration, foam blowing and aerosols because they are less active ozone depletors.

Heavy metals - a loose term covering potentially toxic metals used in industrial processes, eg arsenic, cadmium, chromium, copper, lead, nickel and zinc. Heavy metals may be discharged to the environment and be found as suspended particulate matter in the atmosphere.

Herbicide - substance used to control weeds.

HFCs - hydrofluorocarbons; halogenated carbons, similar to HCFCs, but not containing chlorine and not therefore, ozone depletors.

Hydrocarbons - compounds of hydrogen and carbon which react in the presence of sunlight and oxides of nitrogen to produce photochemical oxidants.

Hydrogen chloride - a poisonous and corrosive gas which, when dissolved in water, forms hydrochloric acid.

Hydrology - the study of water on and below the earth's surface and in the atmosphere.

Insecticide - substance used to destroy or repel insects.

Integrated pollution control - an approach to pollution control in the UK, which recognises the need to look at the environment as a whole, so that solutions to particular pollution problems take account of potential effects upon all environmental media.

Intensive agriculture - a general term applying to agricultural practices which produce high output per unit area usually by high usage of manure and agrochemicals, mechanisation etc.

Invertebrate - any animal that does not have a backbone.

Ion - an electrically charged atom or grouping of atoms.

Ionisation - the process by which a neutral atom or molecule acquires or loses an electric charge. The production of ions.

Ionising radiation - radiation that produces ionisation in matter. Examples are alpha and beta particles, gamma rays, X-rays and neutrons.

Irradiation - the exposure of a body or substance to ionising radiation.

Isotherm - temperature contour, ie line joining points of the same temperature.

Isotopes - variant forms of the same element whose atoms contain the same number of protons and electrons but different numbers of neutrons.

Larvae - immature free-living forms of many invertebrate animals.

Leaching - loss of soluble substances from a solid mass, eg soil, by the action of percolating liquid.

L_{Aeq} - measure of the average sound level, often over a specified period, for example 16 hours.

L_{A10} - recognised measure for the level of sound exceeded for 10 per cent of a given time period.

L_{A90} - recognised measure for the level of sound exceeded for 90 per cent of a given time period.

Leys - land sown to grass for a year or more, grown in rotation with annually cultivated crops.

Lithomorphic soils - shallow soils usually formed directly over bedrock in which the only significant soil forming process has been the formation of an organic or organic-enriched mineral surface.

Loam - soil of medium texture composed of roughly equal proportions of clay, silt and sand.

Man made soils - soils formed in material modified or created by human activity, eg soils containing manures or refuse, or which result from unusually deep cultivation, or soil forming materials for use in land restoration following mining or quarrying.

Marginal land - typically land of poor quality for agricultural use, due to adverse soil, site or climate.

Maximum Potential Soil Moisture Deficit - a measure of the potential use of water by plants.

Medium - eg air, water, soil.

Medulla - the central portion of an organ or tissue.

Methane (CH_4) - the simplest member of the hydrocarbon family. Methane is a potent greenhouse gas.

Microbiology - the study of micro-organisms, eg bacteria, viruses, etc.

Monoculture - repeated cultivation of a single crop on a given area of land.

Nitrate - a major component of many fertilisers

and a natural product of the breakdown of organic matter in soil. It is highly soluble in water and is easily leached from soil.

Nitrogen dioxide (NO_2) - an oxide of nitrogen which arises from fuel combustion in vehicles, boilers and furnaces, which is toxic in high concentrations and contributes to ozone formation and acid rain.

Nitrogen oxides (NO_x) - a range of compounds formed by the oxidation of atmospheric nitrogen. Some of these oxides contribute to acid rain and smog, and can affect the stratospheric ozone layer.

Nitrous oxide (N_2O) - a relatively inert oxide of nitrogen emitted by soils and during the manufacture of nylon. This substance is a potent greenhouse gas.

Noise and Number Index - a measure of noise annoyance from aircraft (not used since September 1990).

North Atlantic Drift - the warm ocean current flowing north-east under the influence of prevailing winds, from the Gulf of Mexico towards north-west Europe and warming its climate. Also called the Gulf Stream.

Noxious - poisonous or harmful.

Nucleus - the positively charged dense region at the centre of an atom, composed of protons and neutrons.

Nuclide - a species of atom characterised by the number of protons and neutrons and in some cases by the energy state of the nucleus.

Nutrient - substance providing nourishment for plants and animals eg nitrogen, phosphorus.

Occult deposition - the deposition of pollutants (in particular acid deposition) on plants and trees from fog and mist droplets.

Oocysts - the environmentally resistant transmissible form of parasites, eg cryptosporidium, excreted in the faeces of an infected host.

Organic - generally any substance containing carbon as part of its chemical make up, eg animals and plants or their constituents.

Organochlorines - any organic compound containing chlorine, eg PCBs and pesticides such as DDT and lindane.

Oxidation - usually a chemical reaction with oxygen, producing oxides.

Oxide - any compound of oxygen with another element.

Ozone hole - significant depletion in stratospheric ozone which has been observed over Antarctica.

Ozone - a naturally occurring gas found throughout the atmosphere. In the lower atmosphere (ground level or tropospheric ozone) it is a secondary pollutant and its formation can be enhanced by other pollutants. In the upper atmosphere (the stratosphere) the ozone layer protects the earth from harmful ultraviolet radiation.

PAH - polycyclic aromatic hydrocarbons, a class of hydrocarbons of high molecular weight emitted by motor vehicles and other processes where there is incomplete combustion. They are toxic in high concentrations and some are believed to be carcinogenic.

Paralytic shellfish poisoning - ingestion of toxic algae by shellfish which are in turn consumed by humans may cause symptoms of this poisoning. The phenomenon may be associated with blooms of algae known as dinoflagellates, although large numbers of the organisms are not needed for toxic effects to be induced.

Particulates - fine, solid particles found in the air or in emissions.

Pathogen - a substance or organism that causes disease.

Pathway - a route by which a substance may come into contact with the human body, eg external exposure, inhalation, ingestion.

PCBs - polychlorinated biphenyls, a group of widely used compounds containing chlorine. PCBs can accumulate in food chains and at high concentrations are thought likely to produce harmful side effects, particularly during the reproductive cycle of some marine animals, eg seals.

Peat soils - predominantly organic soils derived from partially decomposed plant remains that accumulate under waterlogged conditions.

Ped - a naturally formed aggregate of soil particles.

Pelosols - slowly permeable clay soils. They crack deeply in dry seasons.

Pesticides - substances used to kill pests, weeds, insects, fungi, rodents etc. See insecticide, fungicide, herbicide.

pH - the measure of acidity or alkalinity of a substance. A neutral substance has a pH of 7.0 and the lower the pH the greater the acidity.

Photochemical reactions - chemical reactions where the energy is supplied by sunlight.

Photosynthesis - a process in the presence of sunlight, where trees and other plants take in carbon dioxide from the atmosphere which combines with water to produce simple sugars. Oxygen is released as a by-product.

Phytoplankton - free drifting mostly microscopic aquatic plants. The plant constituent of plankton.

Plankton - plant and animal organisms, many of which are microscopic, living in the surface layers of seas or lakes.

Podzolic soils - soils with dark brown, black or ochreous subsurface layers resulting from the accumulation of iron, aluminium or organic matter leached from upper layers. They normally develop as a result of acid weathering conditions.

Pollutant - a substance which is present at concentrations which cause harm or exceed an environmental quality standard.

Primary energy consumption - total consumption of energy including fuel consumed by energy producers.

Radioactive decay - the process of transformation or disintegration of a radionuclide, resulting in the release of radiation.

Radioactive decay product - a nuclide or radionuclide produced by decay.

Radioactivity - the property of radionuclides of spontaneously emitting ionising radiation.

Radionuclide - an unstable nuclide that emits ionising radiation.

Radon - a natural radioactive gas which forms in the ground indirectly by the radioactive decay of natural uranium and thorium in soils and rocks, and which mixes with air and seeps into the atmosphere. It is particularly prevalent in granite areas.

Raw gley soils - soils which occur in mineral material that has remained waterlogged since deposition. They are chiefly confined to intertidal flats or saltings.

Red List Substances - toxic substances, defined by the UK, which cause the greatest potential threat to the aquatic environment. The list includes some heavy metals, certain pesticides, industrial chemicals and solvents.

Rills - see runoff.

Runoff - the flow of surface water from snow melt, rainfall or spring seepage which flows directly into streams, rivers and lakes. Runoff may erode the soil and leave shallow or deep channels (rills and gullies).

Salting - an area of low ground regularly inundated with salt water; a salt marsh.

Selenium - a trace element present in the ambient atmosphere.

Semi-natural vegetation - habitats which, though modified by human activity, support communities of native plants and animals which are similar in structure to natural communities.

Sewage - liquid wastes from communities, conveyed in sewers. Sewage may be a mixture of domestic sewage effluents from residential areas and industrial liquid waste.

Sewage sludge - semi-solid and solid waste matter removed from sewage at sewage treatment plants.

Sewerage - a network of pipes and associated equipment for the collection and transportation of sewage.

Sievert - see dose equivalent, effective dose equivalent.

Silage - crop harvested while green for animal fodder and not dried but preserved by excluding air from storage site.

Slurry - liquid manure.

Smog - smoke fog which contains gaseous pollutants such as sulphur dioxide, soot and ash; often used as a general description for unusually polluted air irrespective of the pollutants present.

Soil moisture deficit - the drying out of soil, occurring when the loss of water by evapotranspiration is greater than rainfall.

Special waste - controlled waste which consists of, or contains substances which are "dangerous to life" as defined in UK regulations.

Stratosphere - upper layer of the atmosphere above the troposphere, approximately 15-50 km above the earth's surface.

Stratospheric ozone - see ozone.

Strontium-90 - radioactive isotope of strontium produced in nuclear reactions, with a half-life of 28 years. It is chemically similar to calcium and tends to be concentrated in milk. When absorbed by the body it is concentrated in the bones.

Sulphur dioxide (SO_2) - a compound of sulphur and oxygen which is emitted into the atmosphere by the combustion of fuels containing sulphur such as coal, diesel oil and fuel oil. It is toxic at high concentration and contributes to acidity in rain.

Surface water gley soils - seasonally waterlogged slowly permeable soils.

TBT - Tributyltin; an extremely toxic substance to aquatic life used as a marine anti-fouling agent. It is a "Red List" substance.

THMs - Trihalomethanes; compounds consisting of one carbon atom, one hydrogen atom and three halogen (eg fluorine, chlorine, bromine or iodine) atoms, some of which are formed during chlorination of water. Trichloromethane (chloroform) occurs most frequently in water supplies.

Tinnitus - persistent ringing or other noise in the ears; it may be caused by exposure to persistent high noise levels or by inner ear disease.

Trace elements - elements which occur in minute quantities as natural constituents of living organisms and tissues. They are however, generally harmful in large quantities. Trace elements include lead, silver, cobalt, iron, zinc, nickel, selenium and manganese.

Tritium - a radioactive isotope of hydrogen produced in nuclear reactions, with a half-life of 12 years. It has a very low radiotoxicity.

Troposphere - lowest layer of the atmosphere, extending to approximately 15 km above the earth's surface.

Tropospheric ozone - see ozone.

Ultraviolet radiation - radiation in the wavelength range between visible light and X-rays, divided into wavelengths bands A, B, and C. The earth is shielded by the ozone layer from lethal ultraviolet-C radiation. The ozone layer also substantially reduces the amount of harmful ultraviolet-B radiation reaching the earth's surface.

Vascular tissue - cells in plants that function as tubes or ducts through which water, and dissolved nutrients move from one part of the plant to another; in animals the term denotes the blood vessels and heart.

VOCs - Volatile Organic Compounds; organic compounds which evaporate readily and contribute to air pollution mainly through the production of secondary pollutants (eg ozone).

Water table - the upper surface of permanent groundwater saturation.

Wetland - an area of low-lying land where the water table is at or near the surface for most of the time, leading to both freshwater and seawater habitats, and waterlogged land areas.

Zooplankton - animal organisms consisting mainly of small crustaceans and fish larvae. The animal constituent of plankton.

ABBREVIATIONS

The main abbreviations used in the report are given below. See also glossary for items marked (*)

ACOPS	Advisory Committee on Pollution of the Sea
ADAS	Agricultural Development and Advisory Service
AONB	Area of Outstanding Natural Beauty
ASSI	Area of Special Scientific Interest (Northern Ireland)
BATNEEC	Best Available Techniques Not Entailing Excessive Cost
BNFL	British Nuclear Fuels
BOD	Biological Oxygen Demand
BRE	Building Research Establishment
BTO	British Trust for Ornithology
CAA	Civil Aviation Authority
CC	Countryside Commission
CCW	Countryside Council for Wales
CFCs	Chlorofluorocarbons*
CCIRG	Climate Change Impacts Review Group
COMARE	Committee on Medical Aspects of Radiation in the Environment
CoPA	Control of Pollution Act, 1974
CPRW	Council for the Protection of Rural Wales
DANI	Department of Agriculture for Northern Ireland
DDT	Dichloro-diphenyl-trichloro-ethane*
DH	Department of Health
DNH	Department of National Heritage
DOE	Department of the Environment
DOE(NI)	Department of the Environment for Northern Ireland
DTp	Department of Transport
DWI	Drinking Water Inspectorate
EA	Environmental Assessment
EC	European Community
EHO	Environmental Health Officer
EIA	Environmental Impact Assessment
EPA	Environment Protection Act, 1990
ESA	Environmentally Sensitive Area
FC	Forestry Commission
GDP	Gross Domestic Product
GWP	Global Warming Potential
HCFCs	Hydrochlorofluorocarbons*
HFCs	Hydrofluorocarbons*
HMIP	Her Majesty's Inspectorate of Pollution
HMIPI	Her Majesty's Industrial Pollution Inspectorate (Scotland)
IACR	Institute of Arable Crops Research
ICES	International Council for the Exploration of the Sea
ICRCL	Interdepartmental Committee on the Redevelopment of Contaminated Land
ICRP	International Commission on Radiological Protection
IEHO	Institution of Environmental Health Officers
IFE	Institute of Freshwater Ecology
IH	Institute of Hydrology
IPC	Integrated Pollution Control
IPCC	Intergovernmental Panel on Climate Change
ITE	Institute of Terrestrial Ecology
IUCN	International Union for the Conservation of Nature and Natural Resources

JNCC	Joint Nature Conservation Committee
LOD	Limit of Detection
LFA	Less Favoured Area
MAFF	Ministry of Agriculture, Fisheries and Food
MLURI	Macaulay Land Use Research Institute
MNR	Marine Nature Reserve
MO	Meteorological Office
MORI	Market and Opinion Research International
NAA	Nitrate Advisory Area
NCC	Nature Conservancy Council
NERC	Natural Environment Research Council
NNR	National Nature Reserve
NRA	National Rivers Authority
NRPB	National Radiological Protection Board
NSA	Nitrate Sensitive Area
NWC	National Water Council
OECD	Organisation for Economic Cooperation and Development
OS	Ordnance Survey
PAHs	Polycyclic aromatic hydrocarbons*
PAN	Peroxyacetyl Nitrate
PARCOM	Paris Commission
PCBs	Polychlorinated biphenyls*
PTWI	Provisional Tolerable Weekly Intake
REHIS	Royal Environmental Health Institute of Scotland
RPA	River Purification Authority
RPB	River Purification Board
RSPB	Royal Society for the Protection of Birds
SCPR	Social and Community Planning Research
SNH	Scottish Natural Heritage
SO	Scottish Office
SOAFD	Scottish Office Agriculture and Fisheries Department
SOEnD	Scottish Office Environment Department
SPA	Special Protection Area
SSLRC	Soil Survey and Land Research Centre
SSSI	Site of Special Scientific Interest
TBT	Tributyltin*
THM	Trihalomethanes*
VOCs	Volatile Organic Compounds*
UKAEA	United Kingdom Atomic Energy Authority
UKSORG	United Kingdom Stratospheric Ozone Review Group
UNECE	United Nations Economic Commission for Europe
WDA	Waste Disposal Authority
WHO	World Health Organisation
WMO	World Meteorological Organisation
WO	Welsh Office
WOAD	Welsh Office Agricultural Department
WRA	Waste Regulation Authority
WSA	Water Services Association
WSL	Warren Spring Laboratory
WWF	World Wildlife Fund (now WWFN)
WWFN	World Wide Fund for Nature (formerly WWF)

INDEX

The references in this index refer to Figure and Table numbers.